'E' Issues for Agribusiness
The 'What', 'Why', 'How'

LEARNING ••••••••••••••• **services**

01579 372213

Duchy College: Stoke Climsland
Learning Centre

This resource is to be returned on or before the last date stamped below. To renew items please contact the Centre

Three Week Loan

2 8 APR 2011		
- 8 MAY 2012		

D1418821

ii

'E' Issues for Agribusiness
The 'What', 'Why', 'How'

by
Kim P. Bryceson
The University of Queensland
Australia

www.cabi.org

iv

CABI is a trading name of CAB International

CABI Head Office
Nosworthy Way
Wallingford
Oxfordshire OX10 8DE
UK

Tel: +44 (0)1491 832111
Fax: +44 (0)1491 833508
E-mail: cabi@cabi.org
Website: www.cabi.org

CABI North American Office
875 Massachusetts Avenue
7th Floor
Cambridge, MA 02139
USA

Tel: +1 617 395 4056
Fax: +1 617 354 6875
E-mail: cabi-nao@cabi.org

©K. Bryceson, The University of Queensland, Australia 2006. All rights reserved. No part of this publication may be reproduced in any form or by any means, electronically, mechanically, by photocopying, recording or otherwise, without the prior permission of the copyright owners.

Every endeavour has been made to contact copyright holders to obtain the necessary permission for use of illustrative material.

A catalogue record for this book is available from the British Library, London, UK.

Library of Congress Cataloging-in-Publication Data

Bryceson, K.P.
  E issues for agribusiness : the what, why, how / K.P. Bryceson.
      p. cm.
  Includes bibliographical references and index.
  ISBN 1-84593-071-1 (alk. paper)
  1. Agricultural industries. 2. Electronic commerce. I. Title.

  HD9000.5.B78 2006
  630.68'8--dc22

                                        2006005878

ISBN   1 84593 071 1
         978 1 84593 071 4   630. 68   BRY

Printed and bound in the UK by Athenaeum Press, Gateshead from copy supplied by the author.

Contents

About the Author

Kim Bryceson BSc(Hons), Grad Cert(HE) METM, PhD

Kim Bryceson is a Senior Lecturer in Agribusiness with the University of Queensland, Australia which she joined in 2000 after working in the private sector developing risk assessment decision support and competitive market intelligence / knowledge management systems for Agribusiness and the Energy sectors. Prior to this, her early career was as a Senior Research Scientist in computer and satellite technology applications in the Commonwealth of Australia's Department of Primary Industries, and the Cooperative Research Centre for Soils and Land Management.

Kim's current research is on the impact of electronic business technologies in agri-industry value chains, the role of the internet and associated technologies in facilitating collaborative commerce, electronic product tracking and biosecurity in agri-industry chains and clusters, and in the development of new domestic and export agribusiness markets. She has research projects in both the international and domestic arenas and has undertaken many business and market analysis consultancies in agribusiness both domestically and internationally.

Kim teaches students undertaking the Bachelor and Masters' degrees in Agribusiness, and the Bachelor of Environmental Management degree and her teaching has a strong focus on 'e' Learning. She is the coordinator of three courses – 'Agribusiness in the eLandscape', 'EAgribusiness' and 'Managerial Decision Support Systems'. Her latest educational research involves developing online Virtual Reality agribusiness environments for immersive learning. In 2004 she won an individual University of Queensland Excellence in Teaching Award – and with Geoff Slaughter and Tony Dunne, she won the 2004 AWB/Agforce Innovation in Grains Education Award for the development of an online Trading Room as part of a course on Commodity Trading. In 2005 she was shortlisted for an Australian Award for University Teaching.

Acknowledgements

The author would like to thank Margaret Cooke, Laura Black, and Judith Millington, who undertook the copyediting of the manuscript and Elana Stokes who has reproduced the figures and created the camera ready copy (all from The University of Queensland). Without this team the book would not have seen the light of day.

I am also very grateful to Tony Dunne, The University of Queensland and Michael Maloney formerly of Incitec Pty Ltd, who provided, through many discussions over many months, insights and advice in the preparation of Chapter 4 and Chapter 8 respectively.

In addition, without input from industry – discussions, interviews, projects, consultancies, etc. much of what has been included in this book would not have had the very real-life flavour that it does. To Nick Evans, Bob Hansen and Wendy Erhart who personally contributed to the Smart Thinkers – thank you.

Finally to Kerry – thank you for your patience while 'The Book' was underway.

Figures and Images

© ANR Communication Services. (p. 189).
© CEN (European Committee for Standardization). Reproduced with permission (p. 94); © Cutter Consortium, LLC. Reproduced with permission (p. 79); © Hay Group. (p. 216); © John Wiley and Sons Ltd, New York. Reproduced with permission (p. 33); © McKinsey and Company Inc. Reproduced with permission (p. 205, 291); © McGraw-Hill Australia. Reproduced with permission (p. 40); © Montgomery Research Inc. Reproduced with permission (p. 147); © Pearson Education Inc. Reproduced with permission (p. 103); © Oxford University Press Inc. Reproduced with permission (p. 69). © Simon and Schuster Inc. (p. 204); © CISCO Systems – image of John Chambers CEO, CISCO Systems. Reproduced with permission (p. 98); © First4Farming Ltd – image of Nick Evans, CEO, First4Farming Ltd. Reproduced with permission (p. 136); © Sally Tsoutas – image of Wendy Erhart, Co-Owner, Withcott Seedlings. Reproduced with permission of Black Communications (p. 169); © The Peanut Company of Australia – image of Bob Hansen, CEO, The Peanut Company of Australia. Reproduced with permission (p. 200); © Dell Inc – image of Michael Dell, Executive Chairman, Dell Inc. Reproduced with permission (p. 281).

Disclaimer

No person, nor any company described or referenced in this book, has endorsed the publication in any way.

Introduction – The Balancing Act

The current harsh reality is that all businesses are under pressure to respond to ever-increasing change in the business environment with the development of new business models and strategies – in fact, it has become a balancing act between keeping up to date and innovatively moving forward! Agribusinesses are no exception.

It has become increasingly clear in the last five years that the internet and associated electronically enabled business practices, are, despite the demise of the dot.coms, business-supporting technologies that are here to stay. Traditional bricks-and-mortar companies are embracing the development of new electronically enabled business channels to accommodate the upsurge in technologically savvy customers – both buyers and sellers. The objectives of this book are twofold:

1. To identify, explain and summarise the concepts, technology, industry and management issues facing agribusinesses today in an electronically enabled business landscape.
2. To create an awareness of current and potential electronically enabled business application areas in the agricultural, agribusiness and rural management areas.

The book was originally conceived as a resource text for professional managers, focusing on the hands-on management issues that need to be addressed in relation to the use of electronic technologies in the conduct of general business today. However, an ongoing operational and research interest in the agricultural and agribusiness sectors by the author – and the progression in the agribusiness world of the use of these technologies worldwide over the last three to four years, has meant the book has morphed to now focus on the agri-industry sector and all the attendant issues that Information and Communication Technologies (ICTs) are bringing to this essentially production driven environment.

In addition, both undergraduate and postgraduate student demand in the agricultural/agribusiness sectors for knowledge in this area has meant that the book has become a combined student text *and* a resource for business practitioners. The main aim however remains the same – and that is to provide an appreciation of operational issues that have become keys to agribusiness success within an electronically enabled business environment, as well as addressing in more detail the theories, concepts and research behind the practical management options.

The underlying theme of the book is 'What, Why, How'. That is: 'What' is electronically enabled agribusiness?, 'Why' would an agribusiness want to embrace it?, and 'How' does one go about doing it? The book pulls together a number of the major issues facing people moving into an electronically enabled agribusiness environment. It is both an introduction to electronic business issues and a comprehensive guide to more detailed business process and strategic planning matters associated with the technologies involved. In particular the following key areas are addressed with specific reference to the agri-industry sector:

- **Concepts** – the underlying conceptual issues associated with running an agribusiness, the difference between eCommerce and eBusiness, and why is electronically enabled agribusiness the way forward for the future?

- **Technology in business** – the underlying technology challenges and systems supporting electronically enabled agribusiness processes, supply chain management and value chain organizational structure requirements.
- **Electronically enabled business models and 'E' strategies** – how businesses work and the effect this has on the development of electronically enabled business. What are B2C, B2B, ESCM, EDM? What was the dot.com crash and why did it happen? Are some models better suited to an electronically enabled agribusiness than others?
- **Management concepts** – the action items for managers in an electronically enabled world such as decision making, creating competitive advantage, risk management, corporate eGovernance, etc.

There are nine chapters – each addressing key concepts and each assuming little or no knowledge of the subject matter to start with, but then progressing quickly into more detailed discussion. Each chapter is heavily illustrated with examples but each also contains four or five 'Byte Ideas' which are more detailed descriptions of companies that have done something particularly spectacular in their business using electronic technologies. These are mainly agribusinesses although some classic electronic businesses such as Amazon.com, eBay etc. are included. Similarly, each chapter introduces and provides a brief biography of a 'Smart Thinker' – an individual who has driven a business in an innovative fashion in the 'E' world. Finally, review and critical thinking questions are provided for each chapter for revision purposes but also to stimulate further thought on the content of each chapter. Final Thoughts flags a general decline in the numbers of young people in the agricultural/agribusiness sectors and proposes that electronic enablement should be used as a mechanism to 'pull' or entice young people into the industry and as a 'push' mechanism through innovative education. A 'Useful URL' section and a Reference Chapter conclude the book.

A very brief overview of each chapter follows – although it should be noted that each chapter has its own introduction and set of chapter objectives.

- Chapter 1 (*An Agribusiness in the eLandscape*) is an introductory chapter discussing the fundamentals of business and of agribusiness. It also introduces the technologies that have created the 'eLandscape' of today's business world.
- Chapter 2 (*Creating Value in the 'E' Agribusiness World*) focuses on a discussion of value adding in business and agribusiness - particularly in relation to how electronic enablement of the business can facilitate this.
- In Chapter 3 (*The KID Triangle – Knowledge, Information and Data in the 'E'World*) a data, information and knowledge framework called 'The KID Triangle' is introduced with most emphasis being placed on a discussion of knowledge creation and knowledge management in an electronically enabled business. A brief overview of Portals and Enterprise Information Portals is included with this chapter.
- Chapter 4 (*Agri-food Chains in the 'E'World*) is an important chapter pushing the point that a business and particularly an agribusiness, is not an island – individual businesses can not exist without other enterprises to either supply them or buy from them. As such, this chapter spends time on fleshing out the supply/value chain concepts introduced in Chapter 1, introduces the 'Lean' chain, and characterizes agribusiness competition as a battle of supply and value chains. The automation of supply/value chain through electronic enablement is discussed in relation to building internal capability first before attempting to extend up and down the chain and therefore EDI, ERP and CRM are addressed as technical components.

- Chapter 5 (*Managing Uncertainty in the 'E' World*) discusses two extremely important issues in any business – trust/privacy and risk management. In addition, disaster management and business continuity planning are also introduced as key requirements in an electronically enabled world.
- Chapter 6 (*Food Tracking and Traceability in the 'E' World*) or, 'we are what we eat', is a chapter that looks at two of the major emerging issues in agribusiness worldwide and the implications for technology use to address these in different agri-industry sectors. The first issue discussed is biosecurity – what it is and what it is not. The second issue is traceability of product (particularly food products and ingredients) through the agri-food chain along with identity preservation.
- Chapter 7 (*Is the Agribusiness 'E' Ready?*) covers an extended discussion about the issues associated with creating and implementing an 'E' strategy, including an evaluation of the objectives of getting into electronically enabled business (Think Big, Start Small and Scale Fast). eReadiness to electronically enable is also discussed covering issues of business capability and the alignment of business strategy and processes.
- Chapter 8 (*The Practicalities of 'Making it Happen'*) and Chapter 9 (*EGovernance and Legal Issues in the 'E' World*) are both about the bottom line in different ways – Chapter 8 pulls it all together in terms of the practicalities of implementing electronic enablement in an agribusiness and looks at issues of organization, project management, outsourcing, resource utilization and organizational change. Chapter 9 deals with Corporate Governance and 'E' Governance including management of the internet and how to legally govern activities on the internet – and the type of bottom line (potential jail time!) that can happen if businesses and/or company directors do not look after corporate governance properly. Cyber law is also discussed as it pertains to the paperless environment and the impact of electronically enabled business on the intellectual property system.

What emerges is that to work effectively in agribusiness in the 21st century you not only need to understand and be familiar with your own specialty area – be it production, processing or retail – but you also need to be aware of the interdependencies between businesses, the technologies that have evolved to facilitate 'doing business', innovative strategies and processes to keep you in front of the competitive pack – and the law associated with the electronic environment....definitely a challenge and a real balancing act!

Chapter 1

An Agribusiness in the eLandscape

Don't panic! it's just about business...

If you believe the hyperbole in the newspapers and other media, it would appear that there has been an 'E' revolution going on in the business world in recent years fuelled by the development of the personal computer and the proliferation of the internet as a means of communication. At one stage during the late 1990s and early 2000, the development of businesses based solely around the internet led to the dot.com era which was so pervasive that if you were not part of it in a business sense, you were regarded as being a lost cause.

Obtaining venture capital (VC) and aiming for an initial public offering (IPO) was on everyone's minds and the talk was all about buying and selling online and the bright young things who were making their fortunes driving this immense change in the way business was being done. Classic business development and management strategies seemed to disappear and talk of the 'New Economy' taking over from the 'Old Economy' was rife. If you were 'in' to IT, you were where it was 'at' man; if you were not – well, you lined up at the employment offices!

However, in April 2000 this all changed with what has become known as the dot.com crash (discussed in Chapter 2). Sense finally came to the world financial markets and companies that could not cut it in terms of providing shareholder value dropped out of the marketplace with many companies going bellyup and fortunes being lost overnight. Some important questions which arise from this experience are why and how did the whole dot.com scenario develop as it did, and what, if any, have been the ongoing effects in the business world and the agribusiness world in particular?

This first chapter takes a very broad look at these questions by looking at the concepts behind a *business*, introducing *agribusiness*, including a discussion on the various stakeholders involved, and discussing the points associated with *doing business*. The chapter also introduces the main electronic technologies available to businesses today using the concept of the *eLandscape* with the emphasis being on technology as a business *enabler* rather than as a revolutionizing force.

Chapter Objectives

After reading this chapter you should understand and be able to describe:

1. What a business is and in particular what an agribusiness is
2. The core drivers and critical responses of 'doing business'
3. The concept of an agri-industry chain and the difference between supply and value chains
4. The practical requirements associated with maximizing profitability
5. The core building blocks of a business
6. The concept of the 'eLandscape' and the functionality of the technologies involved
7. The difference between eCommerce, eBusiness and mCommerce.

What Is a Business or Firm?

A business or firm is a unit that can commit itself to a contract (the word 'firm' is from the Italian, *firma* – a signature). 'Doing business' includes activities involved in the purchase or sale of commodities, or in related financial transactions. 'The business' is thus the decision maker in supplying goods and services (Jenson, 2003).

Founding a business, deciding on the structure of ownership, and determining the core business activities are all critical strategy decisions and should be undertaken with great care. This chapter does not intend to look in great detail at these issues. It simply introduces the concepts to ensure a contextual framework is in place in the reader's mind for the rest of the book. If you are interested in following up on this subject you could read further – perhaps starting with the 'Business Owner's Toolkit' found at www.toolkit.cch.com/text/P01_0000. asp

Understanding the main structures associated with business ownership is useful in a book dealing with eBusiness issues because the structure of the business you are working for – or indeed own – will quite possibly be reflected in the type of electronic enablement you might experience.

Essentially, there are three kinds of firms in modern market economies: **proprietorships**, **partnerships**, and **corporations**.

A *proprietorship* (or proprietary business) is a business owned by an individual, the 'proprietor', who is almost always the decision maker for the business e.g. The Tenterfield Saddler (www.tenterfieldsaddler.com). Some proprietorships are too small even to employ one person full time. At the other extreme, some proprietary businesses employ many hundreds of workers in a wide range of specializations e.g. Glenlogan Park Stud (www.glenloganstud. com.au/) (Partridge, 2004).

A *partnership* is a business jointly owned by two or more persons e.g. The Peanut Van (www.peanutvan.com.au) – see Byte Idea this chapter. In most partnerships, each partner is legally liable for debts and agreements made by any partner. There are also 'limited partnerships' in which some partners are protected from legal liability for the agreements made by others, beyond some limits. In many cases, one partner is designated as the managing partner and is the main decision maker for the business.

A *corporation* has two characteristics that distinguish it from most proprietorships and partnerships:

1. *Limited liability* where the owner of shares in a corporation cannot lose more than a certain amount if the company fails
2. *Anonymous ownership* where the owner of the shares can sell them without getting the permission of anyone other than the buyer.

In a typical corporation, e.g. IBM (www.ibm.com) or Dell (www.dell.com) – see Byte Idea this chapter, the shareholders formally elect a Board of Directors, who in turn select the officers of the company. One of these officers, often called the 'President' or 'Chairperson', will be the principal decision maker for the firm, and will be expected to make decisions in the interest of all the stakeholders in the business, but most particularly the shareholders (Jenson, 2003).

Stakeholders (Fig. 1.1) are all those people inside and outside the business who have a vested interest or 'stake' in the business and include suppliers, buyers, distributors, customers, employees and shareholders. While all stakeholders are important to a business – indeed a business could not 'do' business without its stakeholders – the shareholders are regarded as particularly important because they actually own shares in the business and thus have significant interest in how a business is being run to maximize their investment in it.

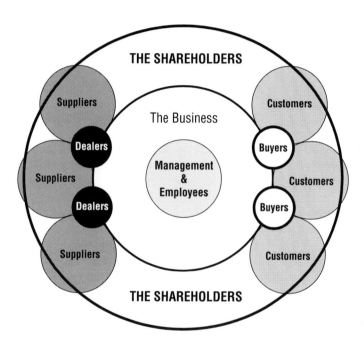

Fig. 1.1. Stakeholders in a business

Byte Idea – The Peanut Van
www.peanutvan.com.au

Electronic Enablement and
a New Business Opportunity

The Peanut Van is a well-known landmark in Kingaroy, Queensland – the caravan has been selling locally grown peanuts on Kingaroy Street on the southern side of town for over 30 years. In 1999, Rob and Chris Patch, who are local peanut growers, bought the business.

Major goals of the new owners were to maintain the quality of the product (only 'Jumbo' grade peanuts are used which are bought on contract from the Peanut Company of Australia on a weekly basis after aflatoxin testing, cleaning and blanching has been achieved), to maintain strict environmental controls on the product (all processing, e.g. roasting, flavouring and packaging, is done in house), and to grow the business. Currently an average of 40 tonnes/year of peanuts are sold from the van.

To supplement their successful mail order business, The Peanut Van website was launched in May 2000 as an information and transaction based website. This has developed into a major marketing tool and a successful eCommerce site that has allowed The Peanut Van to expand its markets to Brisbane (guaranteed delivery in 24 hours), other domestic markets, and to deliver overseas (America in particular).

Why has The Peanut Van been so successful as an eBusiness? Basically the owners have used a mixed mode of business (offline and online) based around product quality and speedy and timely service. The established offline presence of The Peanut Van and the new online website presence which makes use of The Peanut Van's established brand name, combine to extend the reach of their message about a quality product to an international market. The basic business principles involved are backed up by attention to quality product, timely service and good governance of the website with clear online security and privacy policies that have allowed confidence and trust to be developed in potential customers.

In essence, The Peanut Van has vertically integrated every component of its industry agri-food chain (supply, processing, packaging and delivery) and, using new technologies, has extended its markets and is growing the business very successfully.

Information sourced from author interview with Rob and Chris Patch and The Peanut Van website.

What Is Agribusiness?

The term agribusiness is a generic one that has encompassed a number of different definitions reflecting the changes that have taken place in the agricultural sector over the last 50 years. The term has moved from referring to *a large-scale farming enterprise* through *mass production of agricultural goods and services through mechanization* to the modern day terminology where agribusiness includes *the agricultural input sector, the production sector and the processing-manufacturing sector*. In the USA the term can have negative connotations to some in that it has been used synonymously with the term 'corporate farming' used to describe the demise of the family farm (Wikipedia, 2005). In this book the term agribusiness is taken as a generic term that refers to the various businesses involved in food and fibre production.

An important issue for agribusinesses is that they deal in low margin commodities where competitive market forces have typically resulted in the cost of production being very close to the value created, thus leaving relatively thin profit margins (Boehlje, 1999). Further, production is directly affected by climate and the resulting weather which is notoriously fickle and very often results in a variable supply of raw product. Ensuring constant volume and high quality product at the right time and price is a driving business force in domestic and international agri-industry chains. As a result, supply and value chain analysis of agri-industry chains has become a valuable tool in determining where added competitive advantage can be generated for these industries (O'Keefe, 1998; Holsapple and Singh, 2000; Dunne, 2001).

What Is an Agri-Industry Chain?

Businesses do not exist in isolation. Every business has suppliers of goods it needs, and buyers of the goods it makes/sells – each having the same driving forces and critical responses. The grouping of these businesses is called a 'chain' of companies and tends to reflect the industry the businesses are involved in. This concept is discussed in more detail in Chapter 4, but is introduced briefly here for the sake of completeness.

The agri-industry sector is a large, multifaceted industry sector that exists worldwide, and involves a range of businesses that create industry-specific (e.g. grains, sugar cane, timber, dairy, cattle/meat, fruit and vegetables, cotton, wool, soybean to name a few) agri-industry chains that often exist across international boundaries. Agribusinesses in such chains include:

- *Input suppliers* (agricultural chemical and fertilizer companies such as Incitec/Pivot, Bayer and Cargill)
- *Service providers* (banks, R&D organizations, governments, consultants)
- *Producers* (growers of grain, meat, dairy, cotton, fruit, nuts, etc.)
- *Traders* (companies such as Mitsubishi, Australian Wheat Board, Bohemia Nut Company, ConAgra, etc.)
- *Processors* (companies such as Peanut Company of Australia, Flour Mills, Parmalat, etc.)
- *Manufacturers* (companies such as Kraft, Mars, Cadbury Schweppes, Unilever)
- *Retailers* (supermarkets such as Tesco, Carrefour, Woolworths and Coles as well as smaller retail outlets)
- *Logistics support companies* (transport and storage companies such as FedEx, TNT).

Generically, industry chains are classified as either *supply* or *value* chains. In this book the following definitions within the general term 'industry chain' are used.

Supply Chains

A **supply chain** (Fig. 1.2) is taken to mean the physical flow of goods that are required for raw materials to be transformed into finished products. A related term **supply chain management** is about making the chain as efficient as possible through better flow scheduling and resource utilization, improving quality control throughout the chain, reducing the risk associated with food safety and contamination, and decreasing the agricultural industry's response to changes in consumer demand for food attributes (Dunne, 2001).

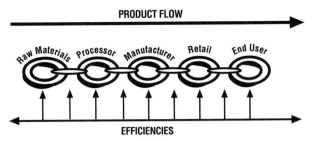

Fig. 1.2. A generic supply chain

Value Chains

A **value chain** (Fig. 1.3) is taken to mean a group of companies working together to satisfy market demands which involves a chain of activities that are associated with adding value to a product through the production and distribution processes of each activity. An organization's competitive advantage is based on its products' value chains. The goal of the company is to deliver maximum value to the end user for the least possible total cost to the company thereby maximizing profit (Porter, 1985). Subsequently, **value chain management** is about creating the added value at each link in the chain and a sustainable competitive advantage for the businesses in the chain.

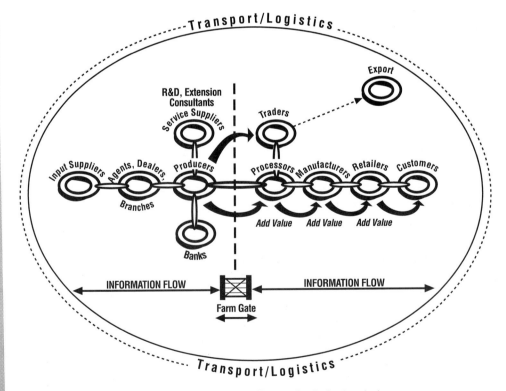

Fig. 1.3. A generic Australian grains industry chain

What Are the Core Drivers of Doing Business?

The core concept of doing business is about buying and selling goods and in the process making money. To make money you have to minimize the costs of doing business and maximize the profits associated with it. From a theoretical standpoint, this philosophy is represented in the 'Theory of the Firm'.

The Theory of the Firm

The traditional Theory of the Firm is an economic one based around the core concept that supply and demand for a product or service are rational decisions by a (primarily) self-interested individual or organization (Clarke, 1907). The firm's or business's goal under this theory is to *minimize costs* and *maximize profit*. In a seminal paper on 'The Nature of the Firm' Coase (1937) stated that a company will tend to expand until the costs of organizing an extra transaction within the company become equal to the costs of carrying out the same transaction by means of an exchange on the open market, or the costs of organizing the transaction in the company. In essence this still holds true in today's global economy although companies have had to adapt to the changing nature of the marketplace – and ways of making a transaction.

Over the years, a number of different theories have been developed to provide a perspective for thinking about organizational objectives. The Resource Based View (RBV) of the firm is one of the most recent and widely recognized theories, as it encompasses the notion of competitive advantage and what is needed to maintain competitive advantage in an ever changing world (Montgomery, 1995; Fahy and Smithee, 1999).

Resource Based View of the Firm

The Resource Based View (RBV) is primarily a theory of competitive advantage (Barney, 2001). The assumption is that the desired outcome of managerial effort within the firm is to create a sustainable competitive advantage which itself allows 'above average' returns compared to the company's competitors. RBV focuses on how firms achieve and sustain competitive advantage and contends that the possession of certain key resources which create 'value' or 'barriers to entry', as well as how effectively the firm deploys these resources in its product markets, is what gives a competitive advantage. Resources can be classified into three broad categories (Collis and Montgomery, 1997):

- *Tangible assets* – appear on the company's balance sheets and include real estate, production facilities and raw materials. Although intrinsically important they are not normally regarded as sources of competitive advantage.
- *Intangible assets* – include brand names, cultures, technological knowledge, patents, trademarks, logos, etc. These assets can play an important part in competitive advantage and are important because they are not 'consumed' by usage.
- *Organizational capabilities* – these are complex combinations of assets, processes and people which are used to transform inputs into outputs.

Not all resources are of equal importance or possess the potential to be a source of sustainable competitive advantage. Advantage creating resources are those that meet four conditions:

1. *Value* – measured by total revenue
2. *Rareness* – resources that are hard to come by
3. *Inimitability* – resources that are difficult to copy
4. *Non-substitutability* – resources that cannot be substituted by others.

How these resources are manipulated and managed will either make or break a business, and is discussed in various chapters of this book as they relate to electronically enabling a business. The driving forces and critical responses of a business are summarized in Table 1.1.

Table 1.1. The driving forces and critical responses of a business

Driving Forces	Critical Responses
• Buy and sell goods • Make money • Minimize costs and maximize profits • Create a sustainable competitive advantage	Monitor and manage three key sets of resources: 1. Intangible assets 2. Tangible assets 3. Organizational capabilities

Maximizing Profitability

As discussed earlier, the primary intent of investors in a business is to earn a return on their investment (ROI) and it is the pursuit of these returns that commands the attention of business managers. The task, in principle, is quite straightforward: deploy the resources available to the business to maximize the profitability of the business for the minimum investment of capital.

In simple terms, business profitability arises from the management of the business to ensure that the revenue generated through sales of products and services is higher than the costs incurred in providing those products and services to customers. The revenue is dependent upon the volume of products and services sold and the price of those items sold. The costs consist both of the direct cost of supplying those products and services and also the overhead costs of running the business.

The other part of the return on investment equation is the capital deployed or invested in the business. This is manifested as fixed assets such as plant and equipment and working capital invested in inventory and receivables.

A measure that business managers use to bring all this together is *return on assets* (ROA). This is an operational measure that focuses on the returns achieved by the business independent of financing decisions and taxation. Throughout the financial year the performance of the business on this measure (and its component measures) is monitored against a previously defined budget or even last year's performance. ROA is calculated as:

$$\text{ROA} = \text{EBIT} / \text{Total Assets}$$

EBIT or Earning Before Interest and Tax is a measure of the operational profitability of the business and, as Figure 1.4 demonstrates, reflects the combined performance of the business in generating sales revenue, containing the cost of goods sold, and managing overhead costs.

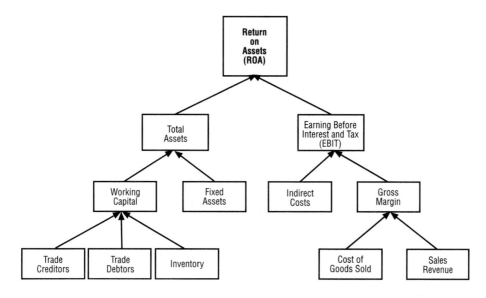

Fig. 1.4. Illustration of the flow of returns involved in obtaining return on assets (ROA)

Practical experience, however, reveals a rich tapestry of complex issues that at times makes the apparently straightforward pursuit of getting a good return on investment quite challenging. The following sections look at some of the main issues involved.

Value Propositions and Customer Satisfaction

Generating sales revenue and hence business profitability relies on providing customers with products and services that allow them to achieve their desired outcomes at a relative cost that is perceived to provide superior benefit for that cost. This is the essence of a successful customer value proposition and it relies heavily on a detailed understanding of the market and customers the business is serving.

Value is not narrowly defined in financial terms but more broadly to include all aspects of the interaction of the customer with the business. Attributes of value include product quality, time taken in presales and sales transaction activities, after sales service and product utility. With a clearly defined value proposition to be offered to customers, management's challenge is to deploy the resources of the business to deliver customer value.

Resource Allocation

In a business context, resources refer to both human and economic capital. The criteria for the allocation of these resources are once again straightforward in theory and complex in practice.

Theoretically, business resources should be deployed on those activities that will contribute the most to generating profit and therefore value for shareholders. In practice constraints such as capital availability, cash flow management, timing of expenditure and value creation all serve to complicate the decision process. By their very nature, these decisions are made with a view to an improved outcome for customers or shareholders or both in the face of an uncertain future. It is at these decision points that senior managers really earn their salary as they combine financial and market assessments with interpretation of the business strategy to determine the deployment of company resources. It is the cumulative effect of these resource allocation decisions that will ultimately determine the success or otherwise of the business.

Supply Chain Optimization

Supply chain and logistics managers are the business world's aficionados of KPIs (key performance indicators). They live by the mantra 'you can't manage what you don't measure'. Their task is to simultaneously ensure that customer service levels are maintained as agreed with the marketing and sales team, while also minimizing the cost of moving and storing products as agreed with the accountants, while minimizing inventory levels to ensure working capital targets are met as agreed with senior management! They often apply sophisticated scientific quantitative methods to plan and execute this difficult balancing act and measure the parameters critical to the processes they control. Their challenges arise from the vagaries of the real world, like customers changing their orders, suppliers having transport problems, the sales team inaccurately forecasting sales demand, and, in agribusiness in particular, the climate playing havoc with raw material production.

Variability is a supply chain manager's enemy and their chief weapon in the battle is information. Without good quality information – the creation, capture and manipulation of which has seen the development of some very powerful software systems over the last ten years – their job in an increasingly information-centric business world would be impossible. The internet provides supply chain managers with the vehicle to extend their reach outside of their own organization up and down their supply chain. With information from suppliers, distributors and customers these managers can continue to improve their operation – see Byte Idea on Dell this chapter.

Cost Management

The revenue and customer value propositions have been discussed as key drivers of profitability. The other component of the profitability equation is costs. Although regarded by some as less virtuous and unexciting, cost management is fundamental in the creation of shareholder value. It is more common than you would imagine to find businesses achieving success in the marketplace and yet under-performing at the bottom line, due to a lack of focus on and control of costs. This is discussed further in Chapter 2.

Byte Idea – Dell
www.dell.com

Built to Order

Dell was founded in 1984 by Michael Dell who gets all the credit for inventing the build-to-order model that everybody else, from car companies to cosmetics makers, now imitates on the internet. This direct business model is based around the concept that by selling computer systems directly to customers, Dell can best understand their needs and efficiently provide the most effective computing solutions to meet those needs. The model allows the company to build every system to order by letting consumers create their own configuration for a PC at Dell.com, (and for the company's corporate customers, PremierDell.com) as well as offering customers powerful, richly-configured systems at competitive prices since it eliminates retailers (middlemen) that add unnecessary time and cost. Dell also introduces the latest relevant technology much more quickly than companies with slow-moving, indirect distribution channels, turning over inventory in less than four days on average.

Dell led commercial migration to the internet, launching www.dell.com in 1994 and adding eCommerce capability in 1996. The following year, Dell became the first company to record $1 million in online sales and by 2000, company sales via the internet reached $50 million per day. Today, Dell operates one of the highest volume internet commerce sites in the world, with the company's website receiving 650 million page requests per quarter at 86 country sites in 21 languages/dialects and 40 currencies and with a revenue for the last four quarters to mid 2005 of US$52 billion.

Dell sells its products and services worldwide and the company is increasingly realizing internet-associated efficiencies throughout its business, including procurement, customer support and relationship management. At www.dell.com, customers may review, configure and price systems within Dell's entire product line; order systems online; and track orders from manufacturing through shipping. At Valuechain.dell.com, Dell shares information with its suppliers on a range of topics, including product quality and inventory.

Dell's corporate headquarters are in Round Rock, Texas as it is also for Dell Americas, the regional business unit for the United States, Canada and Latin America. Dell has other regional headquarters in Bracknell, England, for Europe, the Middle East and Africa; in Singapore to serve the Asia-Pacific; and in Kawasaki, Japan, for the Japanese market.

Dell is a publicly traded company on the NSADAQ under DELL and in 2005 ranked No. 84 on the Fortune Global 500 and No. 1 in the U.S. and No. 3 globally on the Fortune 'most admired' lists of companies.

Information sourced from Fortune Global (www.fortune.com) and the Dell website

Commercial Relationships

Porter (1985) characterizes business as a battle of value chains. Underpinning these value chains are fundamental relationships between suppliers, distributors and customers. Ultimately the success of a particular value chain will be determined by the quality of the relationships that exist between participants in that value chain. The efforts of highly sophisticated supply chain managers can be undone by a lack of trust or willingness to cooperate.

The objectives of value chain partners can at times be divergent. While the objective of any value chain is to create value for the end customer, there is a struggle over how much value each of the participants can capture for their shareholders along the way. Further, each of the individual value chain participants may be part of multiple value chains that can have differing priorities, which lead to sub-optimal outcomes for the value chains that are less profitable for that business.

The Building Blocks of Business

There are three basic building blocks for any successful business, whether they are 'Old' or 'New Economy' businesses. These are the *processes, people* and *technology* of the business. Processes are sometimes called work practices or operations – these are the tasks and the sequences by which work really gets done. People are the lifeblood of any company – people work within defined or de facto business processes to meet customer needs. The technology of a company enables work to be accomplished more effectively once it is known *how* work is to be done, and how the people within the processes can be best used.

Managers need to be aware that all three are linked and that the task is to ensure compatibility of needs, wants and requirements across all three to create a useful business model which defines how the organization will conduct its business. In addition it should be noted that technology *enables*, rather than drives, business operations.

Processes

Processes are the plans, tasks and sequences by which work really gets done. They include the development of a business model and business processes that are specific to that business as well as the undertaking of tasks associated with ensuring a good return on investment for stakeholders.

The Business Model

Optimizing organizational operations requires a model of the business. This model must represent the composition of the business and reflect the competitive issues of timeliness, cost and quality. Its components must be able to be adjusted and the model analysed to examine the effect of changes. There is no formal definition of a business model but Peterovic *et al.* (2001) show that it is taken as an organization's core logic for creating value in a sustainable way – that is, as Hannon and Atherton (1998) point out, for making money over a long period of time, not just the next three to twelve months.

The business model is more than just an organizational structure, a set of pricing policies or the creation of a value proposition – it is the structural design which interlinks all aspects of creating value and revenue for the company and for how the company will appropriate that revenue – see Figure 1.5.

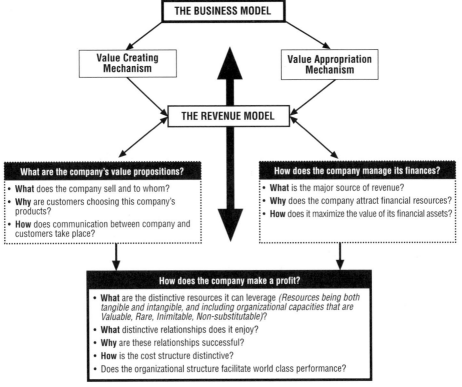

Fig. 1.5. A generic business model flow chart

The process of modelling systems such as a business allows the organization to experiment with change, and to see the effects of success or failure, before implementing change (Sterman, 2000). This capability is key for organizations considering significant transformations such as becoming an electronically enabled business. A formalized business model helps managers to:

- Identify and understand the relevant elements in a specific domain and the relationships between them (Morecroft, 1994)
- Easily communicate and share their understanding of the business among other stakeholders (Fensel, 2001)
- Facilitate change by mapping and using the business models as a foundation for discussion
- Identify the relevant measures to follow in a business, similar to the Balanced Scorecard Approach of Kaplan and Norton (1992).

Business models themselves do not offer solutions, rather it is how the business is run which will determine whether competitive advantage and success can be achieved. Further, business models must and do change over time, particularly in the present-day rapidly evolving business environment. Redefining a business model is fraught with peril, however, particularly if technological innovation is being forced upon the business arena as is the case with internet-enabled business, and the process must be addressed with care.

Historically, business models have been a hierarchical representation of the organizational structure, which related little to timeliness, cost and quality issues. However, the reality today is that rather than looking at the organization vertically (using a hierarchy), an alternative view is to look horizontally, by analysing the business processes that flow across organizational boundaries.

Business Process Modelling

Every company has its own business processes, but what exactly are they? A business process is a specific ordering of work activities across time, with a beginning, an end, and clearly identified inputs and outputs. A process is essentially a structure of action which is initiated by an event, consumes resources, performs activities, and produces goods or services.

The event component of the business process allows for an analysis of timeliness, the resources consumed allows for the analysis of associated costs, and the activities performed leads to analysis targeting process quality. Finally, the resulting goods and services lead to the analysis of product quality. True business improvement is driven by optimizing the processes that a business performs.

Business process modelling is a business process in its own right – it defines the business process/s and subdivides it/them into activities and sub-processes which can be studied and improved iteratively. In Figure 1.6 note that the business process is defined in a *process definition* stage and is managed by a *workflow management system* and that activities can be manual (require human intervention, e.g. perform an onsite service) or automated (does not need human intervention, e.g. generate a statement).

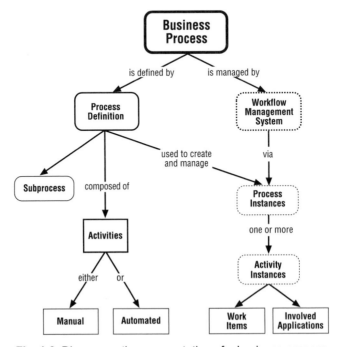

Fig. 1.6. Diagrammatic representation of a business process

Business process modelling is a method of documenting the business architecture of the organization and techniques include:

- *Data or work flow diagrams* – represent the flows of data/work between different processes within a system and describe what users do rather than what computers do. Context diagrams are used to scope the system within which the processes are modelled.
- *Flow charts / data flow charts* – diagrams expressing the sequence and logic of procedures using standardized symbols to represent different types of input and output, processing and data storage.
- *Structured English* – a way of representing the precise logic of a procedure by writing out that logic using a few limited sequences of words, iteration and if-then-else formats.

Business Process Re-Modelling for Electronic Enablement

Typical objectives for wanting to move a company into the electronically enabled business environment are: to reach new markets, eliminate intermediaries and redundancy, lower transaction costs, provide convenience to external parties, reduce delivery times, and personalize and improve customer service.

To electronically enable a company and improve customer, supplier, business partner and employee interactions, a business's existing relationships with internal and external parties need to change. This change requires a redesign of the business processes supporting these relationships, and, in particular, system interfaces must be redesigned to provide straightforward support for new and untrained users. Typical electronically enabled business applications that have significant underlying business process redesign include:

- *Image management.* Creating a web presence allows an organization to define its image in the eLandscape marketplace.
- *Information and order management.* Perhaps the most desirable of all electronically enabled business capabilities is the more efficient management of company information as well as the management of electronic transaction. Chapter 3 discusses this in more detail.
- *Customer relationship management.* Various interactions with the customer can be greatly enhanced by providing customers with tailored information when they require it. This is discussed in more detail in Chapter 4.
- *Supply chain management.* A fully electronically integrated supply chain provides efficiencies and transaction cost reductions at every link in the chain. Chapter 4 discusses this in more detail.
- *Human resource management.* Enterprise information portals (EIPs) enable employees to access appropriate information associated with the company or themselves. EIPs are discussed in more detail in Chapter 3.
- *Finance management.* Many of these types of transactions are handled through electronic data interchange (EDI) and electronic funds transfer (EFT) which are discussed in Chapter 4.

Essentially, however, in moving a business towards electronic enablement the business process modelling scenario is no different from any other business process remodelling situation (see Fig. 1.7) and involves:

- Defining the scope and objectives of the project
- Defining and documenting the current processes
- Reviewing and verifying the current processes
- Defining the eBusiness requirements
- Doing a gap analysis on what is currently in place and what is required for electronic enablement
- Developing new processes
- Undertaking an impact analysis of the proposed new processes in terms of timeliness, resources and costs.

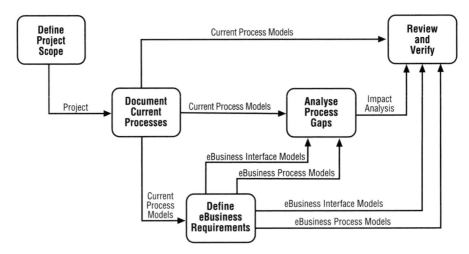

Fig. 1.7. A generic 'E' enablement process remodelling flow chart

People

Without a motivated, empowered and focused team, business success is illusive. Leadership and change management are perhaps the most important issues to deal with and while outlined briefly here for completeness, they are dealt with in Chapter 8 in more detail, particularly in relation to electronic enablement of businesses.

Leadership

The role of quality leadership at all levels within an organization cannot be understated. The leader's ability to present the vision for the future and explain effectively the impact of pursuing that vision on individuals within a business is critical to ongoing success. Without that clarity of purpose and alignment, organizational performance will suffer in the face of new challenges like electronic enablement of the business.

It is important to note that the challenges that electronic enablement of the business presents to leadership and employees are the same as the challenges of the introduction of any new technology or for that matter any other change. The fundamentals of good leadership remain the same.

Change Management

Change management is a term used often and seldom considered. IT project managers will tell you it's about the introduction of new processes and the training of people. Management consultants will tell you it's about leading people as they deal with endings and beginnings and experience the emotions of a 'grief cycle' (Owen, 1994).

Whatever your perspective, effective change management is critical in dealing with the impact of electronically enabling a business. It's about leaders providing clarity and direction and it is about providing structured mechanisms to allow people to 'get on board' with new processes and work practices. In the end, people are still people and electronically enabling the business is just another change to be dealt with.

Technology

Essentially, what has been discussed in previous sections is not new – 'doing business' has been going on for centuries and people have both made and lost money in the process. Why then is it that today 'doing business' seems to have become extremely complicated and time consuming?

The answer will be different depending on who you ask, but almost certainly will include the fact that doing business in general in the 21st century not only means adhering to the basic 'minimize cost and maximize profit' maxim outlined by Porter (1985), but in addition businesses need to address the issues of doing business in a world that is rapidly becoming more electronically enabled and dependent. In short, businesses are now doing business in an electronic or *eLandscape*.

The eLandscape

It has become increasingly clear that the internet and associated electronically enabled business practices are, despite the demise of the dot.coms, business-supporting technologies that are here to stay. They form, along with other sophisticated electronic communication devices and associated systems (Table 1.2), an electronic landscape or 'eLandscape' (Fig. 1.8) within which business is carried out (Ryan, 2000; Porter, 2001; Bryceson and Pritchard, 2003; Schiefer, 2003; Bryceson and Kandampully, 2004).

Table 1.2. Technologies associated with the eLandscape in which businesses find themselves in the 21st century

Technology	Function
Telephone, fax, mobile (SMS messaging)	Communication
Personal digital assistants (PDAs, e.g. palm pilots), PCs, computerized office systems and software, data warehouses	Data storage and manipulation, word processing, spreadsheets, accounting systems
Business information systems (e.g. transaction processing systems, management and executive information systems, decision support systems)	Managerial information sources and decision making
Internet, intranet, WANs, LANs, email	Communication, electronic business information flow
Electronic transaction processing (e.g. electronic funds transfer, EFTPos, BPay, internet banking, web pages	Electronic money movement, online transaction and marketing
Electronic data interchange (EDI)	Movement of information via electronic networks – primarily used in ordering and purchasing
Material resource planning (MRP) and electronic resource planning (ERP)	Complete enterprise wide data and information manipulation resource
Customer relationship management (CRM)	Not a technology per se but more a business strategy associated with the collection and use of customer information
Electronic demand forecasting and supply chain management	More accurate forecasting of demand and supply based on real time electronic information flows
Vendor managed inventory (VMI)	A means of optimizing supply chain performance where the manufacturer or retailer is responsible for maintaining suppliers' inventory levels by having access to the suppliers' inventory data and is responsible for generating purchase orders
Electronic marketplaces	Bring together suppliers and buyers into a virtual trading space. Public eMarketplaces have not been successful in the agri-industry sector. Private eMarketplaces are more so.

A number of the technologies listed in Table 1.2 are well known to people (e.g. telephone, fax, computerized office systems and the internet), but others are less so (e.g. intranet, extranet, LAN and WAN, plus some of the more complex business information systems). More detail on these particular technologies and the business information systems listed in the table can be found in Appendix 1. Technologies and concepts such as electronic data interchange (EDI), electronic resource planning (ERP), customer relationship management (CRM), electronic demand forecasting and supply chain management are explained in greater depth in Chapter 4.

Fig. 1.8. The electronic landscape or eLandscape within which business is carried out
Graphics: Chris Hall, 2005

Electronic Device Connectivity

A key issue in terms of the eLandscape is connectivity – that is, what is connected to what, why and how, to ensure that business can be accomplished easily and efficiently. Computers, entertainment systems and telephones along with their various parts and systems make up a community of electronic devices. These devices communicate with each other using a variety of wires, cables, radio signals and infrared light beams, and an even greater variety of connectors, plugs and protocols. The bringing together of more and more complex systems is an important part of successful electronic enablement, particularly when dealing with supplier and customer businesses. Franklin (2000) outlines three main issues you need to be aware of in relation to electronic device connectivity:

1. *The physical requirements*. For example, will the devices talk over wires or through some form of wireless signals?

2. *The data transmission requirements*. For example, information can be sent 1 bit at a time (serial communications), or in groups of bits (usually 8 or 16 at a time and known as parallel communications).

3. *The protocol used*. All of the parties in an electronic discussion need to know what the bits mean and whether the message they receive is the same message that was sent. In most cases, this means developing a language of commands and responses known as a protocol.

DTN – see Byte Idea this chapter, is a company that has seen a niche market in this information delivery arena in the agricultural sector.

Byte Idea –
Data Transmission Network (DTN)
www.dtn.com

Agmarketing Information via mCommerce

DTN is a private business to business provider of on-demand, real-time information services in the USA, and serves a variety of industries that are weather sensitive – including agriculture, commodity trading, energy and public safety.

The weather affects everything in agriculture (e.g. when to plant crops, apply pesticides, irrigate fields and harvest produce) as well as the supply and market price. Thus a service that provides real-time location-specific weather information as well as market prices and agri-industry news is invaluable. With over 120,000 subscribers through its four brands – DTN Ag, DTN Energy, DTN Market Access and DTN Meteorlogix – DTN is cornering the market.

In particular DTN Ag with the development of DTN ProduceTM provides location-specific weather, marketing and pricing information to help maximize operations and control costs, and with the development of DTN MobileTM enables the access and download of real-time market quotes, weather and news via a WAP 2.0 compatible cell phone or mobile device such as a PDA when using a carrier service plan that provides access to the mobile web (i.e. a producer can check weather radars and forecasts, commodity quotes, market commentary, and agriculture news via the phone and download that information right in the cab of their truck or tractor).

DTN advertises DTN MobileTM as being compatible with any mobile device that has a wireless web service and a WAP 2.0 web browser and as a mobile companion to its existing online and satellite services.

Information sourced from AgriMarketing 1 October 2001, DTN web page (www.dtn.com)
and DTN MobileTM and DTN ProduceTM brochures

The next section briefly outlines the basic principles of electronic device connectivity including infrared and WAP technologies. There are many specialist books and articles available in this area. If you would like more in-depth analysis you should access these. A good starting point is www.findarticles.com

Physical Requirements

This is about how the communication will happen between two (or more) devices – how will the electronic devices be connected and thus be able to talk to each other? Connection implies a medium such as air, wire or fibre of some sort that provides the communications system with a channel resource (e.g. bandwidth) and a systems architecture that can ensure the signaling between the devices takes place.

Most stand alone systems are wire connected (computer to printer, and/or scanner, TV to video/DVD player, etc.) but as soon as you get many devices linked together – particularly computer systems in different rooms and perhaps in different buildings – then wires can become an issue. There are two main ways of getting around connecting devices with wires: the first is to carry information between devices via **infrared light** and the second is the **wireless application protocol or WAP.**

Infrared is currently used in most television remote control systems and it is also used to connect some computers with peripheral devices. Infrared communications are generally reliable and are inexpensive to build into a device, but there are drawbacks – for example, infrared is a 'line of sight' technology (i.e. you have to point the remote control at the television to make things happen). The second drawback is that infrared is almost always a 'one to one' technology – that is, you can send data between your desktop computer and your laptop computer, but not your laptop computer and your personal digital assistant (PDA, e.g. palm pilot) at the same time.

Wireless application protocol (WAP) is a set of open, international communication protocols that define a network architecture for content delivery over wireless networks (Yuzdepski, 2000). The protocol was originally developed by four companies (Ericsson, Motorola, Nokia and Unwired Planet – now Phone.com), to facilitate mobile device users in accessing online (web-based) services. It has revolutionized the personal connectivity market by providing freedom from wired connections – enabling links between mobile computers, mobile phones, portable handheld devices, and providing connectivity to the internet.

WAP includes the concepts of browsers, servers, URLs and gateways similar to the online environment. On a mobile device, a WAP works by replacing a web browser with a WAP browser (Oak, 2002), which can request data from a website and which can translate the HTML (hypertext markup language) data of the website into wireless markup language (WML) viewable text and graphics. The WAP browser requires a WAP gateway to function as an intermediary between the mobile and internet networks. When placed between a WAP browser and a web server, this takes care of the necessary binary encoding of content and can also translate WML to and from HTML.

In the near future, almost every mobile phone and many PDAs on the market will be WAP enabled. The only limitation is that as long as mobile phones are tiny, practical and easy to carry in your pocket, the WAP display has to be small. This means that you can't read through the same amount of information as on a computer, and the screen quality is not as good. You can send pictures with WAP, but they are not of a high quality. Currently, you can use your WAP phone to read news, order tickets, find schedules for trains and buses, book a plane ticket, read email, play a mobile game and share information as well as to access other useful WAP services like maps, dictionaries, city guides and business pages. Bluetooth™ – see Byte Idea this chapter, is one of the biggest players in the wireless technology arena

Data Sending Requirements

As previously stated, information can be sent 1 bit at a time (serial communications), or in groups of bits (usually 8 or 16 at a time – parallel communications). A desktop computer uses both serial and parallel communications to talk to different devices: modems, mice and keyboards tend to talk through serial links, while printers tend to use parallel links.

Byte Idea – Bluetooth
www.bluetooth.com

The Standard of the Future

Bluetooth is an industrial standard which provides a way to connect and exchange information between devices like personal digital assistants (PDAs), mobile phones, computer gear and even digital cameras via a secure, low-cost, short range radio frequency. The specification was first developed by Ericsson (www.ericsson.com) and was later formalized by the Bluetooth Special Interest Group (SIG) which was established in 1999 by Sony Ericsson, IBM, Intel, Toshiba and Nokia, and later joined by over 1000 other companies.

Bluetooth gets around the problems that come with both infrared and cable synchronizing systems by communicating on the licence-free frequency of 2.45 gigahertz which has been set aside by international agreement for the use of industrial, scientific and medical devices (ISM). Bluetooth was primarily designed for low power consumption, with a short range 10 cm, 10 m, 100 m or up to 400 metres (depending on power class) and with a low-cost transceiver microchip in each receiving device.

One of the ways Bluetooth technology may become even more useful in the future is in Voice over Internet Protocol (VOIP – also called Internet telephony). When VOIP becomes more widespread, Bluetooth may then end up being used for communication between a cordless phone and a computer listening for VOIP.

The latest version of Bluetooth technology currently available to consumers is 2.0.

Information sourced from SIG website and Wikipedia

The Protocol to Be Used

All of the parties in an electronic discussion need to know what the bits mean and whether the message they receive is the same message that was sent. In most cases, this means developing a language of commands and responses known as a **protocol** (see Appendix 1 for more details). Some types of products have a standard protocol used by virtually all companies so that the commands for one product will tend to have the same effect on another. Modems fall into this category. Other product types each speak their own language (e.g. printers), which means that commands intended for one specific product will seem gibberish if received by another.

Doing Business in the eLandscape – eBusiness, eCommerce and mCommerce

In today's global marketplace, the impact of technology has pervaded almost all fields of business. In the manufacturing industries, technology has gone beyond its initial role as a substitute for humanpower to become a value-adding component. In service industries it is a unique tool to enhance services in all aspects of the service business's endeavours, whether these be operations, marketing or human resources. The emergence and adoption of internet technology as an electronic medium of communication and commerce is bringing about changes not only in how firms operate and conduct business but in how they compete and offer greater value to all stakeholders.

The 'unique selling point' of the internet is that it provides a timely and efficient method of communication and information flow 24 hours x 7 days a week (24 x 7). The quantum advances in technology, communication and digital science that have taken place over the last five to eight years have enabled firms to transcend the challenges previously posed by distance and time, and to convert them into potential opportunities. The creative utilization of distance and time, referred to as the 'death of distance' through the impact of technology, constitutes a major economic force that will shape many businesses and countries around the globe by proffering an entirely new set of value enhancement options (Holsapple and Singh, 2000; Phan, 2003).

In addition to the impact of new communication technologies on individual businesses in any industry sector, the potential impact on the supply and value chains of those businesses and the whole industry sector in which they trade, is a major consideration for the likely future success or otherwise of those industry sectors (Anderson and Lee, 2000, 2001). Gaining both customers and partnering businesses' acceptance and loyalty as well as focusing on how their needs can be met is one of the primary preoccupations of managers today. This is discussed further in Chapter 4 with particular attention being paid to what this means for agribusiness as more and more companies look to extend their reach in terms of market and product development.

eBusiness versus eCommerce versus mCommerce

One of the legacies of the dot.com boom is not only the new tools for conducting business it left behind – but also the jargon it generated. There was a period in 1999 when it was just about impossible to keep up with the new phrases, terminology and TLAs (three letter abbreviations) that the dot.com revolution was generating. Driving this wave of reengineering of the English language was a remarkable explosion of business models and technological innovations that continue to create opportunities for business years after the hype and excitement have all but disappeared.

However, the jargon (Table 1.3) that remains can create confusion, particularly when the terms eCommerce and eBusiness are used where there is a significant difference in both the meaning and electronic enablement complexity associated with the two terms. The definition used in this book for **eCommerce** is:

> **The buying and selling of products over the internet – initiatives are focused on the business to consumer (B2C) model.**

The definition used in this book for eBusiness is:

> **eBusiness embraces all aspects of electronically enabling a business – initiatives are focused on the organization's core competencies and the business to business (B2B) model. The buying and selling of goods does not have to be via the internet but will be facilitated by electronic technologies and systems within and between businesses.**

Table 1.3. Common eBusiness and eCommerce classifications. A comprehensive set of 'E' jargon definitions can be found at: www.wiredmedia.co.uk/docs/64.html

Classification	Definition	Example
Business to consumer (B2C)	Retailing transactions with individual shoppers	Amazon.com (www.amazon.com)
Business to business (B2B)	Interorganizational information flow and electronic market transactions between organizations	Cyberlynx (www.cyberlynx.com.au)
Customer to customer (C2C)	Selling personal services, advertising property, etc.	eBay (www.ebay.com) Ray White Real Estate (www.raywhite.com.au)
Customer to business (C2B)	Selling knowledge products or services to organizations	Accenture (www.accenture.com.au)
Non-business eBusiness	Non-business institutions	Academe www.uq.edu.au or not-for-profit organizations (www.bluecare.com)
Intrabusiness eBusiness	All internal activities via intranets	
Peer to peer (P2P)	P2P is not a technology or a business model – P2P networks allow any internet-connected computer to talk to any other without having to go through centralized servers	Napster (www.napster.com)

The essential characteristics of any electronically enabled business are that the dealings between two parties (business to consumer or business to business) are *electronic*, the key commodity being traded is *information*, and that it is a transaction that may, but does not necessarily, lead to the delivery of a physical product – in effect a *gateway* to a deal. Successful electronically enabled business requires (Norris *et al.*, 2001):

- Robust technology – hardware, software and networks
- A sound business model – how the business works and interacts with others
- Secure transaction processing
- The establishment of trust in the absence of personal contact
- The handling of intangible, information-orientated products
- Integration of presentation, billing and fulfillment of the requirement.

mCommerce or mobile commerce is business that is done on the move using mobile devices such as PCs, PDAs and in particular the now ubiquitous mobile phone (cellphone or cell in the USA, Canada, South Africa, Pakistan) (see Appendix 1).

Currently mCommerce is mainly used for the sale of mobile phone ring-tones and games and for mobile messaging, although it is increasingly used to enable payment for location based services such as maps, as well as video, music downloads, and for news and sports headlines. According to Juniper Research (2005) mCommerce is currently one of the fastest growing business models with revenues from mobile sports video-clips and text-updates to double in 2005 to more than US$1.3 billion; from lotteries, betting and casino gambling to generate over US$2 billion in 2006; the sale of mobile music is tipped to reach US$9 billion by 2009; and its use in a worldwide tracking service, primarily in food safety (see Chapter 6) is projected to go over the US$1 billion mark in 2007 in Europe alone. Mobile games are another source of revenue, pundits suggesting that more than US$4.5 billion will change hands in this sector alone by 2009.

Banks and other financial institutions are also exploring the use of mCommerce to broaden and retain their business by allowing their customers to not only access account information (e.g. bank balances, stock quotes and financial advice) from anywhere, but also the possibility to make transactions (e.g. purchasing stocks, remitting money) via mobile phones. Stockmarket dealings via mobile are also increasing in popularity as they allow subscribers to react to market developments in a timely fashion, irrespective of their physical location. Corporations are also using mCommerce to expand services and access a new marketing and advertisement channel.

While there are currently very few regulations on the use and abuses of mobile commerce, this will change in the next few years with the increased use of mCommerce requiring a commensurate increase in the need for good security.

Summary

Without knowing or understanding what a business is, why it exists or what creates value for it, the importance of electronically enabling a business as a new component of business models for the future would be difficult to understand. This chapter has covered the basic concepts behind a 'business', an 'agribusiness' and associated 'agri-industry chains', 'doing business', and the 'eLandscape', including the basic components of an electronically enabled business environment.

Remember, however, when confronted with electronically enabling your business – don't panic – it is still really just about doing business. The next chapter further explores the issue of creating 'value' in the business as well as how doing business in the eLandscape creates a better value proposition for the business. Marc Andreessen was one of the early net wizards to demonstrate value creation through the net – and is the Smart Thinker for this chapter.

Smart Thinker – *Marc Andreessen*

Mosaic and Netscape Developer, Chairman of Opsware

Marc Andreessen at the tender age of 21 revolutionized the internet by creating the first prototype of the Mosaic browser (the first popular graphical user interface to the World Wide Web) along with a team of students and staff at Illinois University. Mosaic was a breakthrough because it gave the internet a friendly, usable face. Highly popular, Mosaic gained an estimated 2 million users worldwide in just over one year.

Andreessen moved on to join former Silicon Graphics Chairman Jim Clark to launch Netscape Communications. As Chief Technology Officer and later Executive Vice President of Products for Netscape, Andreessen played a critical role in guiding the company's hypergrowth based on the company's first product, Netscape Navigator, which quickly became the browser of choice for the millions of people using the internet.

Netscape's IPO in 1995 propelled Andreessen onto the cover of People magazine and other publications and he rapidly became the poster-boy of the dot.com generation – a hype that would repeat itself over and over: young, twenty-something, high-tech, ambitious, and worth millions of dollars overnight. Netscape's success attracted the attention of Bill Gates – and Microsoft then went on to license the Mosaic source code from the University of Illinois, turning it into Internet Explorer. The resulting battle between the two companies became known as the 'Browser Wars'. Microsoft eventually won, and Netscape was bought in by AOL in 1999 for US$4.2 billion. Andreessen stayed on as Chief Technology Officer at AOL for a short time before moving on to co-found Loudcloud which focused on providing new solutions for outsourcing global internet operations for enterprises. Loudcloud underwent IPO in 2001 and in 2003 changed its name to Opsware (www.opsware.com), where Andreessen continues to serve as chairman. Andreessen has always maintained that:

- Business success depends on a strong and reliable internet infrastructure in an increasingly web-centric economy.
- Building and managing the infrastructure for reliable internet operations is difficult and time consuming.
- The complexity of managing infrastructure will continue to grow exponentially.
- At the end of the day, the Net has everything to do with communications and nothing to do with computers. What's important is that everyone gets connected either one-to-one or one-to-many, and with highly targeted communications of different kinds.

Source: FastCompany (www.fastcompany.com) and ibiblio (www.ibiblio.org),
Wikipedia (www.wikipedia.com)

The Review and Critical Thinking Questions below have been provided for you to revise and reflect on the content of this chapter.

Review Questions

1. What is a business or firm and what are the driving forces and the critical responses of a business?
2. What does the term 'agribusiness' encompass – why does it have negative connotations for some?
3. What are the core components of doing business?
4. What creates value for a business?
5. What do we mean by the eLandscape?
6. What is the difference between eCommerce, eBusiness and mCommerce?

Critical Thinking Questions

1. What was so innovative about Marc Andreessen's work? What did Microsoft 'see' in Mosaic?
2. Why has The Peanut Van been so successful, and what business model would best describe this success?
3. Explain why Dell Corporation took it to IBM in a business sense and won.
4. What is meant by 'technology being a business enabler'?
5. Research and discuss the issues associated with WAP and mCommerce enabled agribusiness.

Chapter 2

Creating Value in the 'E' Agribusiness World

Recreating the old from the new

Following on from Chapter 1 where the key concepts associated with 'a business', an 'electronically enabled' business and 'doing business' were introduced, this chapter discusses *creating value* in a business by electronically enabling it. Electronically enabling a business inevitably involves costs associated with, among many things, hardware, software, training, strategy development and implementation. The critical questions are: what value will be conferred to the business by electronically enabling it and, if there is some advantage, how can it be sustained?

This chapter will look at the fundamentals of creating value in a business, in particular an agribusiness. It will also look at the drivers of the 'Old' and the 'New' Economies in relation to value creation including some of the lessons learned from the dot.com crash. The more common B2C and B2B business and procurement models will be outlined with a discussion on how these can be used in an agribusiness to generate value.

Chapter Objectives

After reading this chapter you should understand and be able to describe:

1. What competitive business advantage means
2. Why sustainable competitive advantage is so important, and the issues associated with obtaining this in agribusiness generally
3. How the eLandscape can confer added value which can confer a sustainable competitive advantage
4. The difference between the terms 'Old' and 'New' Economy and their respective drivers
5. What the dot.com hype was about and why the dot.com crash occurred
6. B2C and B2B internet business models and the applications involved
7. The critical success factors of electronic marketplaces.

Competitive Business Advantage

Unless a business has a product or service that is completely unique, there will always be competitors for any potential customers that exist for that business. Indeed, in a series of famous books on competition and competitive advantage in business, Michael Porter (1985, 1990 and 1998) states that competition is fundamental to the success or failure of a firm's activities because it determines whether the product or service that a company provides is appropriate in terms of the marketplace.

Porter further goes on to outline how the development of an appropriate strategy to both deal with competition and *generate* competition will lead to a competitive advantage for a business in the marketplace. While competitive advantage generation and maintenance – particularly as they apply to electronically enabled business – are discussed in some detail in the next sections, this chapter in no way attempts to paraphrase Michael Porter's work or other large bodies of work in this area. For further in-depth study, readers are advised to access other literature on the subject.

Competitive Advantage

Competitive advantage was referred to in Chapter 1 as the major component of the Resource Based View of the firm and is fundamentally important to the success – or otherwise – of a business. It is defined by Porter (1985) as:

> **the differential among firms along any comparable dimension that allows
> one firm to compete better than its rivals.**

Competitive advantage in an industry can be measured by the business's long-term return on capital relative to its competitors in that industry. A business's competitive advantage – for example, Coke's brand or Microsoft's control of the personal computer operating system market – largely determines its ability to generate excess returns on capital and links the business strategy with fundamental finance and capital markets. In the end, it is a business's competitive advantage that allows it to earn excess returns for its shareholders (Porter, 1985). Without a competitive advantage, a business has limited economic reason to exist – its competitive advantage is its lifeblood and without it the organization will wither away.

How Is Competitive Advantage Created and Maintained?

Traditionally, there have been two major sources of competitive advantage: the behaviour of a business's *costs*, and the business's **differentiation strategy** – either for a product or components in the whole-of-business value chain (Porter, 1985).

A business has a **cost advantage** if the cumulative costs of undertaking the business and creating value in the business (i.e. across a product or service value chain) are less than that of its competitors. If this cost advantage is difficult for a competitor to replicate, then it becomes a sustainable cost advantage. Creating and maintaining a cost advantage is a major if not *the* major goal of most managers.

A business differentiates itself from competitors when it provides something unique (at any point in the value chain – not just at the final product level) that has value to buyers over and beyond a low price. Differentiation is usually costly, the costs reflecting the cost drivers of making the product or service unique. Costs drivers associated with differentiation include:

- *Economies of scale* – if these exist they will reduce costs
- *Interrelationships and linked activities* – sharing costs across organizations or business units within a business can reduce the costs of differentiation
- *Timing* – development of products or services on time and under budget can be delivered through better coordination of quotations, procurement and inventory management, when undertaken under the auspices of strategic relationships with other members of the value chain
- *Knowledge* – a knowledgeable workforce is a valuable commodity but will cost more to obtain and maintain.

More recently, competitive advantage has been regarded as a more complex situation than simply being defined by costs and differentiation. It has been suggested by Ma (2000) that there are two key aspects of competitive advantage that must be taken into account – **positional advantage** and **kinetic advantage**.

1. *Positional advantage* is a status-defining position that leads to better company performance, and which comes from a business's unique resources, market positions, established accesses, and other traits that are relatively static in nature. It is based on the business's status, social or economic, actual or perceived, in the eyes of customers, competitors, partners, regulators and other stakeholders. Cost structure and differentiation strategies of the business contribute to positional advantage.
2. *Kinetic advantage* is an action-oriented ability that allows a firm to function more effectively and efficiently. Kinetic advantage typically arises from a firm's knowledge, expertise, competence (Prahalad and Hamel's 1990 core competency work picks this up particularly), or capabilities and skill in conducting business activities. It includes the company's ability to identify market opportunities, its knowledge of customers, technical know-how and capability, its speed of action and response in the marketplace, and its efficiency and flexibility of business or organizational processes.

Positional and kinetic advantages often reinforce each other providing a more complete set of interactions to produce competitive advantage – particularly in today's eLandscaped business world. With high-speed changes in the current business environment, including globalization and advances in information technology, many positional advantages traditionally banked on by leading firms have become less durable or irrelevant. Constantly operating at an industry frontier with a host of kinetic advantages – however temporal they may be – could prove to be a necessary task in the future in order to succeed (Ma, 2000).

Sustainable Competitive Advantage

A sustainable competitive advantage is the prolonged benefit of implementing some unique value-creating strategy not simultaneously being implemented by any current or potential competitors, and which competitors are unable to duplicate.

However, once a business is successful, how does it remain so? In reality, the fierce competitiveness of the current market system is good for consumers, but bad news for companies that seek to make extraordinary profits over long periods of time. Sustainable competitive advantage is generally derived from a combination of the following:

- A unique competitive position created by combining the firm's resources – financial, physical, legal, human, organizational, informational and relational – in unique and enduring ways
- Clear tradeoffs and choices vis-à-vis competitors
- Activities that are tailored to the business's strategy
- A high degree of fit across activities (it is the activity system, not the parts, that ensures sustainability and rivals will get little benefit unless they successfully match the whole system)
- A high degree of operational effectiveness.

Competitive Advantage in Agribusiness

Agribusinesses are about producing food and related products – products that as humans, we cannot do without. However, at the production end of any agri-industry chain agribusinesses usually produce commodities where few, if any, unique value creating strategies are possible. Very often, the returns on investments at this end of the chain have not met market expectations (Ginns, 2002). If businesses at the production end of agri-industry chains are not realizing full value, how does the industry sector as a whole survive?

Since the 1980s, the agribusiness sector globally has been subjected to continuous structural change as a more dynamic and demanding consumer base has thrown down the gauntlet in relation to expanded expectations (Wilkinson, 2003). Businesses at all stages in all agri-industry chains have consolidated dramatically (Heffernan *et al.*, 1999) leaving a smaller and continually decreasing number of increasingly larger and more powerful players. Agribusiness is a heavily regulated industry sector and many of the issues that it faces are global – from environmental impacts and quarantine issues, to free trade agreements. Sustainable value creation in such a situation is a moving window that requires the participation of all parties involved in a particular agri-industry chain – and then with a keen eye on the global markets and political and economic arenas involved. Innovation and agility are the keys to success. Electronic technologies are one of the potential tools to facilitate such value creation (Boehlje *et al.*, 2000; Thompson *et al.*, 2000; Leroux *et al.*, 2001; Ng, 2005; van Hemert, 2005).

Competitive Advantage Generation in the 'E' World

Doing business in an electronically enabled world is producing clear changes in the competitive landscape; however, the main issues that must be confronted by electronically enabled businesses are the same as for doing business in the non 'E' world – and those are:

- The need to create value within the business
- The need to minimize costs and maximize profits
- The need to maintain a competitive edge in the marketplace.

Value Creation in the eLandscape

To remain successful in today's internet driven marketplace, established companies have to be able to function as extended electronic business in the eLandscape. The process of creating 'value' in an electronically enabled world or 'E' world is no different than in a non-electronically enabled business environment, other than that the technologies involved in the electronic landscape – particularly those associated with the internet as a communication and information transfer medium – have created some additional vehicles that should be taken into account (Porter, 2001). Amit and Zott (2001), working on this premise, proposed four sources of value creation for electronically enabled business (Fig. 2.1) including:

- *Efficiencies.* The internet reduces information asymmetries between buyers and sellers which in turn improves the speed of information flows and timeliness of information availability on which decisions can be made.
- *Complementarities.* Companies can leverage value creation for their own products when they bundle them via the internet with complementary products from other suppliers.
- *Lock-in.* The ability of a business model to prompt users to engage in repeat transactions – for example, in relation to revisiting a web page and engaging in online transactions.
- *Novelty.* The internet offers limitless possibilities to innovate in the manner in which transactions are enabled.

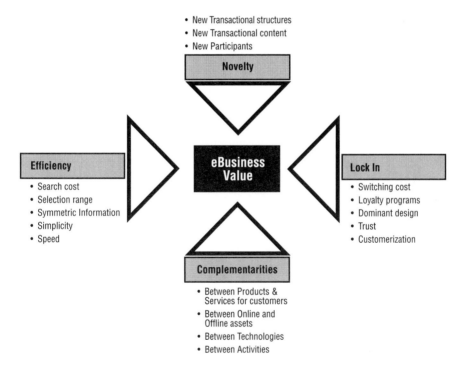

Fig. 2.1. The four components of creating value in an 'E' enabled world
(after Amit and Zott, 2001)

Lock-in, novelty and complementarity while applicable to internet-based online trading enterprises are less important value creation mechanisms for most agribusinesses which essentially remain offline in their business trading models other than to participate in some online marketing and information dissemination activities via a web page (Theuvsen, 2003; Stricker *et al.*, 2003; Fritz and Schiefer, 2003; Ng, 2005). Creating efficiencies, however, is another matter and Bryceson and Kandampully (2004) give a more specific list of value-creating activities involving electronic technologies to facilitate the business processes in and across the agri-industry sector. These include:

- Ensure best practice implementation of technology to better effect traditional business strategies (Bhatt and Emdad, 2001)
- Develop an agile business strategy and supply chain that enables the business to adapt to new business pressures quickly and smoothly (Martin, 1999)
- Develop good business intelligence capability using the internet for real time market and competitor information (Allee, 2000)
- Service different customers and launch new products to address new markets that develop through the extended reach provided by the internet
- Rationalize suppliers and form new partnerships that are collaborative and 'learn' utilizing the internet as a 24 hours x 7 days x 365 days communication tool (White, 2000; Collins *et al.*, 2002; Bryceson and Pritchard, 2003)
- Change delivery frequency and routes to maximize logistic efficiencies
- Alter distribution/warehouse strategies to minimize inventory
- Electronically integrate the supply/value chains
- Improve the information flow (speed, timeliness, accuracy) up and down the chain.

Of these, the full electronic enablement of supply and value chains resulting in an improvement in the speed, timeliness and accuracy of the information flow up and down the chain is perhaps the most important and is discussed in detail in Chapter 4. It involves the complete integration of internal business enterprise resource planning processes and business systems, as well as developing good technology links to customer and vendor systems and innovative logistics systems (Boehlje *et al.*, 2000; Chapman *et al.*, 2002; Bryceson and Pritchard, 2003).

In agri-industries, electronically enabled supply chains can enhance established relationships amongst supply chain members as well as reducing transaction costs through the efficient flow of information both up and down the chain (Tapscott *et al.*, 2000; Bryceson, 2003). This latter outcome also enables more accurate and timely demand forecasting and supply chain management as well as providing the infrastructure for food traceability and product tracking to source (Vaxelaire, 2004), an ever increasing regulatory requirement of the food industry. See Byte Idea on the Shenzhen Agricultural Products Wholesale Markets in this chapter.

As can be seen, some of the above points are difficult to quantify in the traditional sense – indeed quantifying 'value' has become more difficult in the current business environment as corporate value is increasingly tied up not in physical assets but in intangible ones (Read *et al.*, 2001; Dunne, 2004). Reconciling this with what shareholders expect as creating value is becoming a significant skill requirement for senior managers.

Byte Idea –
Shenzhen Agricultural Products Co Ltd
www.ap88.com

EBusiness in Chinese Agricultural
Wholesale Products Markets

The Shenzhen Agricultural Products Co. Ltd is a major Chinese enterprise that has grown from its establishment in 1989 with registered capital of 5.17 million yuan into a listed company worth 4 billion yuan today.

The company's core business is in managing the production, processing, packing, storage, transportation, wholesale, direct sale, retail chain (e.g. the company owns 100 Minrun retail stores), import and export, logistics and distribution of agricultural products. The company is also involved in managing fresh product relationships with the main overseas retail chains (e.g. Carrefour, Metro, WalMart, etc.) and owning a subsidiary company (ap88) that deals in eCommerce transactions associated with retailing fresh agricultural products via online auction.

The company has had an active strategy of growth to position itself as the leader, not only in China but also in Asia, in handling and distributing fresh agricultural products. To achieve this the company has founded a network of 14 wholesale markets throughout southern China (a basket of food from all over the country). Furthermore, the company's goal is to efficiently combine wholesale markets with retail chain stores, direct sales with material distribution, and tangible markets with network (eCommerce) markets.

The company initially boosted the progress of office automation and internal electronic enablement by introducing an advanced logistics management system and international logistics technology which were aimed primarily at accelerating the integration of business information flow. Then came additional development to facilitate the network connections with suppliers and with processors using electronic ordering, transfer and bill payment. Finally in setting up a high tech electronically enabled distribution centre, Shenzhen Agricultural Products Co. Ltd hopes to provide a state-of-the-art facility to ensure a streamlining of all functions in accordance with the strictest possible international food standards.

Information sourced from author interview with Mr Chen Xiaohua,
Director/Vice General Manager of Shenzhen Agricultural Products Co. Ltd

Minimize Costs, Maximize Profits in the eLandscape

In relation to minimizing costs and maximizing profits, it is still difficult to effectively evaluate B2C eCommerce, and B2B online trading companies since, even now, some years after the dot.com crash (discussed in a later section of this chapter), there are still very few of either that have actually generated any profit. However, there are some basic principles of competitive advantage generation when doing business in the eLandscape that can, according to Laseter *et al.* (2000), be followed, including:

- *Recognize that the internet is the grain of sand and not the pearl.* The internet simply provides the infrastructure for providing better communication – the pearl is still the product. Managers must not become so technology focused that they neglect their core business outcomes. Michael Porter (2001) says it all:

  'virtual activities do not eliminate the need for physical activities, but often amplify their importance'.

- *Foster entrepreneurial and innovative capability.* Through its entrepreneurial and innovative capability, a company can effectively align itself with its environment – which in this instance is the eLandscape – and in so doing broaden its business space. It will enjoy an edge over its rivals if it can forecast trends and changes in its environment more accurately (McGrath and MacMillan, 2000; McGrath, 2003).
- *Understand the customer.* Develop strong customer relationships – if a company possesses superior intimate knowledge of its customer base and is good at targeting and selecting customers (e.g. those who will form the backbone of the customer base and who will be loyal, predictable and with high profit-generating potential), or if the company has a keen sense of how to identify and create new market opportunities from environmental stimuli, then it has a firm competitive advantage. Attracting the wrong customers for a quick profit is a disservice to the firm.
- *Gain insight into how the company can use the internet.* Be aware of and take advantage of the forces that drive industry profit potential – for example, the internet's 'unique selling point' is 24 hour, 7 days per week communication and information flow capability, thus increasing customer convenience.
- *No advantage is sustainable on the internet.* The current pace of change in the business environment is rapid and the internet facilitates this by its 24 x 7 availability, thus almost eliminating time and distance issues relating to communication and information flow. To keep ahead of the pack, however, a company needs to be constantly innovating and developing new strategies to give itself a competitive advantage.

First Mover Advantage in the eLandscape

Although the term 'first mover advantage' became common during the internet glory days of the late 1990s when it was a label so powerful it could instantly win a start-up millions of dollars in financing, it has actually been a part of the marketing and management literature for a long time. The concept, which originally came out of the oil and gas industry 100 years ago with the saying then of: *'He who builds the first pipeline controls the flow of hydrocarbons'* (Maney, 2001), can be defined as:

The advantage a company gains by being first to market with a new product or service.

The first mover is often an innovator with a spark that lights up a niche or even an entire industry. For example, Amazon.com was the first big online bookstore. EBay was the first online, auction-based flea market. Yahoo was the first web directory. Unfortunately it takes more than a spark to make it through the long haul and that is where most first movers falter (Waide, 2001). Practically, what is very often missing in first movers are the skills to manage a product or company that is growing at a great speed since these skills are quite different from those required to start a market (Tellis and Golder, 1996). To maintain the thrust, constant innovation is necessary as is the ability to control costs and the strategic vision to know when, or when not, to buy other companies and assimilate them successfully. Two examples of well-known companies demonstrate these points:

1. *IBM (Big Blue)* was the first company to sell PCs in huge numbers and it virtually created the market for corporate PCs. Little did the first movers at IBM dream that some day a Houston based upstart called Compaq would beat them to the number one position in the PC market. Or that a company started by a university student named Michael Dell would also edge them out leaving IBM very much a has-been in a market niche it invented.

2. *Amazon.com* was a first mover in every sense of the word – struggling in 2001 to keep its founder's promise to turn a profit in its oldest lines, books, music and videos (see Byte Idea this chapter). The trouble for Amazon is that it has competitors such as Barnes and Noble who are more focused and have the benefit of Amazon's experiences in avoiding the pitfalls that the first mover fell into. However, Amazon itself has taken advantage of what could be called a 'second mover's advantage' by overwhelming first mover CDnow in the online music market. In Amazon's first full quarter selling music CDs it sold US$14.4 million worth of music titles beating CDnow which had a two-year advantage.

In fact, Henry Ford once said that *the best strategy is to be the first person to be second* (Webber, 2000) – a modern success story built on that maxim is Microsoft. Bill Gates's vision of a personal computer on every desk and in every home was revolutionary and not only made technology a powerful tool for all of us, it also created a new industry that changed the world – yet despite this, his career delivers the message that it can be wiser to follow than to lead. Let the innovators take the losses; if you hold back and follow, you can clean up in peace and quiet. For example: Microsoft's first product was a version of the programming language BASIC for the Altair 8800, arguably the world's first personal computer. BASIC, invented by John Kemeny and Thomas Kurtz in 1964, was someone else's idea. So was the Altair. Gates merely plugged one into the other, and voila! – the whole was greater than the sum of the parts.

Until recently, in most businesses, investing in electronic enablement was viewed as tactical, as it would probably have to be changed or replaced in a relatively short space of time. The current approach is more along strategic lines with the use of information and communication technologies being developed and implemented as part of the fundamental business infrastructure – infrastructure that is now regarded as crucial to simply doing business, as well as for developing efficiencies and competitive advantage within businesses.

The real opportunity associated with electronic enablement of the business and particularly for agribusiness is for the development of collaboration throughout the entire supply chain (Chapter 4). This kind of collaboration requires partnerships with material suppliers, customers, logistics providers, and the most sophisticated technology and service companies. The early adopters have a clear advantage in establishing tight relationships with their industry's leading buyers and sellers and securing the assistance of the most knowledgeable technology experts.

Byte Idea – Amazon.com
www.amazon.com

Books and Other Goodies Online

Amazon.com, Inc. is an electronic shop where customers can find almost anything they might want online. The company lists millions of items in categories such as books, music, DVDs, videos, consumer electronics, toys, camera and photo items, software, computer and video games, tools and hardware, garden and kitchen products, as well as wireless products.

In addition to the US-based website, www.amazon.com, the company operates from websites in Canada, the UK, Germany, Japan, France, Austria and China. The company also operates the Internet Movie Database (www.imdb.com), a comprehensive and authoritative source of information on movies and entertainment titles, and cast and crewmembers.

Amazon Auctions allows buyers and sellers to conduct transactions with a wide variety of products in an auction format. Amazon zShops allows individuals and businesses to offer popular as well as hard-to-find items to Amazon.com's customers. The participants in Amazon Auctions and zShops can use the company's Amazon Payments service, which allows individuals and small businesses to accept credit card payments through Amazon.com's 'one-click' payment feature, thus eliminating the inconveniences associated with cheques and money orders.

In 2000, the company expanded its selection with the launch of Amazon Marketplace, which enables customers to buy and sell used, collectible and rare merchandise alongside the corresponding new product, and Amazon Outlet, where customers can find year-round bargains on thousands of products across the company's product lines.

The company has a number of strategic partners including Toysrus.com, a toy company; drugstore.com, an online retail and information source for health, beauty, wellness, personal care and pharmacy; Audible, which delivers spoken audio over the internet; Ashford.com, an online retailer of luxury and premium products; NextCard, an online issuer of consumer credit cards; Sotheby's.com, an online auction site devoted to art, antiques, jewellery and collectibles; Ofoto.com, an online photography service; and CarsDirect.com, an online source for auto purchasing in partnership with local dealerships.

Amazon.com corporate headquarters are in Seattle, Washington, USA and it is a publicly traded company on the NASDAQ under AMZN.

Information sourced from CIO Magazine (www.cio.com), NASDAQ (www.nasdaq.com) and Amazon.com

The Old and New Economies

The New Economy is a high-tech, services and office-based economy that is being driven by, and places a premium on, innovation, information and knowledge (Varian and Shapiro, 1999). Social capital (networks, shared norms and trust), as fostered in collaboration and alliances, may be as important as physical capital (plant, equipment, etc.) and human capital (intellect, character, education and training) in driving innovation and growth (PPI, 2002).

The internet, with its enormous potential to increase communication capability, and improve efficiency, is a critical component of the New Economy (Baily and Lawrence, 2001). Internet commerce includes consumer retail and business to business transactions; online financial services; media; infrastructure; and consumer and business internet access services (Lawrence *et al.*, 2000). Internet development has also meant that it is now a competitive requirement that businesses invest all over the globe to access markets, technology and talent. Increased competition is being driven by many factors, including the emergence of a global marketplace, the increased number of firms, new technology that makes it easier for firms to enter new markets, and ever-increasing pressure from securities markets to raise shareholder value.

Essentially, the difference between the 'Old' and the 'New' Economies is that while the Old Economy was fundamentally organized around standardized mass production, with manufacturing and mechanization at the core, the New Economy is organized around flexible production of goods and services in a digital and dynamic networked environment (Martin *et al.*, 2003). Competitive advantage increasingly stems from customization, design quality and customer service, and more of this 'value-adding' is produced in offices by 'knowledge workers' – people who work primarily with information or who develop and use knowledge in the workplace (Burton-Jones, 1999).

The ability to innovate and get to market faster is also becoming a more important determinant of competitive advantage. In some sectors, such as information technology, the pace of innovation causes such rapid obsolescence that firms have to run just to stay in place. This situation is clearly illustrated in Stace and Dunphy's (2001) book 'Beyond the Boundaries' which discusses the reorientations in markets that have led to major changes in what, why and how organizations exist in the current business environment (Fig. 2.2).

The New Economy is also characterized by constant churning of job creation and destruction which has undermined the predictability and stability of old economic arrangements and has increased the insecurity faced by workers. It should be noted, however, that churn has always been present and is not something related solely to the New Economy. Churn is essentially the process of innovation and competition that recycles labour into new jobs. New ideas, new products, new technologies, new forms of business organization and new markets upset the status quo, rerouting demand from existing companies and industries to new ones. As less innovative and efficient companies die or contract, more innovative and efficient companies take their place. Such turbulence is a major driver of economic growth although it increases the economic risk faced by workers, companies and even localities (Federal Bank of Dallas, 2005).

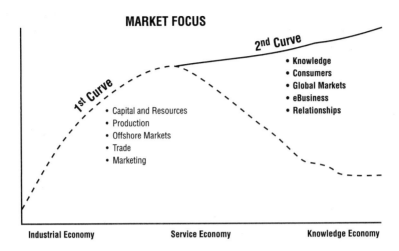

MARKET FOCUS

2nd Curve

• Knowledge
• Consumers
• Global Markets
• eBusiness
• Relationships

1st Curve

• Capital and Resources
• Production
• Offshore Markets
• Trade
• Marketing

Industrial Economy   Service Economy   Knowledge Economy

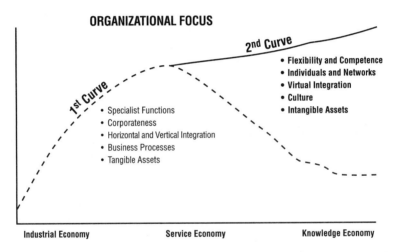

ORGANIZATIONAL FOCUS

2nd Curve

• Flexibility and Competence
• Individuals and Networks
• Virtual Integration
• Culture
• Intangible Assets

1st Curve

• Specialist Functions
• Corporateness
• Horizontal and Vertical Integration
• Business Processes
• Tangible Assets

Industrial Economy   Service Economy   Knowledge Economy

Fig. 2.2. The changing enterprise over time (adapted from Stace and Dunphy, 2001, *Beyond the Boundaries*, 2nd Edition, McGraw Hill, p. 38)

This is nowhere more obvious than in the agribusiness sector where a fundamental change from production driven to demand driven agri-industry chains (Ballenger and Blaylock, 2003) with associated business strategies has led to changes in the numbers of people and the types of jobs involved. The flow-on effect of population movement from the rural areas to the towns has had a major negative impact on the economy and social fabric of rural communities (Pritchard and McManus, 2000; Cavaye, 2001; Collits, 2003).

Some agri-industries are only slowly coming to grips with the consumer driven market focus particularly as it relates to the need for electronic enablement. However, in industries where there have been major food safety scares in recent times – such as the beef (mad cow and foot and mouth diseases) and peanut (aflatoxin) industries – the need for full traceability from paddock to plate is getting to the point where regulations demand a level of record keeping and tracking that can only be accomplished using electronic technologies (see Chapter 6).

The implications for agribusinesses at all levels in their respective agri-industry chains from the point of view of human resource management, knowledge creation, capacity and skills development, as well as business strategy development, are immense. Along with changes to the businesses themselves, these issues will almost certainly require a change away from the slow, plodding, farmer-on-a-tractor image of what 'agribusiness' involves, to an image of an industry sector incorporating dynamic technological innovation requiring substantial business acumen, in order that the right people are attracted to the right organizations so that agribusinesses not only remain viable but are sustainable in this new era.

The Ten Drivers of the New Economy

The actual drivers of the New Economy are not essentially any different from those of the Old Economy; however, the issues associated with them are, and these issues taken together could set the scene for a change in the rules of business.

1. *Physical presence.* While having a product to sell is still what counts, intangible assets are the key to the New Economy – that is, people, ideas and the strategic aggregation of key information-driven assets are providing the innovation and flexibility necessary in a constantly changing business environment.
2. *Location.* Where a company is located in the world is less important than it once was – the world is now both the customer and the competitor. Geography has always played a key role in determining who competed with whom. Now a business can connect instantly with customers all over the globe. The opportunity – and the threat – has never been greater. Distribution of the product remains an issue with new models of offshore warehousing to facilitate international distribution and delivery being developed (e.g. FedEx – see Byte Idea Chapter 8).
3. *Time.* 24 x 7 is the mantra (24 hours, 7 days a week) because the instant interactivity available via the internet is breeding accelerated change. In a world of instantaneous connection, there is a huge premium on instant response and the ability to learn from and adapt to the marketplace in real time. Winning companies accept a culture of constant change and are willing to constantly break down and reconstruct their products and processes – even the most successful ones. For example, Dell has revolutionized PC sales by offering machines built directly from buyers' requests. With very fast inventory and purchase cycles and its ability to analyse customer orders almost on the spot, Dell adapts to trends ahead of the pack.
4. *People.* Humans have become the 'knowledge capital' of a company because knowledge is the prime factor driving the New Economy and it can't be quantified in the accounts (thus the term intangible asset). More than ever before in history, huge value is being leveraged from smart ideas – and the winning technology and business models they create. People who can deliver them are becoming invaluable, and methods of employing and managing those people are being transformed. A good example is that of Microsoft who early on recognized this and successfully 'locked in' one of the world's most talented workforces by giving them stock options worth billions of dollars.

5. *Growth/Scalability.* The internet can dramatically boost the adoption of a product or service by viral marketing which is essentially, network-enhanced, word of mouth (Wilson, 2000). Communication is so easy on the internet, product awareness spreads like wildfire. Once a company reaches a critical mass, it can experience increasing returns leading to explosive growth. Conversely, of course, criticism of a product or company can similarly be explosively fast and create major threats.

6. *Value.* Value rises exponentially with market share. For products that help establish a platform or a standard, the network effect is extremely pronounced and the more plentiful they become, the more essential each individual unit is. This is a striking exception to the economic rule that value comes from scarcity. Network effects were experienced historically in the adoption of telephones and fax machines – the difference today is that because everyone is linked, far more products and services gain their value from widespread network acceptance.

7. *Efficiency.* The vast amounts of information available which are constantly being pumped into businesses these days need to be managed otherwise gridlock develops as a result of information overload. Information and communication technologies (ICTs) are actually designed for this very task (Chapter 3). In addition, outside of the business, traditional distributors and agents can be seriously threatened by a networked economy in which buyers can deal directly with sellers. A new brand of middleman, the **infomediary,** has been created who offers aggregated services, or intelligent customer assistance to help in navigating through the mass of information at our fingertips. A good example is Hoovers. com (www.hoovers.com) which specializes in business intelligence gathering and filtering for all industry sectors or Agribusiness Online (www.agribusinessonline.com).

8. *Markets.* The internet provides buyers with dramatic new power, and sellers with new opportunity. It's no longer necessary for a customer to walk down the street to compare prices and services. A competitor may just be a mouse-click away and intelligent software will help buyers find the best deal. The most famous of all online markets is eBay, the online auction site (see Byte Idea, Chapter 7). An interesting agribusiness example is Sea-Ex International Seafood Trading Board (www.sea-ex.com/trading/) which is a subsidiary of Labiche Corporation.

9. *Transactions.* Information is easier to customize than hard goods and the information portion of any good or service is becoming a larger part of its total value. Thus, suppliers find it easier and more profitable to customize products, and consumers are now demanding this sort of tailoring. In other words it has become a one-on-one game. A good example is the Australian Wheat Board's grower website – see Byte Idea this chapter.

10. *Competitive advantage generation.* While internet technology and the electronic business environment offer too many potential benefits to be ignored, the following questions need to be asked if an organization is looking to deploy internet technology to access innovative competitive advantage gain:

   • What are the potential electronic technologies that are relevant to the business?
   • How are the competitors using these technologies?
   • How can these technologies change customer priorities, the efficiency of the supply chain, communication across the organization, and across the supply chain?
   • How can a business design and supply chain be constructed to meet these new priorities?
   • What technology investments must be made in order to take advantage of these priorities?

Byte Idea – AWB
www.awb.com.au

The Australian Wheat Board

AWB Limited is Australia's major national grain marketing organization and is one of the world's largest wheat management and marketing companies. Its core business function is to serve Australian grain growers, and specifically wheat growers, by marketing and financing their grain. In particular, AWB has the responsibility for the management and marketing of all Australian export bulk wheat (i.e. it is the single desk export trader for Australian grain).

Around 90% of the grain managed by AWB is wheat, although AWB also competes in the trading and management of non-wheat grains including barley, sorghum, oilseeds and pulses with two other bulk handlers – GrainCorp and ABBGrain. AWB manages the national wheat pool which represents an efficient mechanism for managing risk through the use of grain market diversification and the creation of economies of scale.

AWB has five business units all of which are heavily information intensive and electronically enabled:

- National pool management services
- Finance and risk management products
- Grain acquisition and trading
- Grain technology
- Supply chain and other investments.

In 2000 a General Manager (eBusiness) was appointed to steer the company through the development of full electronic enablement including ECRM and ESCM. This is an ongoing task.

In 2003 AWB bought the company Landmark, the rural input supply business arm of Wesfarmers Landmark, making AWB the most powerfully vertically integrated agribusiness in Australia.

AWB is a publicly listed company on the Australian Stock Exchange (ASX) under AWB.

Information sourced from the ASX (www.asx.com.au) and the AWB website

Internet Business Models

Electronic business in one form or another has developed at a very rapid rate over the last few years. Each business model that has been thrown up has seemed viable only for a very short time. In truth there is no one model that is 'right' – as with the 'Old Economy' there are endless combinations of models and themes (Kuhn *et al.*, 2000; Afuah and Tucci, 2003).

The classifications commonly used to describe the 'type' of business carried out using the internet were listed in Chapter 1. This section looks at internet business models associated with the business to consumer (B2C) and business to business (B2B) classifications (Hoque, 2000; Norris *et al.*, 2001; Turban *et al.*, 2004, Ng, 2005).

The B2C Environment

In a B2C buying environment individual consumers purchase goods from a virtual marketplace. The actions of the buyer drive the process and typically (Norris *et al.*, 2001) the buying processes consist of four phases: product identification, catalogue search, product comparison, and purchase. Figure 2.3 illustrates these phases.

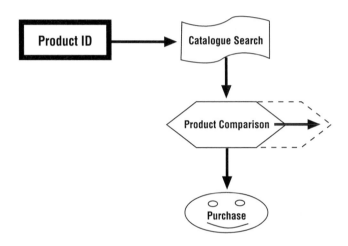

Fig. 2.3. The B2C buying process

1. *Product identification.* This involves a combination of online advertising and one-to-one marketing in a virtual marketplace. To provide personalized service which improves over time e-tailers have to rely on individually targeted product promotions and advertisements for repeat customers. If possible most retailers try to customize content from the very beginning by tracking a consumer's browsing and customizing content to fit suspected tastes.
2. *Catalogue search.* This tends to be focused on the product catalogue itself. The virtual custom catalogue is an essential component of any system that hopes to provide goods and services to either consumers or other businesses.
3. *Product comparison.* When a catalogue search phase crosses into the product comparison phase, features that enable consumers to navigate catalogues, compare similar offerings, and make product selections are required.
4. *Purchasing.* When a customer is ready to make a purchase the system must provide authentication and encryption services to ensure that the transaction is confidential and accurate. Advanced virtual marketplaces can also be connected directly to back office legacy applications and data sources in order to automate the vendor's fulfillment process and trigger third party service providers.

While there was a slow start to using the B2C environment in agribusiness – due in most instances to a mix of conservative culture in the 'way business is done' and the lack of technology in many rural environments in the world (Amanor-Boadu *et al.*, 2000; Hooker *et al.*, 2001; Bryceson, 2003) – in the last two or three years there has been a greater uptake.

B2C Business Models

B2C business models used during evolution from Brochureware to eCommerce (Hoque, 2000) include: the Content Business or Connect-Time/Revenue Split Business Model, the Advertising-Driven Business Model, the Transaction Business Model or eCommerce, the eCommerce Monetize Business Model and the Portal Business Model.

1. *The Content or Connect-Time/Revenue Split Business Model.* In this model, people who supplied content to online services got credit for helping keep users online. Since users paid by the minute or hour, this generated connect-time revenues that were allocated according to a negotiated split between content providers and online services. As the number of users increased these deals initially generated riches. However, as more service providers came online offering their content 'for free' this model became unviable. Content-based models, however, are taking off again in 2005 (Blossom, 2005) with the emerging interplay of content, technology and people in the online environment creating value for businesses. Delivery of mobile-based content is another area that is gaining rapid ground in the value creation stakes.

2. *The Advertising-Driven Business Model.* This model was conceived as a way to generate revenues based on the sheer volume of traffic that some sites could achieve. The model consists of offering a free service and collecting revenue from advertising. Users of these services see banner ads while accessing the website. Clicking on an ad usually sends the surfer to that advertiser's web page or displays relevant product information. Appendix 2 defines some common advertising terminology associated with the internet. There have been three main difficulties with this model:

    i. Determining appropriate CPM (cost per 1000 impressions) rates (segmented according to banners, buttons and interstitials)
    ii. Measuring traffic with integrity (definition of 'hits', 'page views' and 'unique users')
    iii. Estimating efficacy of messaging (whether impressions or click-throughs).

    While the advertising model has declined in value in terms of revenue production particularly as online users got tired of banner ads, another model that remains current is the *Affiliate and Syndication Network Model* which developed from the advertising model. This model combines the content-based business model with the advertising-driven model. The underlying premise is that any website can leverage the content of other websites to create new value for itself by linking itself to that content and using a click-through application to bring the reader of the original content back to where they started. Of course appropriate revenue-sharing must be agreed on. In other words, this model views the web as an interconnected network of value in that by using a network of resellers, you increase volume and share the revenue (Laplant and Levine, 1999).

3. *The Transaction Business Model or eCommerce.* The strategy associated with this model is well known – that is, selling real products for real money through online channels. The idea is that the internet provides a lower cost channel structure resulting from the 'disintermediation' of middlemen such as distributors, wholesalers and bricks-and-mortar

retailers. The subsequent reduction in transaction costs rewards new intermediaries, such as web-based retailers, with fatter margins. Unfortunately such rewards have not eventuated for most e-tailers using this model primarily because the transaction costs associated with doing business – via any medium – are in fact embedded so profoundly into the process of creating the product that they do not get 'dropped off' the cost of selling something e.g. Amazon.com (www.amazon.com). However, there is no doubt that e-tailing is becoming more popular and there are a number of agribusinesses that have developed websites with an e-tailing function e.g. eharvest.com (www.eharvest.com) and Raptis and Sons Pty Ltd (www.raptis.com.au).

4. *The eCommerce 'Monetize' Business Model.* Here the strategy revolves around the idea of never making a profit selling real products for real money! In other words, Freebies (e.g. affiliate programs) for customers are given in exchange for market relationships. The monetizing concept argues that online businesses must first capture large audiences of users or shoppers and then 'monetize' those audiences through subscription fees, advertising and eCommerce.

5. *The Portal Business Model.* As discussed in Chapter 3, a portal is generally a single, web-based interface that offers a gateway to a broad array of resources and services that would otherwise be a series of disconnected and incompatible information sets spread across numerous separate applications. Services include email, forums, search engines (e.g. Yahoo!, Excite and Lycos) and online shopping malls. There are three key characteristics of a portal which make them valuable to users:

   i.   *One stop shopping* that provides simplified access to content, applications and collaborative or community building functionality
   ii. *Personalized views* that provide this access in the context of an individual's preferences and business rules
   iii. *Flexible navigation* including both pre-defined roadmaps and powerful free-form search.

   Agribusiness specific portals are quite well represented – in Australia, Aglinks is an agricultural directory (www.aglinks.com.au) and AgriGate is a portal to ongoing agricultural research in Australia (www.agrigate.edu.au). In Canada, Farms.com is an information portal which is one part of a multifaceted 'E' business in Canadian agriculture (www.farms.com). This also exists in India (www.agriwatch.com) and in Holland (www. agriholland.com). In the UK, Agritrading net (www.agritrading.net) is an agricultural trading portal and AgriFor (www.agrifor.ac.uk) is a free portal to agricultural and food resources. Similar resources can also be found in Latin America (http://lanic.utexas.edu/la/region/agriculture/), China (www.Farmchina.com) and the European Union (http://agritrade.cta.int).

B2C Application Models

The applications that ensure that the business model works are important – the vast majority of B2C electronic application initiatives are e-tailers and consumer portals that aim to add value to business and consumers by breaking down barriers caused by time, distance and form. B2C applications can:

- Reduce the cost of selling goods and services and contribute directly to the enterprise's bottom line (e.g. by empowering consumers with customer self-service which reduces customer service costs and improves the efficiency of the buying process)
- Have the ability to build personal relationships with customers (e.g. by greeting customers by name as they enter the site or by tracking customer behaviour and preferences in order to present completely customized interfaces and content such as 'my Yahoo' page)
- Enable retailers to target customers for individual promotions with one-to-one marketing.

There are four distinct vertical B2C application models: E-Tailing/Consumer Portals, Bidding and Auctioning, Customer Relationship Management (CRM), and Electronic Bill Payment (EBP).

1. *E-Tailing/Consumer Portals.* These e-tailing or consumer portal applications aim to aggregate consumers, market goods and services, and make transactions from simple static catalogues with fixed prices. Examples of activities supported in such an application environment include:

   - *Catalogue management.* E-tailers must integrate product and service information from multiple vendors into a single cohesive catalogue that is intuitive to navigate and that presents products in an attractive manner.
   - *Parametric searching.* Consumers often search for product categories (e.g. 'detergents') rather than specific product names (e.g. 'Morning Glory'). Parametric searching enables buyers to search product categories incrementally to determine product selections that meet their needs.
   - *Personalization and profiling services.* Consumer portals should identify visitors individually, wherever possible, in order to collect personal information about the types of products that they research and ultimately buy.
   - *Advertising or targeted marketing.* E-tailers can dramatically increase the instance of impulse buying by targeting advertising banners and special promotions to consumers with identified interests and buying habits.
   - *The shopping cart.* Most e-tailers use a virtual shopping cart metaphor.
   - *Payment processing.* The site must include some form of electronic payment processing to enable users to make purchases using credit cards, online payment methods, etc.

   It is unlikely that consumer portals will take the place of conventional retailers because they will never be able to deliver the hands-on instant gratification that the customer receives when making a purchase and taking home a new product from a bricks-and-mortar store. This is particularly the case with products such as food.

2. *Bidding and Auctioning.* Bidding and auctioning sites sell products and services with non-traditional, flexible pricing models. Often the host of the auction or bidding application is a third-party intermediary who operates as a link between the buyer and the seller. Successful bidding and auctioning applications must support a common core functionality in order to provide the basic features and services that customers expect in an auction environment – these include:

   - *Catalogue integration and management.*
   - *Chat capability.* All bidding and auction sites include some sort of chat functionality to allow potential buyers to discuss the relative merits of available products.

- *Bid boards*. Consumers must be able to place their bid information on a public bid board to keep other potential buyers up to date on the auction process.
- *Personalization buyer/seller account management.* Many auction sites concentrate on matching buyers with multiple sellers in a buyer cyber-mediary supplier model rather than a simple buyer/supplier model.
- *Notification system*. Auctions are driven by a series of events (i.e. new item posted for bidding, new higher bid received, auction closed). To ensure that all potential buyers have access to up-to-date information about auctions, the application often utilizes an event notification component to deliver updates by email, fax, web page or text message.

The most famous auction site is eBay but agribusiness auction sites include Farms.com in Canada (www.farms.com) which is a general auction site for the agricultural and agribusiness sector, AuctionsPlus from Australia (www.auctionsplus.com.au) featuring livestock and machinery, and Rangeland (www.rangeland.com.au) which specializes in online auctioning of livestock from around the world.

3. *Customer Relationship Management (CRM)*. CRM applications aim to provide services to customers that add value, either to transactions that have already taken place or long-term customer relationships between the end user and the host of the application. They support a wide range of consumer processes and functionality to deliver a robust relationship building experience. For example:

- *Profile management*. Typically includes accessing information that is unique to individual customers. The application must keep track of individual user profiles to ensure that each user has access only to information that pertains to them.
- *Customer content delivery.* Applications should be able to identify and deliver support information directly to the user for whom it is intended.
- *Account management product feedback*. E-tailers should aggregate user experiences with products in order to deliver real-time feedback to design teams about each customer's use of a product. Ultimately this information can be used to automatically provide customer product configurations for individual buyers.
- *Information gathering*. The site should attempt to aggregate a knowledge-based user experience in order to enable customers to engage in self service and self support.
- *Interactive community building*. Customers who have experienced similar problems are the best source of solutions or advice so services such as online bulletin boards, FAQs and chat rooms are all expected.

4. *Electronic Bill Payment (EBP)*. EBP applications streamline the process of collecting, presenting, and paying repetitive consumer charges such as credit card, telephone and utility bills. They include:

- *Bill consolidation*. To offer a complete solution, EBP sites must be able to integrate, present, and collect payment for bills from multiple billing organizations. Typically this involves customizable data translations through standards such as XML.
- *Payment processing*. EBP applications must be able to accept multiple payment methods including credit cards and electronic cash.
- *Analysis and reporting*. Consumers should be able to automatically analyse and summarize information in charts, graphs, and other formats.

- *Integration with bill accounting systems*. When a consumer makes a payment it can be automatically integrated into the internal accounting systems of the biller to reflect the latest, up-to-date information at all times.

In addition to these applications, B2C applications that support the consumer buying process (e-procurement) need to address and provide functionality consistent with the four phases of the B2C buying process outlined at the beginning of this section (product identification, catalogue search, product comparison and purchasing).

B2C Critical Success Factors

Every B2C platform relies on three strategies to bring value to online consumers: branding, one-to-one marketing and online community building.

Branding

In the past branding has almost exclusively been about positioning and product differentiation in advertising. In the old advertising mode the consumer was always the passive recipient of the message (e.g. on billboards it takes quick messages and appealing images to grab the eye of the passers-by). The internet, however, is a two-way medium. Dot.com branding is more about listening to customers than telling them what to believe (Figs 2.4a and b).

Fig. 2.4a. One-way advertising   **Fig. 2.4b**. Two-way information exchange

One-to-One Marketing

One-to-one marketing enables e-tailers to replicate the intimacy of shopping at the corner general store. It is about personalized service, about knowing a customer's preferences and buying habits. Benefits include:

1. *Economies of scope* – it is more important to own a single customer's whole business than to have one-time relationships with many customers
2. *Close long-term relationships* – enable retailers to listen to the needs of their customers and respond directly as well as to prevent defection to competitors
3. *Manage customers individually* – consumers have individual and highly variable tastes. Remember though that no single product can please everybody
4. *Targeted customers* – these are consumers who are highly subject to impulse – don't wait for them to come to you; take your product to them.

Online Communities

The idea of building communities of interest in order to generate consumer interest is not purely an internet phenomenon (Coat, 1993). However, because it is a two-way medium the internet can enable consumers to easily generate their own content and can turn interviews into conversations, articles into discussions, and editorials into debates. There are five requirements for successful online community building:

1. *A shared space* – a website
2. *Shared values* – community members must have some sense of common appreciation and values in order to establish the foundation for discussion
3. *Shared language* – if the community members do not speak the same language it will be difficult to talk
4. *Shared experience* – communities are normally based around a common experience that members share
5. *Shared purpose* – there must be a common purpose either as a commerce oriented consumer site or as a community based consumer site.

A good example is the Australian agribusiness Incitec/Pivot's Big N Community (www.bigN.com.au). The website was developed around an existing offline and specialized user base for the fertilizer anhydrous ammonia – branded as 'BigN'. Anhydrous ammonia is the most concentrated nitrogen fertilizer available and is stored, transported and applied directly to the soil as a liquified gas. As a product, it is potentially explosive and thus has some very stringent and specialized working conditions associated with it that users have to be aware of and capable of dealing with. The website, which was inexpensively set up (approx. A$20,000), has become the major space for active dissemination of information and updates on the product, as well as acting as a focus for general discussion and news (M. Maloney, Brisbane, 2003 and 2004, personal communication).

The Dot.com Enterprise

A dot.com organization is one where the majority or all of the business of the enterprise is conducted online via a website on the internet. In the mid to late 1990s it was the quintessential B2C type of eCommerce business and was much vaunted as part of the 'New Economy' way of doing business. As with most businesses, the dot.com itself could (and those that remain still can) be seen to have a development life cycle – the number of stages of which vary according to how they have been classified. IBM Global (2000), for example, suggests three stages – emerging, midlife and mature – whereas Accenture (2001) suggests five. Below is a simple five stage life cycle based on website development.

Stage 1. Secure an online identity (simple non-transactional sites). These companies are identified by domain name registration and perhaps some rudimentary web home page development (1–3 pages). Many of these names employ 'website forwarding', which is simply a tool for channeling traffic from many domains to one primary website.

Stage 2. Establish a web presence (simple transactional sites). The companies are represented by fully developed websites offering a one-way flow of information from the site to the audience. These sites are referred to as *billboard* or *promotional* sites and are known collectively as Brochureware. They present information on products, companies and topics of interest. Transactions are not conducted, customer information is not collected and they do not promote a continuing customer relationship.

Stage 3. Enable eCommerce (complex selling sites). These true eCommerce sites are websites that have moved beyond a one-way stream of information to engaging their audience in a transactional relationship. About half of the sites in this category feature some sort of 'shopping cart' application to streamline the shopping experience.

Stage 4. Provide eCommerce and customer relationship management. These companies have gone beyond the realm of simple electronic commerce to applications designed to build a relationship with the customers. Account management features, email newsletter subscriptions, customizable interfaces, and rewards programs are functionalities that are added to the online presence to make the customer experience more fulfilling and to maintain a relationship with customers. Sites in this stage have web-enabled key business processes on an intranet or extranet which, for instance, may allow distributors to check product inventory online. It is in this stage that the true power of the internet as a sales channel becomes clear. When a company can offer consumers the efficiency and convenience of an online transaction, and at the same time preserve the customer/vendor relationship that is so easily derived out of storefront, person-to-person interaction, they have captured the best of both worlds.

Stage 5. Use a service application model (ad-orientated sites). These sites use the most advanced internet technologies to offer real-time business processing through the application service provider (ASP) model, online communities and trading consortiums (e.g. the auction site). The real target of the Stage 5 website is the B2B eMarketplace.

The Dot.com Rise and Fall

The dot.com rise of the late 1990s was vigorously marketed by venture capitalists, market analysts and the media. It is impossible to avoid the story of Amazon.com, which despite having never made a profit, has often been proclaimed as the 'wonder' of the dot.com world. The dot.com growth phenomenon is best explained through the National Association of Securities Dealers Automated Quotations (NASDAQ) Stock Market – see Byte Idea this chapter. The NASDAQ is dominated by IT companies, including Dell, Microsoft, Cisco, Intel, Oracle, ETrade, Sun Microsystems, Yahoo!, Amazon.com and many other dot.coms. In 1999:

- The top 15 initial public offerings (IPOs) were dominated almost exclusively by technology, telecommunications and dot.coms, including names such as Priceline.com, eToys.com and MP3.com.
- Market capitalization reached US$5.2 trillion.
- The NASDAQ composite market index reached an all-time record of 5,133 on 10 March 2000.
- The growth of stock prices of dot.coms and internet companies over the late 1990s was unbelievable (e.g. Amazon.com's stock price went from US$2.90 in May 1997, to over A$780 just 19 months later, and Commerce One Inc, a major developer of business to business software, had an IPO price of US$7.00 in late 1999, rising to US$10.43 at close of first day's trading, then finishing the calendar year 1999 at US$196.50).
- Speculation, hype and rumours were driving many stock price increases (e.g. on 28 September 1999, rumours of an announcement to be made by Amazon.com 'significantly affecting the world of e-commerce', boosted Amazon's value by US$1.5 billion).

The Dot.com Crash

The dot.com hype continued in the first quarter of 2000, but came to an abrupt end when the NASDAQ composite index fell over 10% on 14 April 2000 making an aggregate fall of 35% over about a month. More than US$1 trillion in market capitalization was lost in a single day of trading. Dot.com stocks of every description crashed. Companies good and bad lost half, two-thirds or three-quarters of their value in one day. Global technology companies such as Cisco, Microsoft and Sun Microsystems saw their market capitalizations cut by tens of billions of dollars. Striking similarities were made to the speculative bubble preceding the infamous crash of October 1929 which precipitated the Great Depression. Figure 2.5 shows what the NASDAQ crash did to the stock prices of three famous dot.coms.

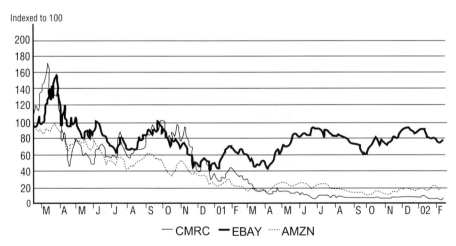

Fig. 2.5. Commerce One, Amazon.com and eBay NASDAQ prices charted over 2000-2002 (Source: NASDAQ Stock Exchange)

The crash was seen as a major event by financial commentators; some called it a 'dislocation', others a 'correction', and even others suggesting it was the 'end of the dot. com era' and the 'end of eBusiness as a separate entity'. An immediate change occurred in the ability of rising dot.coms to raise finance, which quickly became very difficult, investors suddenly realizing that dot.coms did not pave their way to riches.

Byte Idea – NASDAQ
www.nasdaq.com

The Technology Stocks Market

NASDAQ (which is an acronym for National Association of Securities Dealers Automated Quotations) is the world's largest electronic stock market, and is also the most sophisticated. Unlike its traditional trading-floor-based rivals, it is not dependent on one physical location or even one computer system. From the beginning NASDAQ was designed as a 'virtual' global marketplace – a trading system built around a sophisticated computer and telecommunications network that transmits real-time price quotes and transaction data to more than 1.3 million users in 83 countries.

It is this system that defines NASDAQ, ensures its reliability and provides it with its key competitive advantage – the ability to build the world's first truly global securities market unhampered by size limitations or geographic boundaries. NASDAQ's 'open architecture' market structure allows a virtually unlimited number of participants to trade in a company's stock.

Key to NASDAQ's market structure are a core group of financial firms called market makers. More than 500 market making firms trade on NASDAQ, acting as distributors for NASDAQ-listed securities. Also known as dealers, market makers are unique in that they commit their own capital to NASDAQ-listed securities, then turn around and re-distribute the stock as needed. They are required at all times to post their bid and ask prices in the NASDAQ network where they can be viewed and accessed by all participants. By being willing to buy or sell stock using their own funds, market makers add liquidity to NASDAQ's market, ensure that there are always buyers and sellers for NASDAQ-listed securities, and enable trades to be filled quickly and efficiently.

NASDAQ.com is one of the most popular financial sites on the internet. Averaging more than seven million page views per day, NASDAQ.com offers excellent visibility to listed companies. Investors can log on to NASDAQ.com and not only scan the latest news and fund commentary and get quotes for stocks, mutual funds and options on major US markets but can also, via links on the NASDAQ site, see how the Dow Jones and the S&P 500 are performing, along with a wealth of other investor services.

NASDAQ is a publicly listed company operated by the Nasdaq Stock Market, Inc. under NDAQ and was listed on its own stock exchange in 2002.

Information sourced from NASDAQ website

Key Reasons for the Crash

In the early days of the New Economy, entrepreneurs argued that everything then known about business would be obsolete as the information age facilitated by the growth of web-based businesses gained momentum. Further, it was thought that profitability and revenues were irrelevant as compared with building 'traffic' through the website of the dot.com and growing user numbers. After the NASDAQ meltdown in early 2000, this approach was at last recognized as deeply flawed, and suddenly dot.com survival required dot.com profitability!

But why were traditional business valuation tools abandoned by analysts and why were traditional cash flow or earnings-related valuations seemingly unable to deal with the dot.com phenomenon? There were many reasons attributed to the crash, but in reality it was greed for quick and substantial returns. One analyst in the now defunct internet analysis company Webmergers.com wrote:

> 'The speed of ascent was both real and imagined. It was real – the Internet achieved 30% penetration in four years. It was also imagined – or at least enhanced – by an unprecedented frenzy of media attention and market research fantasia. The media infected stock market investors; investors infected markets; markets infected IPO debuts; IPO blowouts infected private equity investors who infected entrepreneurs in an ever-accelerating vicious circle of exuberance.'

It was also recognized that there is no business on the internet that is *not* a service business and that as a service-based industry, customers are a dot.com's most important resource and thus the customer experience – the shopping and buying process and the fulfillment of products – is the key to a dot.com's survival. Service businesses survive if they remember that revenues and profitability are based on optimizing *three* dimensions of customer relationship management:

1. Companies must select customers with whom to do business profitably. As obvious as this sounds, a large number of businesses do business with customers who result in a negative gross margin for the company by developing bad debt situations, or never repeating business with the company.
2. Companies must find customers who value the services offered enough not to just browse but to purchase as well – and to return frequently to the site.
3. Companies must find customers who will maintain a relationship with the business over a long time. Studies show that loyal customers become less price sensitive over time, will refer friends and spread positive word-of-mouth references, and also incur lower costs of relationship management for the company.

Lessons Learned from the Dot.com Crash

The crash affected a great number of people and businesses and there have been significant lessons learned from it for the business world. These include:

- The decline of dot.coms was mostly a failure of business plans, not technology. The winning companies were those who sifted through the technologies and picked products and services that gave them the best bottom line.
- Profitability is important. Companies that don't make money are not only questionable investments, they may not be around that long.

- Source and certainty of revenues is important. Statistics like hits, page views and eyeballs don't necessarily relate to revenue or profitability. Some of the most popular websites visited are notoriously unprofitable.
- Knowing what consumers want is essential, but clever advertising does not guarantee profits or revenues. The novelty of doing something on the web just because you can do it on the web wears off almost immediately. People will use the web for something only if it makes sense, such as for savings or convenience.

The B2B Environment

The online B2B environment essentially involves electronic exchanges or eMarketplaces where businesses register as sellers or buyers with the main aim of communicating and conducting business over the internet. For example, businesses representing each section in a supply chain could join an eMarketplace to transfer information and purchase products.

There are many types of eMarketplaces based on a range of business models. They may operate on a cost-recovery basis by an independent third party (such as an industry association) or be set up as a business offering, with a middle-person providing a value-added function such as transaction services (e.g. www.emarketservices.com).

The B2B buying process is predictably quite different from that in consumer marketplaces since these tend to be either marketplaces revolving around a specific industry sector (*a vertical marketplace*), or marketplaces based on products that form around a supply market cutting across several industries (*a horizontal marketplace*), or marketplaces focusing on functions such as management of a particular business function (e.g. recruitment) (Sculley and Woods, 2001).

B2B procurement phases include: requisitioning, request for quote, purchase order generation, payment processing. Figure 2.6 illustrates these phases.

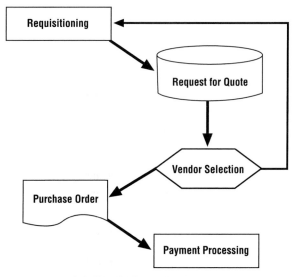

Fig. 2.6. The B2B purchasing process

The B2B procurement purchasing process translates into applications that are more complex and tightly integrated with corporate systems than B2C initiatives.

- *Purchase requisitioning.* This consists of an online form containing product information, quantity and cost filled out by the buyer and submitted over the internet for requisition approval. A workflow component in the form of an email message, fax or pager message is then sent to the appropriate manager. When approval is received the requisition is routed to the purchasing department and the appropriate purchasing agent begins the process of requesting quotes.
- *Request for quote.* This begins when the purchasing agent notifies potential suppliers that the enterprise is interested in making a purchase of specific goods in specific quantities. This request for quote is typically posted at a message board.
- *Purchase order generation.* A purchase order is generated through appropriate online channels. This may involve integration with the business's intranet.
- *Payment processing.* In this phase the data transformation component transmits payment information in the form of a corporate card, smart card information or an EDI transaction to the supplier in a format customized to integrate with internal supplier accounting systems. As more of these processes begin to be automated and integrated across enterprise boundaries in real time, corporations will begin to do business entirely in the virtual world.

B2B Business Models

There are different business priorities at play in the B2B arena compared to the B2C arena. Business professionals spending their company's money want to know how a purchase will add to their bottom line. B2B commerce is dominated by long-term, symbiotic, bio-supplier business partner relationships where stakeholders collectively work for common interests such as lower costs and improved product and service offerings. Successful B2B trading partners will be those who are prepared to collaborate electronically to maximize the new opportunities provided by the internet. Like any strategic business decision, it requires planning and careful return on investment (ROI) analysis.

B2B Application Models

As with B2C, the applications that ensure that the business model will work are important. Successful B2B trading requires a sound understanding of the three application models which are: **virtual marketplaces**, **procurement resource management**, and **extended value chains**. Customer Relationship Management (CRM) which is also an important part of B2B business is discussed in Chapter 4.

1. *Virtual or eMarketplaces.* Applications that enable enterprises to sell goods and services on the internet are referred to as virtual marketplaces. They must have one of the following trading means: auction, reverse auction, bulletin board, catalogue, exchange, or commodity exchange (e.g. www.emarketservices.com). Such markets provide new distribution channels for access to customers worldwide. Some good agribusiness examples are FarmNet (www.farmnet.com.au/html/services.asp), the Chicago Board of Trade which is a global commodity trading company (www.cbot.com), or E-Markets

– see Byte Idea this chapter. The virtual marketplace must have four key processes to be successful. These are:

- *Catalogue aggregation and creation.* An extension of the B2C model, enterprises populate an online catalogue with goods and services that interest and add value for the customer. Once the complete product offerings are in the catalogue, developers can focus on providing an interface to buyers that is consistent with the B2B buying process.
- *Support for complex buying processes.* Although the same see-buy-get cycle of B2C purchases is involved in B2B environments, developing and deploying a virtual marketplace is a continually evolving process of monitoring and responding to marketplace conditions. This inherent variability in the B2B selling process means that a flexible workflow and application design are necessary.
- *Integration with existing enterprise systems.* Virtual marketplaces connect multiple enterprises, each having distinct business architectures and buying processes. To operate seamlessly with these processes, virtual marketplaces must include a component capable of sharing information with internal company systems such as ERP applications, corporate databases and custom developed legacy systems.
- *Support for multiple types of payment.* Payment for purchases made in a B2B environment is typically made with a corporate cheque, corporate payment card, stored value card or smart card.

2. *Procurement resource management.* Traditionally most procurement applications have focused on maintenance, repair and operations (MRO) goods and services because these are often standardized and are thus well suited to automated high volume purchasing (e.g. office equipment such as ball point pens). The procurement resource management applications associated with B2B streamline the buying of both production and non-production goods and services. They facilitate the following processes:

- *Requisitioning.* During requisitioning, employers or purchasing agents identify which products they would like to buy, in what quantity, and at what price. In order to make these decisions buyers must be able to search integrated catalogues of supplier offerings with pre-negotiated prices.
- *Request for quote (RFQ).* Once the requisition has been approved the process moves onto the request for quote phase. By automating this process, buying organizations can reduce human error, shorten the purchasing cycle time, and incorporate suppliers around the globe who otherwise would be unable to participate in the sourcing process.
- *Purchase order generation.* Advanced procurement resource management applications customize electronic purchase orders to automatically integrate with individual supplier order entry and accounting software as well as internal processes.
- *Payment processing.* The receiving process represents the actual hand-off of goods from the supplier to the buying agent. Payment is typically in one of the same common methods as in a virtual marketplace – either corporate cheque or corporate smart card.

In procurement, since application-cost savings go directly to the bottom line, the benefits of these B2B applications include that they:

- Allow long-term relationships with suppliers to be managed and leveraged to create an enterprise-wide most favourable buying environment
- Lower requisitioning costs by automating the internal requisitioning process, reducing buyer personnel costs and time inefficiencies normally associated with requisition approval and order processing
- Reduce supplier costs by integrating with internal supplier inventory management and accounting systems, reducing supply costs and allowing savings to be passed down to the buying organization
- Ensure that appropriate approval is secure before any purchase can be made, thus increasing accountability and control
- Reduce the cycle time of purchases, decrease stocking requirements and lower inventory management costs for buyers with automation and workflow facilities
- Allow purchasing managers to monitor and analyse purchasing data so as to optimize order quantities and to ensure that business goes to selected suppliers.

3. *Extending the value chain.* The notion of the value chain is deliberately vague in order to accommodate and apply to very broad markets. Businesses buy something, add value to it, and sell it for a higher price. Extending the value chain simply means integrating isolated enterprise value chain environments to create a superior collaborative commerce infrastructure. In essence the members of the value chain of the e-enterprise become virtual companies that think and react as one entity and communicate around the clock in real time (Hoque, 2000).

   The result is the deconstruction of the linear value chain into dynamic value groups or nets (Bovet and Martha, 2000) that provide the latest, up-to-date information to all participants, and optimize the transaction of goods and services between partners from suppliers of raw materials to end customers.

   While the most obvious extended value chain platforms often aim to simply extend the functionality of the conventional supply chain management (SCM) systems, the ultimate aim of extended value chain applications is to share enterprise information with suppliers, buyers and business partners so as to enable supply planning, demand planning, production planning, and logistics to occur in real time (Thompson *et al.*, 2000; Boehlje *et al.*, 2000; Porter, 2001; Bryceson and Kandampully, 2004).

   Eventually the traditional isolated value chain model breaks down and new value grids are created where enterprises of all kinds share information to provide each of them with important strategic advantages over their competitors (Chapter 4).

Critical Success Factors of a B2B Environment

As indicated above, eMarketplaces are becoming a web of supply demand relationships with collaborative commerce blurring the line between competition and cooperation (i.e. coopetition – or in other words, it may not be the swiftest or strongest who wins out, but the one with the best partners).

A true eMarketplace should provide both market-makers and market participants with an open, flexible, reliable, highly available and scalable environment. Its functionality should span an array of capabilities that cross business processes delivering the greatest value to the customer, industry, or groups of customers and industries (Hajibashi, 2001). Table 2.1 lists the five critical factors for a successful eMarketplace.

Byte Idea – E-Markets
www.e-markets.com

ECommerce for Agri-Food Chains

E-Markets is a private company first set up in 1997 as an eCommerce applications provider for the agri-food industry. It provided the first online grain production contracting system, the first internet-based agricultural input ordering system and the first internet-based risk management tool for the grain industry. Since 1997, applications have become more targeted at agri-industry supply chains with applications focused on removing barriers between buyers and sellers, enabling information sharing across agri-food chains and streamlining back office processes in the chain companies.

Online solutions created by E-Markets address procurement, ordering and inventory, risk management, trading systems, and enterprise resource planning systems. Demonstrations of some of E-Markets' products are available online via the company website.

Agricultural partners include AgriBiz, Dow AgroServices, Feed and Grain, GSC and CropPlan Genetics with technology partners including Best Software, IBM, Great Plains Business Solutions, Global Exchange Services and Rapid Inc.

Information sourced from the E-Markets Media Kit found at the E-Markets website

Table 2.1. Critical factors for a successful eMarketplace

Critical Factor	Description
Early liquidity	B2B exchanges most likely to succeed are those with the greatest liquidity. The more buyers trade on a marketplace, the more suppliers will be tempted or forced to join them.
Right owners	Companies that bring liquidity to a marketplace are the logical founding partners and have the best chance of capturing the value it creates in the form of lower prices for goods and services.
Right governance	Good governance is required to make marketplaces, and particularly multibuyer marketplaces, work. It is necessary to ensure that buyers agree on the terms of their involvement and commit themselves to supply liquidity. Suppliers tend to reward well run marketplaces with better terms. The management team will also have to make critical decisions regarding technology, payment settlement, buying policies and the resolution of disputes among buyers or between buyers and suppliers.
Open standards	The marketplace must attract as much buying and selling as possible and thus they must operate under 'open standards' technically to ensure full integration between suppliers, the marketplace and the buyers.
A full range of services	Companies becoming involved in marketplaces are not only wanting to use the purchasing model as a way of increasing the competitive advantage but they want to cut their total cost ownership – that is, not just reducing the price they pay for their supplies, but also other supply-related services such as excess inventory and spoilage.

Summary

Competitive advantage is about having that certain 'something' which allows the organization to do better than its rivals in the marketplace. It tends to involve innovation in products and methods of doing business as well as having an underlying infrastructure to enable the business to proceed. Moving to doing business over the internet is both innovative from an infrastructure perspective and from a business strategy point of view – having an innovative and well thought out strategy is a key to future success in a business world that is in a state of constant change.

Lessons learned from the dot.com crash – most particularly that technology does *not* create a successful business, but that having a good revenue generating business model to ensure profitability *does*, along with the need to cultivate and care for the business's suppliers, buyers and customers – are the driving forces for developing the successful electronically enabled businesses of the future. One company that rode out the dot.com crash successfully was Amazon.com – its CEO, Jeff Bezos is the Smart Thinker for this chapter.

The Review and Critical Thinking Questions below have been provided for you to revise and reflect on the content of this chapter.

Review Questions

1. Define competitive advantage and explain why sustainable competitive business advantage is so important for increasing the life span of a company.
2. What has electronically enabling a business got to offer in the way of conferring SCA?
3. What are the value drivers of electronic enablement?
4. Why is the term 'New Economy' used and what are the drivers?
5. What are the main B2C and B2B business models and what electronic applications are involved in ensuring that they are successful?
6. Why has agribusiness been slow to adopt 'E' enablement?

Critical Thinking Questions

1. Analyse the dot.com crash for a company of your choice – what were the major issues? Relate this analysis to an agribusiness that you know of.
2. What are the critical success factors (CSF) of a B2C and B2B agribusiness of your choice? Why?
3. Compare and contrast the NASDAQ and the ASX.
4. What makes Bill Gates so successful?
5. Discuss the issues associated with the use of internet technology in the rural environment – both from a social and a business perspective.
6. Compare the growth strategies of Microsoft and Amazon.com and disucss the pros and cons of both.

Smart Thinker – *Jeff Bezos*

Founder and CEO of Amazon.Com

Jeff Bezos is the founder and CEO of Amazon.com, a company which is consistently ranked as one of the top retail sites on the internet. After graduating from Princeton in Electrical Engineering and Computer Science in 1986, Bezos worked for FITEL, Bankers Trust Company and then D.E. Shaw and Co. in New York where he helped build one of the most technically sophisticated and successful quantitative hedge funds on Wall Street. At Amazon.com, not only has he built one of the most successful e-tailers but he has showed how to build an online customer base so loyal that other e-tailers will pay Amazon big dollars to reach customers on the Amazon.com site.

Amazon.com opened its virtual doors in July 1995 with a mission to use the internet to transform book buying into the fastest, easiest, and most enjoyable shopping experience possible. Today, Amazon.com is a public company and is the place to find and discover anything you want to buy online. Millions of people in more than 220 countries use the online shopping site to peruse a huge selection of products, including free electronic greeting cards, online auctions, and millions of books, CDs, videos, DVDs, toys and games, electronics, kitchenware, computers and more.

What is interesting is that Amazon.com did not return a profit until 2002!! Mind you, since then, it has remained profitable and in January 2004 posted its first full-year net profit (of US$35.3 million on revenues of $5.65 billion in the calendar year 2003). Much of the growth of the company was due to its international division.

A TIME Magazine Person of the Year in 1999, Jeff Bezos's approach to the scepticisms surrounding the concern as to whether Amazon.com would ever turn a profit was to take the view that it would be short-sighted to optimize for short-term profits. He strategized that in the long term, investment in building great customer experience and introducing new customers to the company was more likely to create a sustainable competitive advantage. It seems he was correct – Amazon's share value at IPO in 1997 was US$18.00 – today it is US$41.40...Slow growth and attention to the customer has obviously paid off big time!

Source: Time Magazine (www.time.com), Wired.com (www.wired.com) and
Wikipedia (www.wikipedia.org)

Chapter 3

The KID Triangle – Knowledge, Information and Data in the 'E' World

Knowledge is power

This chapter examines a fundamentally important issue for successful businesses in the 21st century – managing the information coming into and out of the organization as well as that within the organization. The chapter also introduces the concepts of information ecology, the information economy and the information supply chain which have particular implications for electronically enabled business. Finally, the chapter takes a look at knowledge and knowledge management exploring the questions of what is knowledge, how do you (or the business) acquire it, is it really possible to manage it, and why is it important to attempt to do so from a business perspective. Nonaka's SECI model of knowledge creation is discussed as is the idea of knowledge as 'capital' in a business.

Chapter Objectives

After reading this chapter you should understand and be able to describe:

1. The differences between data, information and knowledge
2. What business information and information management in organizations are about
3. What the information ecology approach is and why it is fundamentally linked to good business management practices
4. What the information economy and the associated information supply chain mean
5. The implications for electronically enabled business of managing information online
6. What portals are and their evolution to the enterprise information portal (EIP)
7. The importance of knowledge acquisition and knowledge management to the business.

Distinguishing between Data, Information and Knowledge

Data, information and knowledge are not easy to separate in practice and essentially form a continuum, with the amount of human intervention increasing along that continuum. However, the following definitions adopted from Davenport (1997) simplify the essential question of what is the difference between the terms '*data*', '*information*', and '*knowledge*'.

- *Data* can be defined as a simple observation of states of the world. Data is easily structured, easily captured in machines (i.e. it is inert), often quantified and easily transferred. Examples include the raw results of a soil test such as a pH value or a clay content value or – from a business perspective – an individual share price of a company on the stock market.
- *Information* can be defined as data endowed with relevance and purpose. Information creates patterns in the data and activates meaning, it requires a unit of analysis, is dynamic, involves a lot of human mediation, and is easily transferred. For example, you could create information by using the soil test values noted above to create a spatial variability map of a paddock for, say, pH – or in the stock market example, by creating a graph of a company's share price over time.
- *Knowledge* is valuable information from the human mind and involves reflection on, and synthesis of, information in context. Knowledge is hard to structure, difficult to capture on machines, is often tacit (exists uniquely in a person's mind), is hard to transfer from person to person and forms the basis of intelligent action. For example, using the spatial variability map of pH talked about above to determine application rates of lime to manage the paddock for a particular crop requires knowledge. Similarly, a knowledgeable person will use share price trend information to help make decisions over buying and selling stock in that company.

The **KID** (*Knowledge, Information, Data*) Triangle shown in Figure 3.1 conveys the inverse difference between quantity and value of data, information and knowledge within an organization. The remainder of the chapter will follow the structure and flow of the triangle, dealing first with data and information and then knowledge issues.

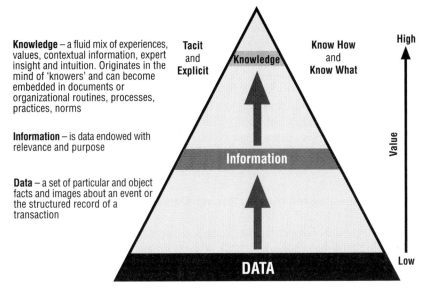

Knowledge – a fluid mix of experiences, values, contextual information, expert insight and intuition. Originates in the mind of 'knowers' and can become embedded in documents or organizational routines, processes, practices, norms

Information – is data endowed with relevance and purpose

Data – a set of particular and object facts and images about an event or the structured record of a transaction

Tacit and Explicit

Knowledge

Information

DATA

Know How and Know What

High

Value

Low

Fig. 3.1. The KID Triangle: information management basics

Types of Data and Information Commonly Found in Organizations

Data can be defined in many ways and includes numbers, characters, images or any other method of recording, in a form which can be manipulated by a human or via computer. Information science defines data as unprocessed information which is then converted into processed information and then into knowledge (Alter, 1999).

There are essentially two types of information in modern organizations: **unstructured information** and **structured information**.

1. *Unstructured information.* This is unorganized information and according to some analysts at least 80% of most organizations' critical information resides in this form. It includes any text-based content such as market research information, information on customer reactions to a new product, information on a competitor's business, or socio-political and legislative environments, along with rumours, gossip, stories, etc. Classically, in an electronically enabled business, it is scattered in various types of file formats across a company's network, thus very often being inaccessible or extremely difficult to find by the individuals who need it.

   Gathering this type of information is largely an *ad hoc* activity and is very labour intensive. Little, if any, value adding goes on as it is gathered, and it is difficult to control and manage – librarians being the only workers extensively trained in this type of information management.

   Intellectual capital is a specialized form of unstructured information since it encompasses the intangible assets and information sources of a company that can be expected to generate revenue, either directly or indirectly (Burton-Jones, 1999). It includes intellectual property (patents, trademarks, copyrights), employee know-how, and business rules incorporated into computer programs or databases.

2. *Structured information.* Structured information includes both paper and computer-based information that is logically and strategically organized to ensure that it is easily accessible. It includes paper files and computer printouts, microfilm, diskettes, computer tapes and CDs, optical disks, audio and video tapes.

This type of information has proliferated with the *Business Systems Planning (BSP)* approach where a top-down identification of business information requirements and a bottom-up identification and detailing of all computerized information items or elements used in business takes place. In such situations, an alignment of key data entities and classes with the processes using them within the organization and a grouping of data class and process relationships into specific computer applications and databases has proliferated.

Organizing and Accessing Information through Databases

There are many texts and books that address data and information management in great depth (e.g. Alter, 1999; Elmasri and Navathe, 2000), and it is not the intention of this text to go into great detail on databases; however, for the sake of completeness, this section briefly outlines the major issues involved in managing business data and information which a manager should be aware of.

Information plays an essential role in business. Maximizing its benefits in the business always requires careful organization and good access methodologies. There are three main questions relating to organizing and accessing information in a business:

1. What information is in the information system?
2. How is the information organized?
3. How can users obtain the information they need?

In traditional information management the focus is very much on managing computerized data and information. The process of identifying the types of entities in a situation (specific 'things' the system collects information about), the relationship between those entities, and the relative attributes of those entities (specific information related to those entities) is called data modelling. Data modelling also includes the decisions involved in structuring the information in computerized information systems or databases (McLellan, 1995). As such, databases today are a component of everyday life. A generalized definition of a database is that it is a collection of related data (facts that have explicit meaning). A more detailed definition includes the following:

* A database represents some aspect of the real world. Changes associated with this aspect of the real world are reflected by changes in the database.
* A database is a logically coherent collection of data with some inherent meaning (i.e. a random assortment of data is *not* a database).
* A database is designed, built and populated with data for a specific purpose. It has an intended group of users and some preconceived applications in which these users are interested.
* A relational database is a set of two-dimensional tables in which one or more of the key fields in each table is associated with a corresponding key or non-key field in another set of tables.

Traditional databases and applications accessed by most people include:

- *Bank accounts* where funds may be deposited or withdrawn from a bank
- *Hotel/airline reservation systems* which enable the booking of a hotel reservation or airline seat
- *Library systems* which enable library catalogues to be accessed online.

Non-traditional database applications include:

- *Multimedia databases* which store pictures, video clips and sound messages
- *Geographic information systems (GIS)* which store and analyse spatial information such as weather maps, land use maps, bushfire hazard maps, satellite images, taxi routes, etc.
- *Data warehouses and online analytical processing (OLAP) systems* which are used in many companies to extract and analyse information from very large databases for decision making
- *Real-time and active database technology* which is used in controlling industrial and manufacturing processes
- *Internet search engines* which have been built up using 'spiders' to scour the web for appropriate (see section later in this chapter on information quality) information to populate the database behind the search engine.

A database may be generated and maintained manually (e.g. library card catalogue), or it may be computerized (e.g. an online library catalogue). A database that includes a database management system (DBMS) is a database system. A database system includes a complete definition or description of the database structure and constraints which is stored in a catalogue. The catalogue describes the structure of the primary database by containing information or *metadata* (data on data) on:

- The structure of each file (number of fields, characters per field, etc.)
- The type and storage format of each data item (e.g. text such as RTF, HTML or graphics such as JPEG, TIF, GIF)
- Constraints to data (e.g. currency, completeness, doubts on collection technique, etc.).

Users obtain the information they need by querying the database via graphical user interfaces (GUI) developed using software to create the 'look and feel' a user encounters when using a system (e.g. Windows).

Text Databases, WWW Hypertext Databases, Search Engines, Spiders and Bots

A *text database* is a collection of related documents assembled into a single searchable unit. The individual documents can be large or small, but they should bear some relation to each other and should be stored on computer so that individual documents and information within the documents can be retrieved. They have become increasingly more important as computers and searching techniques have become fast enough to find information within text documents.

Text database management differs substantially from standard DBMS software, which will include some rudimentary text/text string search and retrieval functions for locating specific records, in that its focus is on searching for, retrieving and categorizing specific words, phrases or combinations of words.

A *WWW hypertext database* is a database in which information is stored as a document written in the hypertext markup language (HTML) which enables it to be displayed onscreen and accessed directly from the display. Very often, these documents contain interactive links that a user may select in order to move directly to other parts of the current document. The use of the World Wide Web (WWW) has focused on this type of access because the information on it is organized as a set of hypertext documents that can be downloaded on request from computers around the world. The WWW is accessed using software called a browser (e.g. Netscape) which provides a user's interface to the web and displays web pages to the user. Browsers either access the WWW via a URL or via a hypertext link within a document.

Search engines are software programs that find documents or web pages that seem to be related to groups of words or phrases supplied by the user (Tyner, 2001). The GUI (graphical user interface) front-end to a search engine allows the user to enter keywords that are run against a database (most often created automatically by 'spiders' or 'bots' – see below). Based on a combination of criteria (established by the user and/or the search engine), the search engine retrieves WWW documents from its database which match the keywords entered by the searcher. It is important to note that when you are using a search engine you are not searching the internet 'live', as it exists at this very moment. Rather, you are searching a fixed database that has been compiled some time previous to your search.

While all search engines are intended to perform the same task, each goes about this task in a different way, which leads to very different results. Factors that influence results include the size of the database, the frequency of updating, and the search capabilities. Search engines also differ in their search speed, the design of the search interface, the way in which they display results, and the amount of help they offer. In most cases, search engines are best used to locate a specific piece of information, such as a known document, an image, or a computer program, rather than a general subject. Table 3.1 lists some of the many search engines that exist with their URLs.

Table 3.1. A selection of popular search engines and their URLs

Search Engine	URL
Google	www.google.com/
Altavista	www.altavista.com/
HotBot	www.hotbot.com/
Fotosearch	www.fotosearch.com.au/
Game Spy Arcade	www.gamespyarcade.com/
Australian Jobsearch	www.jobsearch.gov.au/

Spiders and Bots are software programs that move across and around the WWW automatically (Venditto, 2002). They access the WWW hypertext structure by retrieving a document, and recursively retrieving all documents that are referenced. Some are more 'intelligent' than others – that is, they carry out tasks automatically although you must always remember that 'someone' has programmed them to be so. See the Byte Idea – The Big Four to get an idea of the way in which so much depends on so few in this data and information rich world.

Information Ecology

In 1997, Thomas Davenport proposed a way of looking at information management that takes into account the total information environment within an organization. Davenport argued that the information that comes from computer systems may be considerably less valuable to managers than information that flows in from a variety of other sources. He suggested an approach that puts people, not technology, at the centre of the company's information world and called the management of this approach *'Information Ecology'* (Davenport, 1997).

There are three environments associated with managing information ecologically and all three can be involved in any one project. These are:

- The information environment
- The organizational environment
- The external environment.

A change in one will affect the others – Figure 3.2 shows the interactions between the three environments.

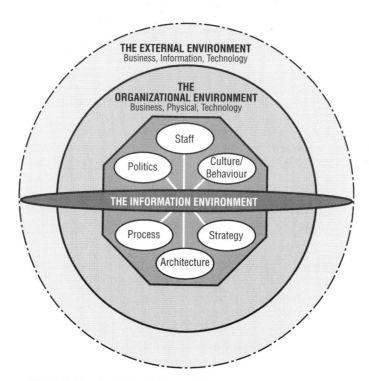

Fig. 3.2. The Ecological Model of Information Management
(adapted from Davenport, T. (1997), *Information Ecology*, Oxford University Press, p. 34)

Byte Idea – Sun Microsystems, Cisco, Oracle and EMC2

www.sun.com   www.cisco.com
www.oracle.com   www.emc.com

The Big Four

In the short-lived but legendary dot.com era, when venture capital grew on trees and the internet held the promise of amazing riches, no companies benefited more than the Big Four that kept (keep) the internet running. The web revolved around Sun Microsystems, whose servers host websites. Dot.com investors worshipped Cisco, whose routers direct traffic to all those servers. Oracle's business blossomed, as web companies used its database software for everything from managing eCommerce to delivering dynamically generated pages and, as the need to store huge quantities of web information grew by leaps and bounds, so did storage provider EMC2.

With the dot.com era passing, so has the phenomenal growth spurt of these companies – and the question is: what strategies will these companies put in place to enable them to weather the downturn? Some current strategies are:

- Sun Microsystems is using the internet to operate a dynamic bidding process in which contracts and purchase orders are auctioned off, slashing costs by 23%. In 2005 they teamed with Google in an agreement to share the two companies' technologies around the world.
- Cisco has become synonymous with a strong service provider channel strategy using the internet and marketing associated with the productivity implications that internet business solutions provide.
- Oracle has taken on board internet-enabled enterprise software across its entire product line: database, server, enterprise business applications, and decision support tools and in 2005 uses the 'Information Driven Enterprise' as its unique selling point – reinforcing this with a merger with Peoplesoft, the ERP software developer.
- EMC2 – offers a complete information management framework including products, services and technologies and is moving towards being the 'ultimate information life cycle company'.

Information sourced from CIO Magazine (www.cio.com),
Network World (www.networkworld.com) and the respective company websites

The Information Environment

The information environment is the core of an ecological approach to managing information. It encompasses:

> **all information services that are accessible to users of the information, together with all available means for coordinating the use of those services that enable users to access, evaluate and use any information that may be extracted from the total information resources available.**

There are six critical components: information strategy, information politics, information behaviour and culture, information staff, information processes and information architecture. These components are discussed below.

Information Strategy

Creating an organizational information strategy (Griffin, 2000) revolves around the question 'what do we want to do with information in this organization?' and must involve senior management (Distefano, 2000). It should encapsulate a set of attitudes in which any information that should be available for sharing (and most will be) is well defined and appropriately accessible (allowing for necessary safeguards); where the quality of information is fit for its purpose (e.g. accuracy, currency, consistency, completeness – but only as far as necessary); and where all staff know, and exercise, their responsibilities towards information. Information strategies are likely to change and will require revision.

Information Politics

Information politics is about the power that information provides and the governance responsibilities for its management and use. Information management can be used to distribute power or to concentrate it and it is a choice issue based around many factors such as organization size, major business and organizational structure. Thus information is affected all day, every day, by power, politics and economics. However, the politics of information remains a subject that is not likely to be discussed in great detail despite its importance, because of the covert nature of power acquisition and maintenance based around information – a dynamic that would be seriously threatened if it were made more transparent.

In terms of managing information politics, there are some key areas that Strassman (1995) suggested can be addressed to reduce the opportunity for covert power gain:

1. *Link business and information technology plans* – it is extremely important that the organizational business goals drive the information technology, information system, information architecture and information management planning processes. By making the plans widely available throughout the organization, the process becomes transparent.
2. *Define information management goals* – understand how to define the goals and principles of information management particularly in relation to the use of standards as a balance between rigidity and chaos.
3. *Promote information sharing* – the need to encourage information sharing within the organization is important. Changing existing cultures which may actually facilitate 'information hoarding' (see below) is a difficult task but must be addressed for a successful business-focused information management culture to be attained.
4. *Organize for information security* – this also encompasses dealing with the issue of privacy of personal information on personal computers. A good security structure facilitates information accountability measures.
5. *Manage organizational knowledge* – knowledge has a tendency to 'walk out of the door' when people leave. Recognize the need to preserve organizational knowledge and promote information management as a core competency of the business. This approach is only now being recognized as truly important for ongoing sustainable competitive advantage of an organization.

Information Behaviour and Culture

Information behaviour and culture are quite probably *the* most important factors in creating a successful information environment and are tightly intertwined. Information behaviour is associated with how individuals handle information – how they search for it, use it, modify it, share it, hoard it or ignore it. According to Davenport (1997), behaviours that should be encouraged include:

• *Information sharing* – as opposed to information hoarding. This should be voluntary to maintain people's trust and goodwill.
• *Handling overload* – the extent and complexity of information content, particularly that available online, requires an emotional engagement of individuals to discern the chaff from the wheat in addition to having physically well structured content, well-managed workflow, clear permissions and security practices (Dalton *et al.*, 2001).
• *Dealing with multiple meanings* – the problem of multiple meanings that can be obtained from information pre-dates computers but the problem has been escalated with their proliferation, the associated jargon and the ready availability of information online from many different sources.

Corporate or information culture is 'the way we do things around here'. It is the pattern of behaviours and attitudes that express an organization's orientation to information. It is invisible and formidable and it provides the context in which business is done. Changing how people use information and, ultimately, how they use it to build a supportive information culture, is the pivotal issue in information ecology – and it does not just 'happen'. Culture is slow, in fact almost impossible, to change. Brooking (1999) outlines a number of knowledge management concepts from which she suggests *'corporate memories may live forever'*, as the most important to remember and understand. This phrase in essence alludes to the tendency that in times of great and speedy environmental/economic change, an organization's culture may well prevent the organization as a whole from moving forward.

Information Staff

Many people provide and interpret information. Of particular importance organizationally are content specialists such as librarians and market researchers, designers and facilitators of information bases along with technical staff such as programmers, systems analysts and database administrators who maintain and implement the IT infrastructure. There are six broad characteristics of a good information person and these include:

1. A broad business understanding and knowledge of the organization's structure and function
2. A knowledge about the diverse sources and uses of information in the organization
3. A familiarity and ease with using IT
4. Political savvy as well as the ability to exercise leadership
5. Strong interpersonal skills
6. A strong orientation towards overall business performance, rather than a narrow allegiance to internal functional goals.

The primary goals of information staff include making information meaningful, ensuring 'clean', accurate and timely information, ensuring that information is easily accessible, and that the user engages with the information. The primary tasks of information staff involve:

- Pruning information to get rid of extraneous material and to improve the quality of the information
- Adding context to information such as source of the information, previous history surrounding the information
- Enhancing the information style, which is defined by wording, emphasis, variation, interactivity
- Choosing the right medium for presentation (e.g. video, overheads, face-to-face, emails).

Information Processes

These define how information work gets done. Ideally information processes involve all activities undertaken by information workers and should be defined by the information management strategy, itself aligned with the business plan.

Information Architecture

Information architecture (IA) is a guide to the structure and location of information in the organization. It may be descriptive, such as a map of where things are found in the information environment, or prescriptive, such as a model of the environment.

The Organizational Environment

A company's information environment is always based in the wider organizational environment which includes the overall business situation, existing technology investment and physical arrangement. The *business situation* includes a firm's business strategy, business processes, organizational structure/culture and human resources. All these have impact on the information environment, particularly the information strategy (and vice versa).

The investment in technology a business makes will influence the information environment and the accessibility of information. Information is best shared when people are in the same space. Studies have shown that physical proximity also increases the frequency of communications, thus the almost ubiquitous presence of desktop computers and laptops connected by local area networks has provided a good infrastructure in most current-day organizations.

However, according to the Gartner Group IT EXPO, Brisbane, (1999) while information management programs perform best when enabled with sophisticated and elegant technology, an emphasis on technology alone will achieve very little towards good information management – indeed it can 'drive out' information and limit creative thinking; conversely, the strongest information management culture that is not supported by robust technology will also falter.

The External Environment

A company's information ecology is affected by external factors, some of which cannot be controlled – for example, new legislation or competitors making sudden moves. However, companies need to interact with, and must have information about, the external environment including the general business market, technology markets and information markets.

- *Business markets* create the general business conditions for firms and thus have an effect on their ability to acquire and manage information and the types of information they need. This includes information on changes in customers, suppliers, business partners, regulators and competitors.
- *Technology markets* are where available technologies for creating and managing the information world are bought and sold.
- *Information markets* are where everything from industry trends to mailing lists are bought and sold.

The Information Economy, Information Supply Chain and 'E' Enabled Business

The *information economy* is essentially the global economy that has emerged from the relatively newly acquired ability to access and transfer information from anywhere to anywhere at any time. In Australia, the Commonwealth Government's vision for the country in the information economy, and a national framework to achieve it, can be found in a policy document called 'A Strategic Framework for the Information Economy' which has been coordinated by the National Office of Information Economy (NOIE, 2002). In this document, originally released in January 1999, with the most recent update being in November 2002, the prediction is that the reach and impact of the information economy will grow enormously and that full engagement with this approach will be essential for Australia's future.

The information economy goes far beyond a simple economic state to encapsulate all aspects of everyday life now and in the future – from government to big business to the home. One of the major changes in this new environment is in the way industries and business organizations are structured and operate (Turner, 2000). As has been emphasized in the early chapters in this book, electronic technologies, particularly the internet, are allowing faster and more efficient methods of exchanging, transmitting and processing information and of networking. From procurement, product design and manufacturing, through to distribution, marketing and customer relationship management, internet technologies are changing the rules, internally and externally. From basic human resources through to the final cost structure of a product, the business systems profile is becoming dramatically different.

This capability has given rise to the term *information supply chain* which contains, according to George Marinos (2005), of PriceWaterhouseCoopers:

> **the full set of elements – technology-based, process-specific and organizational in nature – that are necessary to 1) collect information from discrete processes, 2) transform this information from data into knowledge, and 3) distribute this information efficiently and in a timely manner to the appropriate data consumers.**

In other words the information supply chain has an impact on every aspect of doing business – how, who with, when, where – and thus the basic structures that support those activities. This is discussed in more detail in Chapter 4 in the section on electronically enabling supply chains, but Figure 3.3 below illustrates what the information supply chain is with an example of how it can directly affect business by reducing transaction costs.

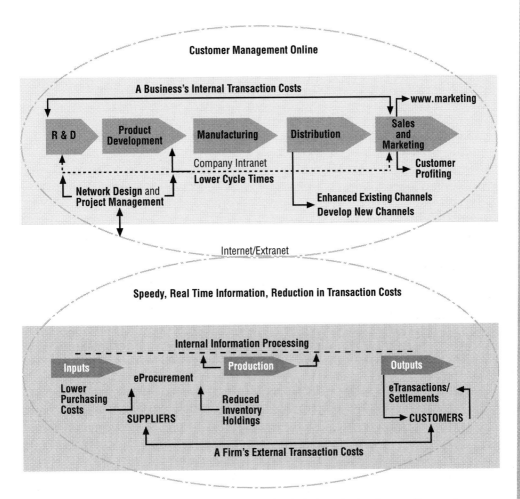

Fig. 3.3. The role of the information supply chain in reducing transaction costs

Transactions are at the core of any business operation, forming one of the most significant components of business systems and overall costs. All transactions are information based, relying on exchanging, obtaining, using and tracking information throughout their life cycle. The costs associated with these transactions are firmly ingrained in the physical world and most industries are arranged into long and complex value chains, the shapes of which are largely organized by transaction costs.

These value chains are often populated with 'middlemen' who prosper by taking a 'cut' out of every transaction they organize within their sphere of influence. Since the process of changing key suppliers and distributors in the physical world is so lengthy and involved, most buyers and sellers simply live with the cut imposed rather than attempting to economize on transaction costs. On the internet, the capability of a free flow of information between anyone means that the transaction costs associated with most forms of commerce can decrease substantially if the scenario is managed appropriately. This does not imply that middlemen have no place in an electronically enabled business environment – many new types of intermediaries will prove quite successful online. The key difference is that these new intermediaries have to find ways to create value for themselves in an environment where certain forms of transaction costs have largely evaporated.

Managing information, information flow and information use are thus significant components of running a successful and sustainable electronically enabled business and must be understood and planned for in managing for the future (Shapiro and Varian, 1998).

Portals

Webster's Dictionary defines a **portal** as a doorway, gate or entrance. More importantly, from an electronically enabled business's perspective, it is defined in the electronic world as an internet, intranet or extranet 'door' or single point of access that users pass through to reach sources of information in their various forms – for example, in databases, documents, within a department, in an application or web-based link to other portals or sites. Using this electronic world definition, there are two main types of portal:

1. *General purpose portals* – a single site where users can find links to various information sources or online businesses (e.g. Yahoo)
2. *Vertical portals* – which provide the same functionality as a general purpose portal but for a very narrow defined area of interest such as travel or business decisions.

The software applications associated with current portals have developed significantly in terms of functionality in enabling access to large and complex volumes of content. Three stages of complexity can be seen across the current portal offerings:

Stage 1. Consolidation. In this phase, relatively non-complex functionality is employed with multiple information sources being brought to a single point of access online.
Stage 2. Customization. In this phase, functionality is increased to allow selected information sources to be used based on personal preferences.
Stage 3. Personalization. In this phase, superior functionality is included in the design of the portal so that specific information, regardless of source, and which exactly meets a set of personal requirements, can be accessed.

Although some of the larger companies and particularly those that do significant business online have portals that display functionality spanning all three phases, most businesses are still at Stages 1 or 2 in developing their portals. There are a number of different portal market segments which are based on the level of complexity a business is after. These market segments include:

- *Information portals*. These automate searching, catagorizing, organizing and publishing intranet-based information and enterprise reporting – for example, the Agribusiness Universe portal which focuses on Asia (www.agribusinessuniverse.com/), or Farming UK (www.farminguk.net), and in Australia InFarmation.com – see Byte Idea this chapter.
- *Application portals*. These are mainly browser front-ends to a software vendor's own application system such as an enterprise resource planning (ERP) or customer relationship management (CRM) system (Chapter 4). They do not connect heterogeneous systems – for example, Bolthouse Farms, one of the largest suppliers/distributors of carrots in the world, has used Ciber software to develop a front end to their company ERP system (www.ciber.com/services_solutions/).
- *Employee portals* (business to employee – B2E). These combine ERP and human resource information systems and the goal is to assemble structured, unchanging processes into role-based routines – for example, Syngenta (www.syngenta.com) working with SAP (SAP, 2003).
- *Collaborative portals*. These facilitate the development of online team rooms, collaborative projects, management, discussion groups and email. A very good example is the AgroWeb Network that has been set up in Central and Eastern Europe (www.agrowebcee.net).
- *Expertise portals*. These provide access to expert networks and markets – for example, the Indian AgMarkNet (www.agmarknet.com) – and may be powered by software such as Ciber.
- *Decision portals*. These are knowledge portals which combine information portals, collaborative portals and expertise portals to automate and improve the decision-making cycle – for example, Agri Beef Co. (www.agribeef.com) which is a privately owned diverse agribusiness grounded in the cattle industry in the USA, has implemented ProClarity software to analyse data and provide decision-making information from its mission-critical production systems for custom cattle feeding, feed supplement manufacturing and distribution as well as high-end beef products distribution and packing (DMReview, 2003).

Byte Idea – Agribusiness Universe
www.agribusinessuniverse.com

A Global Reach from an Asian Perspective

Agribusiness Universe is one of the euniverse.com research websites run by Intermix (www.intermix.com). The Intermix Network has more than 30 websites providing a wide variety of content and is also a permission-based marketing network.

Agribusiness Universe focuses on providing a gateway to agri-industry information for people in the Asian agri-industry sector, researchers and financial investors (as a subsidiary of Mulliscapital – www.mulliscapital.com). The portal primarily services the dairy, fruit and vegetable, poultry and meat, rice and grains, sugar and seafood industries.

Divided into 'industry portals' it includes a wide range of interesting industries (including the Tea Auction Portal – www.teaauction.com), specific industry sector research sites, pertinent company websites, and related links to international agribusiness organizations. Agribusiness Universe puts people searching for agri-industry information in touch with that information.

Information sourced from Agribusiness Universe website

Corporate or Enterprise Information Portals (EIPs)

The terms *corporate portal* and *enterprise information portal (EIP)* are interchangeable and refer to a browser-based application that allows employees and/or business partners to gain access to secure business-related information from a variety of sources to foster productive decision making (Maza, 2001).

The EIP concept (Firestone, 1999; Hall, 2000) uses both 'push' and 'pull' technologies to combine different information from the WWW, corporate databases and applications using web browsers and search technology. EIPs should be able to link to information sources outside the corporate firewall such as internet news feeds, real-time events (e.g. stock quotes), calendars, news announcements and internet discussion groups, and should also be able to support a bi-directional exchange of information with these sources. IBM is one of several companies offering advanced EIP software (www-306.ibm.com/software/data/eip/).

Data feeding into an EIP can be structured (i.e. numeric as in corporate databases and applications) or unstructured (e.g. data found in email, eCommerce site log files and the internet) and thus data and information acquired from many different sources can be used for further processing and analysis. A key functionality requirement is that EIPs should provide 'interactivity', which is the ability to question and share information, on user desktops.

EIPs integrate disparate applications (see Fig. 3.4) including content management, business intelligence tools, data warehousing, data marts and data management, along with other data external to these applications, into a single system that can *'share, manage and maintain information from one central user interface'* (Luce, 2002).

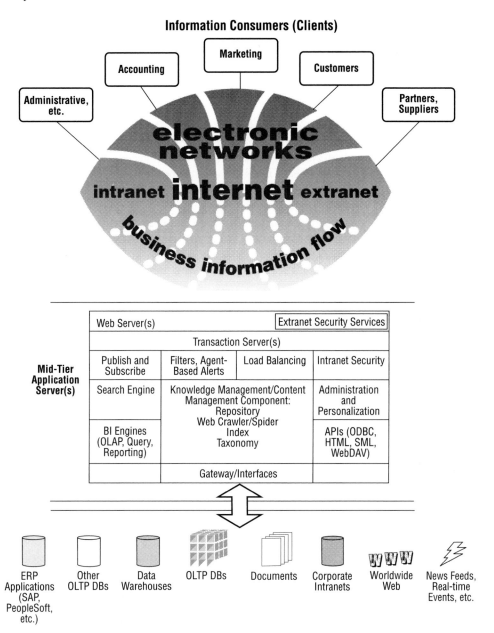

Fig. 3.4. EIP architecture (adapted from Hall, 2000)

Because the EIP concept is based on a functionality that can be tailored around a particular group of users, EIPs exhibit the trend towards 'verticalization' in application software. That is, they are often packaged applications providing targeted content to specific industries or corporate functions (Mercy, 2001).

EIP Key Features

EIPs offer a solution to the problems of distributing and consolidating business information (e.g. reports, documents, spreadsheets, etc.) generated anywhere in the enterprise by any application, and of making them easily accessible, subject to security authorization, to non-technical users via standard browser technology. The key features include (Firestone, 1999; Hall, 2000):

- *Scalability.* The portal must scale to support a huge number of discrete information objects without affecting availability or response time. Portal architecture must be capable of scaling to support a large number of users and applications on a wide range of platforms and devices.
- *Access/Search/Navigability.* EIPs should offer multiple ways to identify potentially valuable information including a search engine for text-based searches, an indexing system that is standard across all information objects, and hypertext linking within documents to enable 'jumps' to associated information objects.
- *Security.* A crucial feature of EIPs is enforcing security for the multitude of information objects that can be accessed. Most objects are assigned security levels when they are generated by the business application. The EIP must be capable of plug-and-play integration with existing authentication schemes. Security enforcement at the portal enables users to log in to access all business information objects they are authorized for.
- *Dynamic execution.* EIPs must provide the ability to execute reports from production systems or databases in real time giving users access to the most up-to-date information – for example, a user clicks on a report object, which updates automatically from the underlying application and displays the most current data.
- *Ease of use/categorization.* Effective portal technology meets users at their skill level, providing a basic interface with simple choices for the novice and a range of interactive business analysis and reporting tools for power users. The portal should also be able to categorize information in a required context as a static report. Interactive features such as drill down and hypertext linking should also be available to users.
- *Ease of administration.* Tools used by administrators and power users to create reports, to file and index business information objects in the repository and to administer the repository (deleting obsolete objects, ensuring standard tagging of new objects) should be easy to use. EIPs must leverage existing data and existing business logic rather than requiring the company to rebuild its data store as with data warehousing technology.
- *Extranet support.* EIPs must function on either side of the corporate firewall to enable a corporation to effectively open its business information object repository to mobile users, customers, distributors, subsidiaries, partners and any other parties who need access to company information (subject to security clearance, of course). This free exchange of information is a key enabler of corporate electronic commerce.
- *Personalization/Customization.* EIPs should permit customization by the end user including the arrangement of the browser (incorporating real-time web feeds, headlines, notification of report availability, etc.). The portal must permit both 'push' and 'pull' report distribution, so that users can 'subscribe' to information based on interest (e.g. all sales reports for the 'X' region) or by 'exception' (e.g. receive an inventory report when levels of item 'Y' dip below a user-defined level) or call up information from the repository upon request.

EIPs and Content Management

Content is the generic term used to cover all the unstructured and semi-structured data and information that an organization owns, in particular the term is used to refer to that material which resides online (Dalton *et al.*, 2001). Such material may be sourced from diverse sources such as web documents, research reports, contracts, government licences, brochures, purchase orders, data warehouses, data marts, and other decision support and executive information systems, legacy systems, enterprise application servers (e.g. SAP, Baan, KDD/Data Mining Servers, Stock Transaction Servers), or any document with content relevant for some corporate interest that resides in an EIP.

   Content management refers to the task of managing a collection of unstructured and semi-structured information that an enterprise owns as well as the tasks of integrating information from upstream content. *Content management systems* process, filter, and refine 'unstructured' internal and external data and information contained in diverse paper and electronic formats, archive and often restructure it, and store it in a corporate repository (either centralized or distributed) (Sullivan, 2001). Because of the focus on such a wide variety of documents, applications and data stores along with the problem of capturing them, and extracting and analysing information contained in them, the technologies involved are quite varied. They include:

- Imaging and scanning
- Workflow and groupware (to get the raw material of articles, photos and other content into the system)
- Document management with permissions-based updating, to let users with different levels of authority update a site to varying degrees and visitor tracking, to provide feedback on how frequently and in what order users view given pages
- COLD storage (a repository of archived material)
- Business process automation
- Search engines providing keyword, phrase or concept-based searching
- Text mining
- Object orientated database management systems
- Video streaming
- Intelligent agents (which have played a role in data warehousing systems are now also important because of their roles in optimizing focused acquisition, retrieval, analysis and transmission of content).

EIPs versus Data Warehouses

Data warehouses (Lambert, 1996) and *data marts* (Brooks, 1997) are integrated, time-variant, non-volatile collections of data supporting decision support systems (DSS) and executive information systems (EIS) applications, and, in particular, business intelligence tools and processes. The differences between a data warehouse and a data mart are shown in Table 3.2.

Table 3.2. Differences between a data mart and a data warehouse

Data Mart	Data Warehouse
Focus is on one subject area or one group of users	Focus is on many subjects and many user categories
May be many in an organization	Tends to be only one per organization
Do not contain operational data store information	Contain operational data store information
Easily navigated and understood	Complex to search and use and contains so much information it can 'overload' a naïve user

EIPs may be viewed as an expansion of current trends in data warehousing systems which now contain, amalgamate and integrate web-based interactive querying and reporting, business intelligence gathering and synthesizing, data warehouses, data marts and data management applications. The main differences between EIPs and data warehouses are shown in Table 3.3.

Table 3.3. The differences between an EIP and a data warehouse

EIP	Data Warehouse
Integrate content management systems	Traditionally have interacted with structured legacy systems dealing with online transactional processing (OLTP)
Emphasize the exchange of data with external data stores and applications	Data resides in the data warehouse. Little or no exchange takes place between external data stores or applications
Connectivity between the EIP and external data stores and applications is important	Connectivity is not important
Emphasize the sharing of data and information among users. Workflow and business process automation technologies are incorporated	Data and information sharing is not a design criterion
Emphasize data mining and analytical applications	Data mining is important but analytical applications are less so
Integrate disparate applications and data sources into a single, integrated application	A high level of integration is not required
Facilitate the flow of information and support a full information supply chain (which is the chain created by 'getting' the information, 'using' the information and 'acting' on the information)	Store information

Data Mining

Data mining is defined as: **a technology that allows the extraction of hidden predictive information from large databases** (Thearling, 1999) and it thus allows questions that traditionally required extensive hands-on analysis to be answered directly from the data quickly.

Data mining automates the process of finding predictive information in large databases. For example, in the case of targeted marketing, data mining of a company's customer information database will make use of past promotional mailings to identify the targets most likely to maximize return on investment in future mailings. Other predictive problems include forecasting bankruptcy and other forms of default, and identifying segments of a population likely to respond similarly to given events.

Data mining tools sweep through databases and identify previously hidden patterns in one step. Discovering the patterns in data could include the analysis of retail sales data to identify seemingly unrelated products that are often purchased together, the detection of fraudulent credit card transactions and the identification of anomalous data that could represent data entry keying errors. The most commonly used techniques in data mining are (Berry and Linoff, 2004):

- *Artificial neural networks.* These are non-linear predictive models that learn through training and resemble biological neural networks in structure.
- *Decision trees.* These are tree-shaped structures that represent sets of decisions. These decisions generate rules for the classification of a data set. Specific decision tree methods include classification and regression trees (CART) and chi square automatic interaction detection (CHAID).
- *Genetic algorithms.* These are optimization techniques that use processes such as genetic combination, mutation and natural selection in a design based on the concepts of evolution.
- *Nearest neighbour method.* This is a technique that classifies each record in a data set based on a combination of the classes of the n record(s) most similar to it in a historical data set (normally 3n x 3n).
- *Rule induction.* The extraction of useful if-then rules from data based on statistical significance.

Byte Idea –
@griculture.com and InFARMation
www.agriculture.com
www.infarmation.com.au

American and Australian Agricultural Information Portals

@griculture Online was one of the first agricultural websites in the world and was launched on 10 May 1995. As with so many of the currently successful 'dot.com' companies, the owners were already successful 'offline' business people, being editors of Successful Farming® magazine (the largest paid-circulation farm magazine in America and the largest paid-circulation business magazine in the US) and the publication that started Meredith Publishing Corporation in 1902.

@griculture Online is an information portal for agriculturalists – it is a community of publishers, organizations, companies and individuals providing the latest information in agricultural news, weather and markets. There is also a 'talk' group where visitors and registered members provide the grassroots voice of farmers and associated individuals. Discussions occur on more than 20 different topics, from livestock and machinery, to agricultural computing and farm humour.

@griculture Online is a good example of a company that has a sound business strategy combined with an in-depth knowledge of their customers and users which has allowed them to stay at the front of the pack.

InFARMation is owned and run by its parent company, InfoChoice Ltd (www.infochoice. com.au) which is an infomediary company specializing in providing consumers with independent, useful, valued information about specific industry sectors.

InFARMation is the agricultural arm of InfoChoice Ltd and is focused on providing information to the Australian agricultural sector. It is designed to act as an agricultural community online, while providing a marketing vehicle for buyers and sellers involved in Australia's rural sector. It delivers customized information such as a range of weather reports and commentary, detailed market prices and analysis for a number of agricultural sectors and the latest industry news.

Information sourced from company websites

Portals and Electronically Enabled Business

As emphasized in Chapter 1, electronically enabling a business is about facilitating trading, particularly in information which may lead to a deal. However, electronically enabling the business doesn't simply mean you build a website, hope your customers and partners will buy into it, share data, allow access to information via a web browser and sell a few items online. That approach is doomed to failure in very quick time. A correct approach requires careful technical architecture design, a well-planned strategy and a commitment to altering the current business model to fit an electronically enabled business (Thompson, 2000). Also, it requires planning and consideration of what partners, retailers and suppliers are using as the technical backbone for their own business to business solutions in an effort to achieve compatibility with trading partners.

To excel in electronically enabled business therefore, the most critical components are *accurate information* and *seamlessly tied systems*. Without accurate information, the integrity of shared data is compromised and without appropriately planned and managed systems, inefficiencies in the availability and delivery of information occur. These two outcomes combine to cause the customer to suffer resulting in the high probability of their taking their business elsewhere, see Byte Idea on The Seam (www.theseam.com) this chapter.

Data warehousing enables the consolidation of data from product, manufacturing and customer databases into operational data stores. Web access via a portal to data warehouses by suppliers and customers strengthens relationships along the supply chain, which potentially allows greater value to be realized. For example, online ordering saves both cost and time – two of the greatest value propositions for any organization and allows *'delivery channel applications'* such as customer relationship management systems access to suppliers, products, inventory and customer data. An effective electronic enablement strategy supported by tightly integrated systems can:

* Make the supply chain more efficient
* Provide the ability to track transactions throughout the supply chain
* Help suppliers understand customer behaviour
* Help more accurately targeted sales and marketing efforts
* Help and automate processes across the enterprise – and with external vendors and partners
* Strengthen the relations between customer and supplier by sharing accurate data among involved parties.

EIPs raise the potential of a company to undertake good business by enabling information flow, analysis and customer centricity. Companies such as Dell use the portal concept to complete the electronic supply chain integration necessary to link directly to their suppliers.

Knowledge and Knowledge Management

Moving towards the apex of the KID Triangle (Fig. 3.1) referred to earlier in the chapter, we need to understand the term 'knowledge' in the context of: a) doing business and b) doing business electronically. There are many texts that cover knowledge and knowledge management but two quotes from the 'gurus' in the theoretical management field – Ikujiro Nonaka and Peter Drucker – can point the way:

> *'In an economy where the only certainty is uncertainty, the one sure source of lasting competitive advantage is knowledge'*
> (Ikujiro Nonaka, 1991, 'The Knowledge-Creating Company', p. 91).

> *'Knowledge has become the key economic resource and the dominant – and perhaps even the only – source of comparative advantage'*
> (Peter Drucker, 1995, 'Managing in a Time of Great Change', p. 271).

The implication of these statements is that knowledge influences everything from a company's strategy to its products, from its processes to the way the business is organized. The statements also highlight some serious issues in the practical arena which can be focused into the following questions:

1. What is knowledge?
2. How is knowledge created?
3. How is knowledge managed?
4. What knowledge needs to be managed to create maximum advantage to the organization?

Byte Idea – The Seam
www.theseam.com

A Global eMarketplace selling Cotton and Peanuts

The Seam was set up in 2000 based upon the old Telcot marketing system owned by Plains Cotton Cooperative (PCCA) – a farmer-owned cotton marketing and denim manufacturing cooperative in Lubbock, Texas (www.pcca.com). Telcot was a computer based system allowing gins to offer their cotton direct to the market through a network of terminals.

In 2000 The Seam was created by PCCA, Cargill Cotton, Allenberg Cotton Co and Dunavant Enterprises. It took the old Telcot marketing system, converted it for web-based operation and took it outside of Texas. Today The Seam shareholders include an electronic title company, other cotton cooperatives, merchants and mills all over the USA.

In February 2003 The Seam launched the International Marketplace. This is an electronic marketplace to allow buyers and sellers to meet and negotiate price and terms for forward delivery cotton from the US, Brazil, Australia, West Africa, Paraguay, Argentina, Greece and Pakistan. West Africa and Brazil represent 85% of volume to date which is just over one million bales. It is currently considered a merchant to merchant system, with no mill or producers actively trading (although in principle they could).

In 2004 the USDA nominated The Seam to auction all cotton that defaults to the CCC loan program (under the Commodity Credit Corporation (CCC) Charter Act, the USDA may make loans to producers to build or upgrade farm storage and handling facilities). In the 2004/2005 season nearly three million bales were traded in the USA – all spot transactions with immediate payment against electronic warehouse receipts. The same services commenced for peanuts in August 2005.

In July 2005 The Seam started running electronic tenders on behalf of the grower association in Mato Grosso, Brazil (AMPA) with over 50,000 metric tonnes being offered by the end of August. This is a new product and allows the growers direct access to the international market (merchants) without going through brokers or agents. Brazilian mills have also started to sign up to it.

It is this last development that shows the power of the internet when it comes to marketing a commodity to the international market.

Information sourced directly from Mr Bill Ballenden – The Seam Management

What Is Knowledge?

Knowledge was defined in the KID Triangle as something that originates in the minds of 'knowers' – a fluid mix of experiences, values, contextual information, expert insight and intuition. A shorthand version of that definition is that:

knowledge is information in context plus understanding.

For all organizations, the cultivation of this knowledge – often an implicit, unreflecting cultivation – is the essence of developing core competencies necessary to maintain the organization and resist its dissolution. However, organizational knowledge that constitutes a core competency is more than just '**know what**' (explicit knowledge that is easily shared with others, e.g. in manuals, reports, etc.). A core competency (defined by Prahalad and Hamel (1990) as the collective learning skills behind a business's product lines) requires the more elusive '**know how**' or in-head tacit knowledge which is an individual's particular ability to put 'know what' into practice (Seely-Brown and Duguid, 1998). While 'know what' and 'know how' work together, they circulate within an organization separately.

- *Know what* circulates relatively easily and therefore can be hard to protect requiring the development of intellectual property laws.
- *Know how* is hard to spread, coordinate and benchmark, but is easy to protect because it tends to be in-head knowledge of individuals who, unless favourably persuaded to release it, will tend towards 'knowledge hoarding' as their employment currency. Tacit knowledge is generally very poorly captured within organizations and is *the* major loss when people leave.

How Is Knowledge Created?

If you accept that there is both personal and corporate value in knowledge, the question is: how do you acquire it?

Knowledge, knowledge acquisition and knowledge management are much studied and discussed subjects across many disciplines, but perhaps the best known knowledge research work is that of Michael Polyani who discussed tacit knowledge in depth in a book called 'The Tacit Dimension' published in 1966, and Ikujiro Nonaka who, in the 1990s, described in a series of papers (Nonaka, 1991; Nonaka, 1994; Nonaka and Takeuki, 1995) that there is a four stage model of knowledge creation called the SECI model or *'knowledge spiral'*.

This model (see Fig. 3.5) describes how tacit knowledge (in-head, unspoken knowledge), through a process of *Socialization*, is *Externalized* (becomes explicit) with the explicit knowledge then being *Combined* via communication and diffusion processes across peers or a group, and finally *Internalized* by group members as learning.

Fig. 3.5. The SECI model and the knowledge spiral

While the SECI model of knowledge creation is highly regarded, there are some difficulties with it. For example, if tacit knowledge really is 'in-head' knowledge that is difficult or impossible to articulate as Polyani (1966) believes, it cannot be externalized – and if it cannot be externalized, how is knowledge ever transferred between people? Indeed, how can knowledge ever be created?

The arguments and discussions about how knowledge is created can be non-ending (even spiral in nature!) but are well captured in a paper by Hildreth and Kimble (2002) who discuss the duality of knowledge drawing on a number of authors and theories of knowledge creation: not least of which is Etienne Wenger's discussion of social participation as a means of learning – and thus knowledge creation (Wenger, 1998). However, what is left at the end is the understanding that learning creates knowledge and that learning arises from within a social context whether that be via a virtual or physical environment. In addition, learning involves combining explicit information with personal experiences and existing knowledge. The question then is how can this type of theoretical work on knowledge creation be related back to the practical nature of electronically enabled business?

Wenger is an advocate of the concept of '*communities of practice*' as a way in which knowledge is created and he defines these as being (Wenger and Snyder, 2000):

> **the process of social learning that occurs when people who have a common interest in some subject or problem, collaborate to share ideas, find solutions, and build innovations.**

Wenger and Snyder further believe that an effective organization has a number of interconnected communities of practice, each dealing with specific aspects of the company's core business. Knowledge is created, shared, organized, revised, and passed on within and among these communities. This is a very useful concept for underpinning organizational learning and innovation development in an online environment where discussion boards, chat rooms, and weblogs are ubiquitous.

How Is Knowledge Managed?

Knowledge management is not new and has been alluded to in subjects such as economics, intellectual capital, engineering, aspects of computing, organizational studies and artificial intelligence to name a few. However, in the academic literature on Knowledge Management over the last ten years, it is apparent that it has emerged as an area of interest in its own right with solid roots in academia and organizational practice. Essentially, Knowlege Management encompasses:

- The identification and analysis of available and acquired organizational knowledge assets and related processes (e.g. those regarding markets, products and technologies involved in generating profits)
- The subsequent planning and control of actions to develop these assets to the benefit of the organization.

When placed in the context of maximizing organizational competitive advantage in a rapidly changing economic environment, Knowledge Management may also be defined (Ruggles, 1998) as:

> **An approach to adding or creating value by actively leveraging know-how, experience and judgement resident within and outside of an organization.**

Using this approach to determining what can be managed about knowledge, there are eight major categories of knowledge-focused activities that can be addressed:

1. Generating new knowledge
2. Accessing valuable knowledge from internal and external sources
3. Using accessible knowledge in decision making
4. Embedding knowledge in processes, products and or services
5. Representing knowledge in documents, databases and software
6. Facilitating knowledge growth through cultures and incentives
7. Transferring existing knowledge into other parts of the organization
8. Measuring the value of knowledge assets and/or impact of knowledge management.

The processes within an organization that need to occur to enable these activities to take place are (NHS, 2005):

- The creation of knowledge strategies to guide the overall approach
- The undertaking of knowledge audits to identify knowledge needs, resources and flows
- Connecting people with people to share tacit knowledge using approaches such as communities of practice or learning events. Currently, communities of practice are regarded as the 'killer application' of knowledge management (Clemmons, 2002)

- Connecting people with information to share explicit knowledge using approaches such as best practices databases, and using content management processes to ensure that explicit knowledge is current, relevant and easily accessible
- The creation of opportunities for people to generate new knowledge, for example through collaborative working and learning
- The introduction of processes to help people seek and use the knowledge of others such as peer assists
- Teaching people to share knowledge in ways that inspire people, for example using storytelling techniques
- Encouraging people to prioritize learning as part of their day-to-day work by reflecting before, during and after the tasks and projects they have performed on what they have learned during the process.

According to the Gartner Group (1999), as with information management, KM programs perform best when enabled with sophisticated and elegant technology, but an emphasis on technology alone will achieve very little towards KM. Conversely the strongest KM culture that is not supported by robust technology will also falter.

Why Are Companies Interested in Managing Knowledge?

Companies are interested in managing knowledge (Table 3.4) as the driving forces in the business world change rapidly on the back of innovative technology development and globalization. For example, prior to the turn of the last century, businesses were based around the premise that:

- The business owned the means of production and the employee could not make a living without it (i.e. the employee needed the business more than the business needed the employee).
- The great majority of employees worked full time for a business. The pay they got for the job was their only income and provided their livelihood.
- The most efficient way to produce anything was to bring together as many as possible of the activities needed to turn out the product, under one management (e.g. the Ford Motor Company).
- Suppliers and especially manufacturers had market power because they had information about a product or a service that the customer did not, and could not have, and in fact did not need if they could trust the brand.
- Each industry had its own technology (i.e. all the technology needed to make steel was peculiar to the steel industry) and, conversely, whatever technology was used to make steel came out of the steel industry itself.
- Every product or service had a specific application, and for every application there was a specific product or material (e.g. beer and milk were sold only in glass bottles, working capital for a business was supplied by a commercial bank through a commercial loan, etc.).
- Competition took place mainly within an industry.

Since the 1970s, things have changed dramatically in the business environment with some or all of the following points contributing towards that change (Drucker, 1999, 2001) (Table 3.4):

- Knowledge now drives production and it is both 'owned' by knowledge workers and is highly portable. That is, knowledge workers provide 'capital' just as much as does the provider of money and the two are dependent on each other making the knowledge worker an equal.
- The majority of employees still have full-time jobs with a salary that provides their only or main income. But a growing number of people who work for an organization are not full-time employees but part timers, temporaries, consultants or contractors or even employees of outsourcing contractors.
- Maximum integration within an enterprise of all tasks associated with product development is no longer a viable business strategy. The knowledge needed for any activity has become highly specialized and it is increasingly expensive and difficult to maintain enough critical mass of knowledge for every major task within an enterprise.
- Communication is so easy, fast and less costly with the internet that the most productive and most profitable way to organize is to disintegrate (e.g. outsourcing the management of an institution's information technology, data processing and computer system has become routine).
- The customer now has access to huge amounts of information via the internet. Information equates to power. Power is thus shifting to the customer, be it another business or the ultimate consumer.
- Virtually no product or service any longer has either a single specific end use or application, or its own market (e.g. cardboard, plastic and aluminium compete with glass for the bottle market).
- Businesses are having to redefine themselves by what they are good at doing rather than by the material in which they have specialized in the past. Merck, America's largest pharmaceutical company, has diversified from making drugs into wholesaling every kind of pharmacy product, including those of competitors.
- The market and the competition is global.

Table 3.4. Reasons behind why companies are becoming interested in managing knowledge

Reasons to Manage Knowledge	
1. The marketplace is increasingly competitive and rate of innovation is rising	• Knowledge must evolve and be assimilated at a faster rate
2. Companies are becoming knowledge intensive, and focused on creating customer value, not capital intensive	• Staff functions are being reduced. The informal knowledge management of staff functions need to be formalized to cope with rapid change
3. Companies are realising how important it is to 'know what they know'	• To be able to make maximum use of the knowledge
4. Unstable markets necessitate 'organized abandonment'	• Target markets are constantly shifting
5. Lets your business lead change rather than change leading your business	• Knowledge creates windows of opportunity
6. Only the knowledgeable survive	• Knowledge has become deterministic – knowledge takes time and experience to acquire – employees have less and less time for this
7. Cross industry amalgamation is breeding complexity	• A proliferation of mergers, joint ventures, multiutilities, etc.
8. Knowledge can drive decision support	• Information and experience combined provide better input into decision making
9. IT and Information Systems (IS) development	• Knowledge requires sharing: IT supports sharing, IS develop collaborative working
10. Tacit knowledge is mobile and there is a need to manage expertise turnover	• More company 'capital' walks out of the door with every employee who leaves the company than in any other way
11. Competitors and markets are global	• Internet technologies have opened up the world market but have also opened up businesses to world competition

What Aspects of a Business's Knowledge Need to Be Managed?

The knowledge assets of a business are the knowledge regarding markets, products, technologies and organizations that the business owns, or needs to own, in order for its business processes to generate profits and add value. There are always conflicting opinions as to what aspects of these assets need managing but the core issues revolve around the management of:

- Innovation and the process involved in generating new knowledge within the business
- The process of accessing valuable knowledge from outside sources
- The use of accessible internal knowledge in decision making
- The embedding of knowledge in processes, products and/or services
- The representation of knowledge in documents, databases and software
- Growing the corporate knowledge by developing appropriate cultures and incentives
- The transference of existing knowledge into other parts of the organization
- The process of measuring the value of knowledge assets
- The process of measuring the impact (value add) of knowledge management.

How Is Knowledge Managed for Maximum Advantage to the Organization?

There are a number of theoretical approaches to managing knowledge in an organization (e.g. Van Der Spek and De Hoog, 1995; Macintosh *et al.*, 1998), but, practically, managing the knowledge of an organization is a complex task. Most traditional company policies and controls focus on the tangible assets of the company and do not manage their important knowledge assets (Van Krogh *et al.*, 2000). More and more, success in an increasingly competitive and changing market depends critically on the quality of information and knowledge which organizations use in their key business processes. For example, the supply chain depends on knowledge of diverse areas including raw materials, planning, manufacturing and distribution. Likewise product development requires knowledge of consumer requirements, new science, new technology, marketing, etc. Thus:

- At the *strategic level* a business needs to be able to analyse and plan its business in terms of the knowledge it currently has and the knowledge it needs for future business processes.
- At the *tactical level* the business is concerned with identifying and formalizing existing knowledge, acquiring new knowledge for future use, archiving it in organizational memories and creating systems that enable effective and efficient application of the knowledge within the organization.
- At the *operational level* knowledge is used in everyday practice by professional personnel who need access to the right knowledge, at the right time, in the right location.

The most comprehensive practical guide for knowledge management can be found in the 'European Guide to Good Practice in Knowledge Management' (2004) the main points of which are illustrated in Figure 3.6 and from which the simple recipe outlined in Table 3.5 can be created.

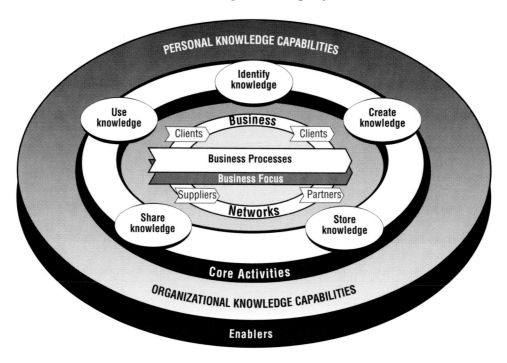

Fig. 3.6. The main issues associated with knowledge management in companies (adapted from the *European Guide to Good Practice in Knowledge Management*, 2004)

Table 3.5. A practical knowledge management recipe

Requirement	Ingredients and Actions
Understand the objectives of the business, department or workgroup	• Annual reports, long range planning and company growth strategy reports, executive interviews with the trade press and media, SWOT studies, etc.
Perform a knowledge audit	• Identify what knowledge assets the company possesses and where they are in the organization • Identify what are their uses, what form they are in and determine how accessible they are • Identify where and how knowledge can add value to the business • Identify the potential increased value to the business of using managed corporate knowledge
Match projects to business objectives (*the key to valuing the intellectual capital in the organization*)	• Identify the success factors of your organization and the relevant business cycles • Match projects to the knowledge context of these cycles (i.e. what are the important points in these cycles where actions must be swift and effective? What is the important content at these points – i.e. what information, currency, quality, etc.)
Know the business content portfolio	• Understand the combination of data, information and knowledge that represent what the company must effectively package • Ensure that there is a formal recognition that data, information, knowledge have relative value (*important and surprisingly lacking*) • Know what people are involved in creating the content and who will use the content
Set up a knowledge architecture	• Success of KM will be based on effective packaging and delivery of content • Identify the scope of the project and the investment to be made • Identify the people, content and technologies involved • Formalize the process
Profile people	• Include employees, key partners, customers • Ask the questions 'what do you do? What information do you use daily? Where do you get the information from? Why is the information of use? Where do processes run smoothly/not?'
Use a personal approach	• DO NOT USE EMAIL

The Problems and Solutions for Knowledge Management

There are many problems associated with attempting to manage corporate knowledge, not least of which is the identification of knowledge assets and the efficient and cost effective use of them. However, the biggest difficulty associated with knowledge management is *changing people's behaviour* to embrace a knowledge sharing philosophy and the biggest impediment to knowledge transfer in an organization is organizational or corporate culture. The solutions include:

- Creating a culture that encourages knowledge sharing
- Having an enterprise-wide vocabulary to ensure that the knowledge is correctly understood
- Identifying, modelling and explicitly representing the corporate knowledge
- Sharing and re-using knowledge among differing applications for various types of users (this implies being able to share existing knowledge sources and also future ones).

Some examples of what are currently considered to be knowledge management tasks include *mapping sources of internal expertise*, for example creating online phone and email directories arranged by knowledge areas, *creating networks of knowledge workers*, for example encouraging informally socially constructed communities of practice, and *establishing new knowledge roles* with a focus on leveraging knowledge, enabling knowledge via training or technology and making knowledge visible by identifying gaps and establishing priorities. In reality this means introducing a change agent who markets the importance of knowledge inside the company.

Portals and Knowledge Management

As discussed earlier in this chapter the amount of information businesses are storing is increasing exponentially with the use of the internet and other communication and IT technologies that are available to them. Customer demographics, accounting data, product information, human resources data, all contribute to the mass of information that companies collect and which is being organized in enterprise information portals.

For knowledge management purposes, EIPS may be further refined as enterprise knowledge portals (EKPs) which can be goal-directed towards knowledge production, acquisition, transmission and management. They may also be focused on enterprise business processes such as sales, marketing and risk management and can provide, produce and manage the information (metadata) about the validity of the information the EKP supplies. Knowledge portals thus provide information about the business, and also supply the meta-information about what information can be relied on for decision making. They distinguish knowledge from mere information and they provide a facility for producing knowledge from data and information.

Summary

This chapter has looked at information management, information ecology, the information economy and the information supply chain, corporate information portals and knowledge acquisition and management – all particularly as they relate to either developing or supporting an electronically enabled business.

Without good information and content management there can be no solid infrastructure to support a business model that is based around good clean, accurate and timely data. Without the ability or knowledge to do something with that information then there is no business. Managers need to acquaint themselves with these facts and act appropriately to steer the business forward. The Smart Thinker in this chapter has certainly thought long and hard about these issues and has applied them with devastating effect in running Cisco Systems – CEO John Chambers.

The Review and Critical Thinking Questions below have been provided for you to revise and reflect on the content of this chapter.

Review Questions

1. What is the difference between data, information and knowledge?
2. What are the major information types in an organization and how are these commonly organized and managed?
3. Why is the information ecology approach so called, what does it entail and why is it fundamentally different from other approaches to information management?
4. What is the information economy and where does the information supply chain fit in?
5. What is a portal and what is its importance to eBusiness?
6. What is the SECI model of knowledge management?
7. How do you acquire knowledge?

Critical Thinking Questions

1. Discuss the benefits (or otherwise) of being able to access information online.
2. Discuss the concept of managing knowledge.
3. What innovations has The Seam introduced into marketing commodities?
4. What are the differences between @agriculture Online and InFARMation and do you think these differences are important to you – the user? Why?
5. Discuss with examples how and why EIPs are useful to organizations.

Smart Thinker – *John Chambers*

President and CEO of Cisco Systems

Cisco Systems (www.cisco.com) is the world's largest supplier of high-performance computer internet routing systems. John Chambers is the third CEO of Cisco and has led the company since 1995, growing the company from US$1.2 billion to approximately US$25 billion in annual revenues thus cementing Cisco's undisputed role as the number one provider of the networking gear that powers the internet and eBusiness.

In the process Chambers has been named 'The Most Influential CEO in Telecommunications' and 'The Most Influential Person in Communications' by several industry publications and he has been lauded by government leaders and countless publications worldwide for his visionary strategy, and his ability to drive an entrepreneurial culture. There is also no doubt that Chambers has created a leadership style that favours personal contact over mountains of paperwork. His public appearances, web casts and broadcasts give a consistently clear message: We never take the efforts of our employees or the challenges of our customers for granted—ever.

Almost more importantly for a book of this nature, John Chambers believes that:

- Technology should force sweeping changes in business processes, rather than just be about bolting new technologies onto existing ways of doing things: *'Investment in IT without process change does not get the results you want.'*
- The internet and education are great levellers: *'They are the two great equalizers in life leveling the playing field for people, companies, and countries worldwide and by providing greater access to educational opportunities through the internet, students are able to learn more..... [and] companies and schools can decrease costs by utilizing technology for greater productivity.'*
- Collaborative supply chains are the way to go: *'Productivity is not just about reducing costs. And reducing costs doesn't always equal increased sales...... a new focus on productivity is emerging based on adding value to the exchange of information. This next horizon for productivity [will be] based on interactions across and between partner and customer organizations.'*

Source: Chief Executive Magazine (www.chiefexecutive.org)
Time Magazine (www.time.com) and the Cisco website

Chapter 4

Agri-food Chains in the 'E' World

A Business is not an island

This chapter tackles perhaps the most important, and at the same time the most difficult, road a business or industry must think about following in terms of electronically enabling to add value, that is: the concept that a *business is not an island*. In other words, the business does not exist and work in isolation from other businesses – either those supplying goods to it or those customers buying goods from it. This concept has already been introduced in Chapter 1, but in the agribusiness arena in particular, where companies in each agri-industry specific chain (e.g. grain, citrus, beef, wool, stonefruit, honey, poultry, cotton, etc.) vary in size dramatically from the sole trader through to the multinational corporates, the electronic enablement of both an individual business and the connections between the businesses that it deals with, need to be thought about and dealt with before the supply chain can function effectively. Therefore, electronic enablement of agri-industry supply chains along with the implications for value adding, cost reduction and differentiation, is discussed in more detail in this chapter under the banners of the *'E' model*, the *supply chain and supply chain management*, the *electronically integrated supply chain* (ESC), *electronic supply chain management* (ESCM) and *collaborative supply chains*.

Chapter Objectives

After reading this chapter you should understand and be able to describe:

1. The issues associated with an electronically enabled business model
2. What electronic enablement for a business and across an industry chain involves
3. What EDI, ERP and CRM are and why they have become so important in today's business world
4. What a supply chain is and what supply chain management (SCM) involves
5. The 'lean' supply chain concept and the potential benefits it can create within an agri-industry chain
6. Why electronic collaboration and supply chain agility are important.
7. The main business and IT integration issues associated with electronically enabled supply chains
8. What demand based management offers as a business tool in the 21st century.

The Electronically Enabled Model in Business and Industry Chains

Before moving into the world of industry chains with all the complexities associated with the multiple individual businesses involved, it is useful to first look at the issues associated with developing the concept of the electronically enabled model of an individual business and the associated systems this approach requires.

As discussed in Chapter 1, a business model is an organization's core logic for creating value in a sustainable way. It features the value creating mechanisms (what the company sells) and the value appropriation mechanisms (sources of revenue for the company) and combines these in a profit-making model. According to Osterwalder and Pigneur (2002), 'E' business models have variously been described as internet models or web-based trading models, but these authors go further and suggest that a true electronically enabled business model describes the logic of a *business system* for creating value in the internet era. Such a model is composed of four main components – product innovation, infrastructure management, customer relationships and financials. In addition, research by Bryceson and Kandampully (2004) in a number of agri-industry sectors indicates that competencies in relation to electronic technologies, and their implementation and use within an agribusiness should also be included (Fig. 4.1).

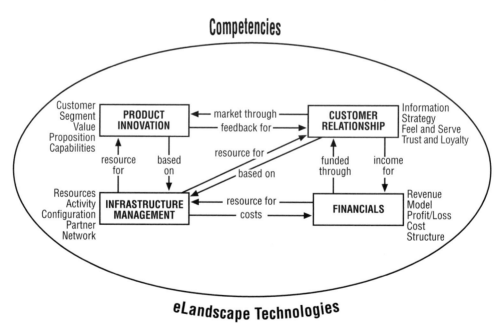

Fig. 4.1. The components of a modern 'E' business model (adapted from Osterwalder and Pigneur, 2002)

An electronically enabled business model not only requires the company to accept electronic technologies and the internet as underlying infrastructure around which strategies and tactics can be developed to conduct business, but also recognizes that it is only the first part of adopting an holistic model around which to run the business and interact with suppliers and customers.

There is no doubt that there are a number of complexities associated with developing an electronically enabled business model. The issues can be clearly defined depending on the developmental stages of the electronic enablement of the business. The following list outlines these stages and the associated levels of 'e' complexity that can be found in businesses – and which can have a major impact on how they are able to relate to their suppliers and customers:

1. *Stage 1.* At this stage, automation of discrete transactions can be found – this predominantly includes accounting processes and online order entry if a company webpage exists.
2. *Stage 2.* At this stage, functional enhancement of activities such as human resource management (HR), sales and product design starts to happen.
3. *Stage 3.* This stage is very accelerated by the internet and involves cross-activity integration within the business using tools such as customer relationship management (CRM), supply chain management (SCM) and enterprise resource planning (ERP) systems – i.e. linking sales activity with order processing starts to happen.
4. *Stage 4.* This stage is where most large scale businesses (e.g. WalMart) are up to and involves the integration of the value chain of the product with the entire value chain system (from raw materials through the industry sector and tiers of customers, suppliers and channels), and product development.
5. *Stage 5.* This stage involves optimizing the workings of the value chain system in real time.

Businesses are under increasing pressures to respond to change. In large part, the internet offers an entirely new set of options for marketing, transaction processing and customer service. While there are many new opportunities, companies must understand how electronic enablement works to capitalize on these. Current key business trends indicate that managers must be aware of a new focus in business on the use of the internet to collapse time and distance from a communication and information flow capability. These trends include the following:

- *Globalization.* This is fostered by the internet and the ability of companies to reach beyond their traditional geographical boundaries.
- *Sophisticated customers.* These demand customized service. Customers want a product quickly and are no longer satisfied with a mass marketing approach from vendors. Suppliers at every level need to be part of the process.
- *Virtual corporations emerge.* Boundaries between companies have become fluid – the internet provides the connectivity necessary for this to work.
- *Speciality skills.* These are developing in the workforce and workers are more mobile – as the market changes, so do corporate requirements of their employees.
- *Information abounds.* Information is out there and available, creating a much more knowledgeable market and customers.

Electronic Enablement and Connectivity

Electronic enablement in business has evolved around the premise that information technologies (IT) can perform six data-based functions (capture, transmission, storage, retrieval, manipulation and display) faster and more efficiently than any other type of technology (Alter, 1999). Each new innovation and technology plays a part in greater efficiencies. Further improvements in the IT used to address these basic data functions continue to take place in four major areas:

1. *Greater miniaturization, speed and portability.* Speed and power is the underlying force of progress to date with the creation of smaller electronic components with greater capabilities being the driving force
2. *Greater connectivity and convergence of computing and communications.* The need and ability to transmit data between electronic devices at different locations has escalated exponentially in the last five years and will continue to be a major driver in IT improvement (See Figure 4.2)
3. *Greater use of digitized information and multimedia.* Information exists as defined data items, text, pictures, sounds, videos. Digitization involves coding data as a set of equivalent numbers or, in the case of a picture, a series of dots. Multimedia is the use of multiple types of data within the same application. Greatest demand in the future is likely to be for better software techniques and interfaces with people and the development of flexibly delivered training programs and online learning.
4. *Software advances.* These have mainly been accomplished by miniaturization of components enabling larger programs using more power and memory to be run on physically smaller and smaller computers costing less and less.

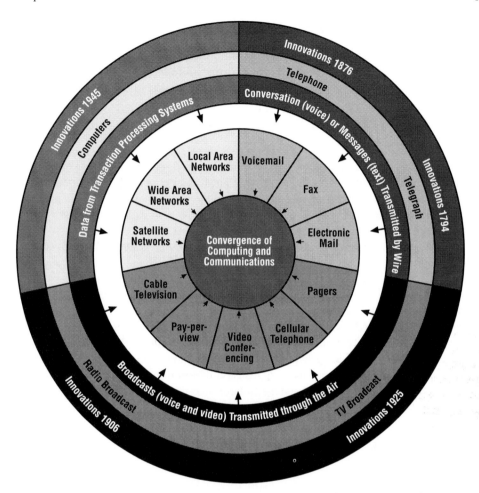

Fig. 4.2. The convergence of innovation, computing and communications (*Graphic Design – Chris Frost, Information Design – Emma Somogyi, The University of Queensland – Source: Alter, S. (1999). Information Systems – A Management Perspective. Addison Wesley Longman, USA)*

An excellent example in the agribusiness arena is the RFID based ear tag traceability systems (see Chapter 6) that are being employed across the globe in an effort to ensure birth to death traceability of livestock. Miniaturization has seen the development of realistically sized ear tags that capture all essential information about a beast for downloading into a database that can be used for individual animal and herd management purposes, as well as traceability of retail products.

Electronic enablement also entails the use of appropriate information systems to capture, transmit, store, retrieve, manipulate or display information, thereby supporting other systems. There are essentially six types of information systems used to support the functional areas of business and these are: Office Automation including Electronic Document Management Systems; Communication Systems; Transaction Processing Systems – including payments, credits, and invoicing; Management and Executive Information Systems; Managerial Decision Support Systems and Knowledge Management Systems. More detail can be found on these in Appendix 1 and through this chapter's references and readings.

Electronic Connectivity

As discussed in earlier chapters, in today's business world, the reliable and efficient access to information has become an important asset in trying to achieve a competitive advantage. With the advent of the internet, time and geographical separation are no longer an excuse for not sharing information. Mind you, the art of connecting components like computers, telephones and personal digital assistants (PDAs), along with their various parts and systems is a real skill of the 21st Century and is best left to the experts.

Our interest in this book regarding electronic connectivity issues is associated with anything that impacts business-based information flows. In the main, this means anything to do with computer networks, particularly the internet – and related technologies. These technologies are described in Appendix 1 under the banner of the eLandscape, but Pidgeon (2000) and Tyson (2000) in two good articles describe the main technological issues and protocols associated with computer networking. Readers are directed to these articles for more detail on these areas but suffice to say, the emphasis of the articles is that computers are connected to the internet via a hierarchy of networks and that the management of these networks is of extreme importance to ensuring good data and information flows. Ensuring the use of appropriate protocols and standards across the business and between businesses as well as the integration of systems (see later in this chapter) is an important management function.

Security

With any form of electronic connectivity – particularly where data and information are being transferred – security is a major issue. This is discussed further in Chapter 5 but the five key elements associated with electronic security from an organizational perspective are listed below for completeness:

- *Authentication* which involves checking that the users are who they say they are
- *Authorization* where the users are empowered to perform the action that they are attempting
- *Connection integrity* where the transactions between the user and the system are not compromised
- *Auditing* when a record of all pertinent interactions between the users and the system is maintained
- *Non-repudiation* where the users cannot deny having performed an action on the system.

Strategic Alignment of Business and Electronic Enablement

Strategic alignment of the business effort and the degree of electronic enablement is a central issue which may actually determine the position of a business in the marketplace. Alignment is a two-way process (Figure 4.3). The business determines the electronic enablement needs, but electronic enablement influences the business. The prime objective of alignment is to ensure that information provision and technology matches the business needs in what it does (*context*), how it does it (*process*), and when it does it (*timing*).

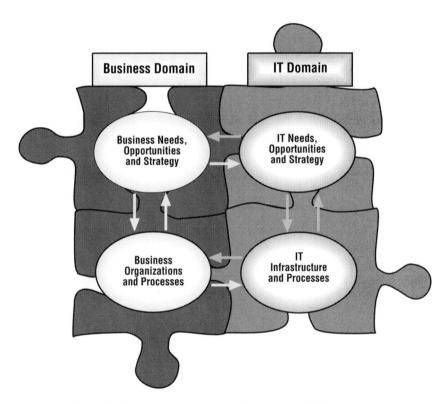

Fig. 4.3. The two-way process of business and IT alignment

As the business needs develop and the stages of electronic enablement progress, the sophistication of technology used increases. The following section deals with the three most significant systems associated with a technology plan aligned with the business requirement of fully electronically integrating the supply chain.

EDI, ERP and CRM

There are a number of specific information systems and applications which make use of the information technologies and systems described above. These include electronic data interchange (EDI), enterprise resource planning systems (ERP) and online customer relationship management systems (CRM).

Electronic Data Interchange (EDI)

Electronic data interchange or EDI is the transfer of data between different companies using networks, such as the internet. As more and more companies get connected to the internet, EDI is becoming increasingly important as an easy mechanism for companies to buy, sell, and trade information (Peat and Webber, 1997; Hildebrand, 2000).

Companies have traditionally used paper as the medium for conducting business. Company records are filed on paper, and paper forms are mailed between companies to exchange information. The advent of the business computer enabled companies to process data electronically – however, the exchange of this data between companies still relied heavily on the postal system. For example data was entered into a business application, a form was printed containing the data, and the form mailed to a trading partner. The trading partner, after receiving the form, re-keyed the data into another business application. Even nowadays, such a process is still being used in many smaller businesses. There are a number of issues inherent in this process including:

- Poor response times – use of the postal system can add days to the exchange process
- Excessive paperwork for both companies involved in the exchange
- The potential for errors as information is transcribed multiple times.

Computer telecommunications for data transmission were adopted in the 1960s, the goal being to eliminate human intervention which adds a large burden of time-intensiveness to the transaction task-at-hand through labour costs, and to reduce error-prone data entry which compounds the costliness of undertaking trading ventures. Today it is recognized that computerization has improved response time, reduced paperwork, and reduced the potential for transcription errors.

Early electronic interchanges were based on proprietary formats agreed between two trading partners. The original formats were only for purchasing, transportation, and finance data, and were used primarily for intra-industry transactions. It remained difficult for a company to exchange data electronically with many trading partners until the late 1970s when work began on international electronic data interchange (EDI) standards (Norris *et al.*, 2001).

Today there are many standards for EDI which define the data formats and encoding rules required for a multitude of business transactions, including order placement and processing, shipping and receiving, invoicing, payment, and many others.

> ***What is now a major concern is that in a globally connected world, all trading partners doing business together need to use the same standards.***

Enterprise Resource Planning Systems (ERP)

Enterprise resource planning (ERP) systems are configurable business information systems packages that integrate information and information-based processes within and across functional areas in an organization including human resources, finance, materials, manufacturing and distribution (Koch, 2002).

ERP systems grew out of the material requirements planning (MRP) systems of the early 1960s. MRPII still forms a core module of current ERP systems. Critical in the step from MRP and MRPII to ERP was the drive towards integration and real time information flow that led to the development of contemporary ERP systems (O'Leary, 2000). Examples of companies that deal in ERP systems include SAP (www.sap.com) and Oracle/Peoplesoft (www.oracle.com/applications/peoplesoft-enterprise.html) both of which are found globally across all industry sectors in most large corporations and a growing number of small to medium enterprises (SMEs).

Ideally ERPs can also be used to better connect the organization to customers and suppliers facilitating the flow of information, goods and services across the supply chain (James and Wolf, 2000). Technically therefore they should be able to address the need to integrate with other company's systems and business processes to transmit information between different firms (Kumar and Van Hillegersberg, 2000). Tasks that inter-organizational systems address include the design of custom products, entering of orders, transmitting invoices, making payments and servicing the product.

Customer Relationship Management

Customer relationship management (CRM) is about creating value for customers (Deck, 2001). It is *not* about technology although technology is required to enable CRM – in essence it is a core business strategy.

Why? Every business has customers; every business needs more of them spending more. While 'looking after your customers' has always been a central concept in good retailing, it is particularly true in today's fluid and highly competitive market where there are a multitude of choices regarding delivery channels, product packaging, loyalty rewards, service levels and ultimately price. Today's customers are transaction-makers that exist not just in the external marketplace but throughout a company's value chain – all are driven by a range of shared needs and expectations including:

- They want to see promises kept with reliable delivery
- They want a seamless, high quality end-to-end experience
- They want service backed by trust and acknowledgement of genuine value.

Maintaining and/or increasing profit and market share is a fundamental tenet of business – developing and maintaining customer relationships to sustain competitive advantage is thus the focus of CRM (Sims 2000).

CRM solutions should manage the total end-to-end customer related process for an organization, optimizing the implementation of marketing, sales and customer care strategies across multiple channels throughout the organization. Electronically enabled CRM applications allow companies to interact directly with customers via corporate websites and Web based storefronts (Say, 2001), the idea being that 'E' CRM will enable a seamless interface between the organization, and existing and potential customers allowing:

- *Integration and consolidation of customer information* providing staff with the relevant customer information and client history to deal with a customer consistently throughout the organization.
- *Provision of consolidated information* and supporting scripts across all channels throughout the organization to assist staff during one-to-one communication with the customer matching their needs with the most appropriate product.
- *Management of customer cases* providing the right person with management control in a planned and transparent manner, ensuring that appropriate action is taken at the proper time.
- *Automation and manual generation of new sales opportunities* prompted by measuring the customer profile against the pre-defined business rules between a customer and staff member.

The components of CRM include a strategy, the technology and its implementation, mobile business (mCommerce) for the enterprise, sales and marketing, business intelligence (BI) and the customer contact centre.

1. *Strategy.* A CRM strategy needs to define where the business benefits will come from, how CRM will deliver those benefits, the priority business areas and the priority customer relationship areas e.g. customer retention or customer servicing. The challenge is to:

    - Improve product sales per customer
    - Reduce customer defections
    - Identify, capture and convert sales opportunities more effectively
    - Improve and integrate with sales process across all channels
    - Have a clear measurement of sales process and performance.

    The fundamental concepts in developing a strategic framework for CRM are based around changing customer expectations, the nature of the customer relationship and the difference between loyalty and captivity.

2. *Technology and implementation.* CRM breaks through the traditional boundaries separating sales, marketing, service, IT and other functional areas. An overall, flexible enterprise architecture plan is required to enable seamless integration of CRM systems, as well as the alignment of 'e' and traditional channels into hybrid technology systems.

3. *Mobile business for the enterprise.* Mobile technology allows field sales, support and service personnel to access critical customer and company information, send and retrieve data, and interact with colleagues and customers. New applications will give birth to innovative customer-facing sales and service channels e.g. wireless application devices connecting CRM and mobile workers to ensure that information is always up-to-date (Edwards *et al.*, 2002).

4. *Sales and Marketing.* Sales is a multi-channel selling system that relies on a combination of field sales, retail, partners, call centres and electronic channels. The goal is to make the customer the focus of sales efforts by integrating customer needs into channel and product strategies via forecasting, push support, personalization and by embedding service into products using networked sensors, microprocessor intelligence and wireless communication.

5. *Business Intelligence (BI).* BI is the intersection of the needs of the business and the available information necessary to make the best business decisions. BI information sources include:

- The internet, the extranet, the intranet
- Online transaction processing systems (OLTP)
- Operational data stores (ODS)
- Data warehouses (DW) and data marts (DM)
- Analytical applications such as data mining applications, statistical analysis applications, predictive modelling applications, reporting systems, and data islands (isolated sets of data such as spreadsheets, desktop databases, etc.).

6. *Customer Contact Centre.* The customer contact centre integrates customer touch-points and provides service through one multi-channel gateway. The customer contact centre, whether it is a help desk, a call centre, or online support via email or chat, is how customers experience the organization. Customers leave the customer contact centre experience with either positive or negative feelings towards your company.

The reality is that EDI, ERP and CRM are all interrelated and interdependent macro processes for shared value creation in a business. Jabali (2003) points out however, that unless there is a core understanding that relationships between people are still fundamentally what drives successful business, then the technology will not be able to deliver what it is capable of.

The Supply Chain and Supply Chain Management

A *supply chain* (Figure 4.4), encompasses every effort involved in producing and delivering a final product or service, from the supplier's supplier to the customer's customer (Dunne, 2001).

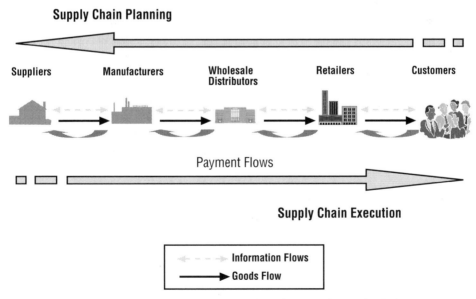

Fig 4.4. A diagrammatic illustration of a generic supply chain

Managing and enabling the supply chain is the single largest cost component of a firm's revenues – typically 55-85% of the sales dollars – and essentially is comprised of two elements:

1. Strategic sourcing
2. Strategic logistics.

These elements are customer-focused and include the flow of information, business processes, and decisions to specify, design, plan, purchase, transport, warehouse, inventory, and deliver materials and services to internal and external customers. Most importantly, the supply chain integrates all decisions, processes, and activities on a lowest-total-cost life cycle basis from raw materials to end-user – it thus integrates core business and customer needs. The efficiency and effectiveness of these processes are seen as the industry-competitive differentiators and help companies achieve price/cost reductions and increased profitability.

Modern supply chains are highly complex and dynamic. They are characterized by constantly changing relationships and configurations, they use a mixture of manufacturing techniques such as *Build-to-Stock, Make-to-Order,* and *Just-in-Time* to fulfill orders, and they involve multiple organizations (Anderson and Lee, 2000).

In addition, the emergence of the internet as a new technology enabler for business (Anderson and Lee, 2001; Lawrence *et al.*, 2003; Forsyth, 2004) has increased the number of customer interactions and product configurations, thereby presenting greater demands on supply chain management and performance (Allee, 2000; Bhatt and Emdad, 2001). The ultimate goal and measure is customer satisfaction: the ability to fulfill customer orders for personalized products and services faster and more efficiently than the competition.

It is critical therefore to focus management attention on the performance of the supply chain as an integrated whole, rather than as a collection of separate processes or companies.

Supply Chain Management (SCM)

SCM is the overall system of coordinating closely and managing supply and demand, the sourcing of raw materials and parts, manufacturing and assembly, warehousing and inventory tracking, order entry and order management, distribution across all channels, and delivery to the customer (Cox 1999).

SCM decisions are both strategic and operational. *Strategic decisions* are normally made over a long time horizon and are closely linked to the corporate strategy. They guide supply chain policies from a design perspective. *Operational decisions* are short term, and focus on activities over a day-to-day basis. Both types of decisions attempt to create a situation that effectively and efficiently manages the product flow in the strategically planned supply chain.

Best practice supply chain management integrates user expectations, commercial requirements, and the flow of purchased materials and services (Beamon and Bermudo, 2000; Hausman, 2002). It wins customers by providing better value and it rewards shareholders by enhancing profitability and providing better returns. Arntzen *et al.*, (1995) describe four major decision areas in supply chain management: location, production, inventory, and transportation or distribution.

1. *Location decisions.* The geographic placement of production facilities and sourcing points is the first step in creating a supply chain. The location of facilities involves a commitment of resources to a long-term plan. Once the size, number, and location of these facilities

are determined, so are the possible paths by which the product flows through to the final customer. These decisions represent the basic strategy for accessing customer markets, and will have a considerable impact on revenue, cost, and level of service. They should thus be determined by an optimization routine that considers a large amount of information on production costs, taxes, duties and duty drawback, tariffs, local content, distribution costs, production limitations, etc. Although location decisions are primarily strategic, they also have implications on an operational level (Arntzen *et al.*, 1995).

2. *Production decisions.* Production decisions include what products to produce, and which plants to produce them in, the allocation of suppliers to plants, and products to customer markets. Such decisions assume the existence of the facilities, but determine the exact path(s) through which a product flows to and from these facilities and thus also have a big impact on the revenues, costs and customer service levels of the firm. Operational decisions focus on detailed production scheduling which includes the construction of master production schedules, scheduling production on machines, and equipment maintenance and quality control measures at a production facility.

3. *Inventory decisions.* Inventory decisions refer to how inventories are managed. Inventories exist at every stage of the supply chain as either raw materials, semi-finished or finished goods. They can also be in-process between locations. Their primary purpose is to buffer against any uncertainty that might exist in the supply chain. Since holding of inventories can cost anywhere between 20 and 40 percent of their value, their efficient management is critical in supply chain operations. The management of inventory is an operational issue and includes deployment strategies (push versus pull), and the determination of the optimal levels of order quantities and reorder points at each stocking location. These levels are critical, since they are primary determinants of customer service levels (Beamon and Bermudo, 2000).

4. *Transportation decisions.* Transportation decisions are closely linked to inventory decisions, since the best mode of choice is often found by trading-off the cost of using the particular mode of transport with the indirect cost of inventory associated with that mode. While air shipments may be fast, reliable, and warrant lesser safety stocks, they are expensive. Meanwhile shipping by sea or rail may be much cheaper, but necessitates holding relatively large amounts of inventory to buffer against the inherent uncertainty associated with either. Therefore customer service levels and geographic location play vital roles in such decisions. Since transportation is more than 30 percent of the logistics costs, operating efficiently makes good economic sense.

The Bullwhip Effect

The term *bullwhip effect* refers to the magnification of demand fluctuations as orders move up the supply chain. It occurs when information about the demand for a product gets distorted as it passes from one firm to another in the chain. (Reddy, 2001). Practically, this distortion means that small changes in demand from the consumer can create large variations in orders placed upstream, leading to excessive inventory throughout the system, poor product forecasts, insufficient or excessive capacities, product unavailability, and higher costs generally.

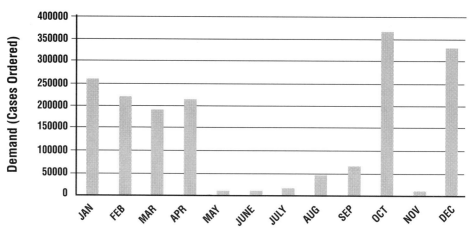

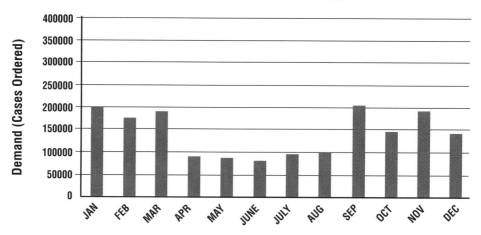

Fig 4.5. Illustration of the bullwhip effect across a supply chain

As is illustrated in Figure 4.5 – Retail orders in the example have followed a relatively stable pattern throughout the year with small drop off in orders mid-year. The distributors, on the other hand, have a wildly fluctuating order situation that has got worse over the year.

Causes of the bullwhip effect include overreaction to backlogs, not ordering in an attempt to reduce inventory, poor communication, delay in passing information and materials through the chain, batching of orders in an effort to take advantage of transport economics, 'shortage gaming' which is when over-ordering in a time of plenty is practiced as a strategy to cover the lean times, and finally – inaccurate demand forecasting.

Improved forecasting techniques at any one level in the supply chain cannot eliminate the bullwhip effect and may worsen it if used improperly. Information flow, coordination of orders, purchase order contracts that specify regular ordering intervals and vendor managed inventory (VMI) strategies across the supply chain offer the only hope of taming the bullwhip effect.

Due to its wide scope, supply chain management must address complex interdependencies; in effect creating an **extended enterprise** that reaches far beyond the production enterprise or factory door. Today, material and service suppliers, channel supply partners (wholesalers/distributors, retailers), and customers themselves, as well as software product suppliers and system developers, are all key players in supply-chain management. Essentially good SCM lets both the firm and its suppliers reap the benefits of smaller inventories, smoother production and less waste.

The Lean Thinking (Lean Chain) Concept

The concept of *lean thinking* or the *lean supply chain* is not new – Japan, in the post World War II era basically rebuilt their car industry around the main principle involved which is streamlined, efficient manufacturing. This is an approach which was further refined by Toyota in the 1970s and '80s to include the elimination of any steps in their production process that did not create value, and the constant pursuit of optimization of those elements that did create value.

Womack *et al.*, (1990) first coined the term in a study of Japanese business methods. They used the word 'lean' because basically, the Japanese used less of everything including resources and time. In other words, they had trimmed the fat out of their systems and reduced waste – thus enabling the components in their business/chain that actually added value, to be commensurately more valuable than previously.

There are six main principles of lean thinking (Womack *et al.*, 1990; Womack and Jones, 1996; Lucansky and Burke, 2003; Barney, 2005), and these are:

1. *Specify what creates value* for an organization/chain.
2. *Define the value stream.* Identify material and information flows (including time taken for each step and between steps) currently required to deliver a product or service which will in turn help identify bottlenecks in, for example, inventory management. It should be noted that few products or services are provided by one organization alone, so identifying and integrating with partners in the value stream is a necessity.
3. *Eliminate waste.* Waste in the value stream is any activity that the customer will not pay for. In reality, in most production operations, only 5% of activities add value, 35% are necessary non-value adding activities and 60% add no value at all and include overproduction, defects, unnecessary movement of items, unnecessary inventory, inappropriate processing, and excessive transportation (Hines, 2004). Thus eliminating as much of the waste as is possible can create spectacular performance improvements. Again, as with point 2, eliminating waste has to be undertaken throughout the whole value stream across all businesses involved in jointly delivering the product or service.
4. *Partner up.* Source partners who can go lean and who can produce quickly and in high volume – it must be noted that partners must have incentives for joining a lean chain such as a long term business relationship.

5. *Good communication.* Ensure this is recognized as the premium requirement between internal and external partners.
6. *Technology adoption.* Technology supports both communication and visibility – as discussed throughout this chapter, the key to good supply chain management is the physical and informational flows necessary to meet consumer requirements and to provide value to the firms involved. Information technology, by offering cheaper and more powerful communications, data collection, and processing, facilitates this as well as improving efficiencies and reducing waste.

The Lean Chain Approach in the Agri-Food Industry

The lean thinking concept can be applied to all industry sectors and although much of the work has been done on the car and manufacturing sectors, recent investigations in the UK and Australia have focused on how the Lean Chain approach can be used in the agri-food industry (Storer, 2001; Simons *et al.*, 2003; NFIS, 2002; Simons and Zokaei, 2005).

To date, it is clear from these studies in the red meat, lamb and pork industries that there are some historical behaviour patterns that have inhibited to some extent the food sector embracing lean thinking approaches to supply chain management in the past. Of particular note is that when compared to other industry sectors, agri-food chains tend to exhibit greater buyer opportunism – that is, there tends to be low contract complexity and poor trust within agri-industry chains with associated poor process information sharing (Simons *et al.*, 2003).

Given that the lean thinking concept is all about a collaborative, whole-of-chain, transparent information flow approach – is the agri-food industry a poor candidate for this important productivity improvement tool? Well, in fact the very nature of most agri-food chains which have a dominant retail sector and a concentrated supply base actually predispose them to the lean way of doing things according to Cox (2001) who states that lean supply can really only occur in situations of buyer dominance. Moreover, in an industry sector where margins are so tight, anything that reduces costs and boosts profits is worth a try!

Electronically enabling the Supply Chain

Supply chain integration and automation makes good sense for most organizations. Standardizing the internal data processing and the data transfer between suppliers and customers is the core strategy of a good SCM strategy because it makes procurement systems faster and more effective which reduces internal costs related to supply chain management. More recently, the real innovation and most of the predicted future growth is in the automation of supply chains to better collaborate and integrate with external partners – i.e. the creation of the *electronic supply chain* (ESC) (Ryan, 2000).

ESCs involve the complete integration of internal ERP and other business systems as well as collaborating with both internal and external supply chain partners and having good technology links to customer and vendor systems. The goal of ESCs is to optimize the established relationships of supply chain participants by streamlining the flow of information between them all. Using the internet as the enabler technology, time and distance collapses and electronic supply chains can connect numerous distributed work environments and facilitate the communication of data dynamically and in real-time across not only countries, but also the world. Lower operating costs, shorter production lead times, fewer process disruptions, and better inventory management are some of the suggested benefits.

Implementing an electronic supply chain is a major and generally irrevocable step for an organization, requiring large investment and a fundamental change in operating processes (Buhr, 2000). Integration between the participants, both of infrastructure technology and of business processes becomes a major issue that either makes or breaks the ESC. Four major issues need to be addressed in automating the supply chain:

1. *The eBusiness supporting technologies required.* These include EDI, ERP and XML (eXtensible Markup Language).
2. *Workflow.* This is the implicit series of transactions in any activity that need to be clearly, accurately and unambiguously defined if they are to be automated.
3. *Direct and indirect goods.* It is necessary to distinguish between general purpose and specialist purchases if they are to be automated.
4. *Multiple suppliers.* Here the need is to instill a single view of the supply options from multiple feeds if they are to be automated.

Agri-industries are not isolated from this scenario. In fact, changes in technology, institutional structures, governance, increasingly tightly aligned supply and value chains that extend from genetics through producers, processors, and consumers, along with the globalization of agricultural markets, are resulting in agri-industries existing as integrated systems with producers increasingly interwoven into the food distribution chain (Todd, 2000; Newton, 2000). As a result, this industry sector needs to move rapidly to update current technologies and business processes to ensure that the various industry chains involved (for example, grains, meat, dairy, cotton, horticulture etc.) remain viable in a rapidly churning world business environment.

EBusiness Supporting Technologies

EDI fits into SCM because it allows transactions that have required paper-based systems for processing, storage and postage to be replaced and handled electronically and with less room for error. Key stumbling blocks to successful implementation of EDI are the cost and the infrastructure involved and the fact that most EDIs must connect over dedicated proprietary or *value added networks* (VANS) (see Figure 4.6).

XML (eXtensible Markup Language). Widespread supply chain automation depends on high volume computer-to-computer interactions which depend on standard formats for the information being exchanged. XML delivers a standard language for supply chain interactions and is regarded as the language of choice for delivering standardized online applications. Appendix 3 provides more detailed information on XML.

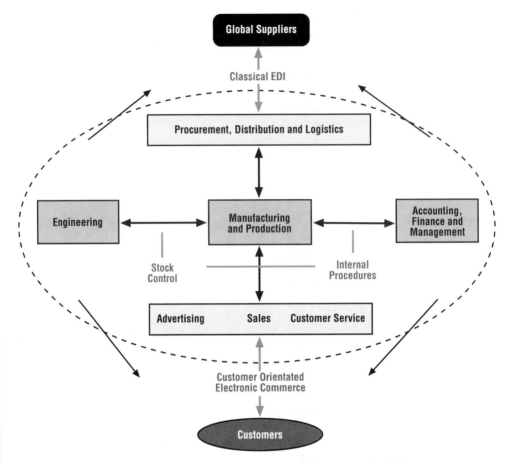

Fig. 4.6. An illustration of where EDI fits into the supply chain

Workflow

Workflow systems improve the access, processing, storage, retrieval and distribution of paper- and electronic-based information. Increased efficiency and considerable cost savings occur when deploying workflow systems into any labour intensive operation. For example large numbers of client instructions, received by fax, email or telex may require action within fixed time periods. A critical challenge is to effectively manage and process the flow of those documents. This not only applies to internal documents but also documents between both other offices in the group and clients, particularly if signatures, comments and notes need to be collected along the way. It may also be necessary to retrieve this paper trail sometime in the future so as to check the validity of the transaction.

Workflows can be simply constructed to automate the flow of messages from receipt to completion, thus ensuring compliance, audit and management oversight throughout the entire process. To automate a supply chain, the links in the chain must be well understood, that is, the following questions must be addressed: What are the steps in the transaction? Who is responsible? What needs to be done from one step to the next? Workflow addresses these requirements by stringing together the processes and creating a clear end-to-end path. The key issues include:

- *Design of Workflow*. A good design should work irrespective of the location of its components and there should be no doubt about where you are in the workflow.
- *Standards*. The accepted step is that you move the business process to match the purchased software rather than customize it – this will facilitate compatibility outside the company.
- *Application of electronic signing*. Using solely username/password protection is not strong enough – personal digital signatures are necessary (Electronic Transactions Bill, 1999).

Direct and Indirect Goods

To automate a supply chain, direct and indirect goods need to be clearly distinguished, because there are differences in the way these two types of good are procured.

Direct goods are goods that you need because of the business you are in, e.g. fertilizers, ball bearings, microchips etc., and the majority of corporate expenditures are related to purchases of these. Sourcing direct goods typically involves complex transactions requiring collaboration and greater information exchange and are thus generally obtained by someone expert in the goods being supplied. The higher value savings come from automating those processes to get products to market faster, reduce the cost of manufacturing products, reduce inventories, and avoid shortages of materials. The implication for an automated system is that the requirement is for a tightly specified, high value, product.

Indirect goods are goods you need simply because you are in business, e.g. PCs, stationery and so on – they are often referred to as maintenance, repair and operations (MRO) goods. Orders are generally initiated by a non-specialist but are within the context of a contract negotiated by a professional buyer. The implication for an automated system is that the requirement is for a loosely specified product, for example the company Grainger Ltd (www.grainger.com) which has 14,000 suppliers accessing 100,000 brands and over five million products.

Multiple Suppliers

One buyer may have many suppliers and several of the suppliers may carry the same line, creating an information overload for buyers wishing to compare information from each supplier. There are two technical solutions:

1. A market site is created that pulls in data from all the separate suppliers and 'normalizes' it so the information is delivered in one single virtual catalogue that can be used by all participants (e.g. E-Markets for the agribusiness sector (see Byte Idea in Chapter 3) and Quadrem for the mining sector – see Byte Idea this chapter).
2. Vendor catalogues are standardized in a common machine-readable way and are presented through an industry standard interface.

ESC networks will eventually replace the electronic marketplace structure as it exists today. Contingent upon an interoperable platform, the eventual integration of marketplaces and online supply chains will result in an unprecedented level of connectivity among participants. Figure 4.7 illustrates the ESC concept.

Byte Idea – Quadrem
www.quadrem.com

An eMarketplace for the Mining, Processed Foods and Oil and Gas Industries

Quadrem is the global eMarketplace for the global mining/mineral/metals industry, the consumer packaged goods and the oil and gas industries with 14,500 suppliers and 436 buyers. It handles more than US$5.9 billion throughput a year. Quadrem commenced trading in 2001 and has been developed around six core principles:

1. **Openness** – Quadrem is open to all sellers and buyers who want to participate.
2. **Global** – Quadrem reaches a global market of sellers and buyers and has established regional offices to provide technical and sales support.
3. **Superior technology that is secure and confidential** – Quadrem provides a highly secure environment for commercial information and transactions with best practice training and support processes.
4. **Neutral** – Quadrem is an electronic transaction platform whose value-added services deliver benefits to both sellers and buyers with a comprehensive digital document set.
5. **Independent** – Quadrem is an autonomous entity. Its shareholder base is comprised of organizations which may also be customers; however, Quadrem's management is completely independent.
6. **Connected and integrated** – Quadrem has a very large trading community of buyers and suppliers many of which have integrated their ERP systems with the eMarketplace to accelerate efficiencies.

Purchase order value has risen from US$0 in 2001 to over US$6 billion in 2005 with a major increase in the number of suppliers joining the eMarketplace between 2003 and 2005 (doubling from 5,000 to more than 13,000 members).

- Quadrem's unique selling point to *buyers* is that through 'E' technologies it streamlines all aspects of purchasing, reducing supply-chain costs, improving operations and increasing revenue opportunities.
- Quadrem's unique selling point to *suppliers* is that through 'E' technologies it delivers transaction speed, order accuracy and new business opportunity.

Information sourced from CIO Magazine (www.cio.com) and the Quadrem website

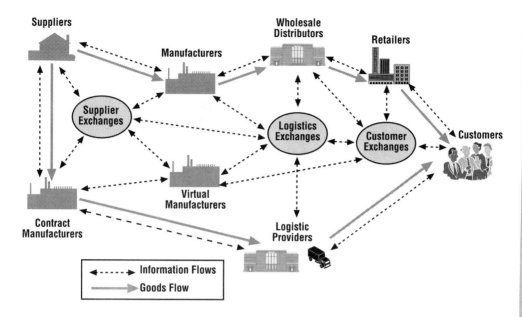

Fig. 4.7. A possible internet enabled supply chain network scenario

Electronic Enablement and the Lean Chain

Lean supply chain management reduces costs and boosts sales and market share for companies by focusing on breakthrough performance improvements wherever and whenever it is possible. Electronic enablement is an important part of adopting a lean approach to supply chain management in that it facilitates:

- The identification and elimination of waste wherever it occurs which leads to reduced lead times, frees up cash, increases customer responsiveness and delivers new products to market quicker. In particular where legacy systems are still being used, many businesses have had to design their processes to fit in with these – and each successive set of changes has added to a patchwork of different fixes which tends to increase waste. Up-to-date technology eliminates this situation and streamlines processes.
- Monitoring the continuous motion that surrounds lean inventory management thereby making a Just-In-Time approach to managing inventory a feasible goal.
- Information flow. Information that is inaccurate or slow will certainly make the flow of materials from suppliers through manufacturing and onto customers a visible problem so speeding it up and making the flow transparent to all concerned, vastly improves this situation.
- Making the right item at the right time and improving the accuracy of product costing by having better and faster access to customer demand information and forecasting data.

A good example is the use of *radio frequency identification (RFID)* technology (AIM, 1999) as a tracking device, both of the obvious issues such as environmental conditions and the history and movement of the item, but also of less obvious issues. Issues such as eliminating human error in monitoring and recording information, controlling shrinkage by improving visibility through the system which also helps in mitigating risk – for example in areas where there are intense regulatory requirements. It should be noted however that RFID technology is an enabler of lean supply chains – not the creator (Brooke, 2005).

Optimizing ESCM through Collaboration

Successful collaboration, in the business sense, means that two or more groups or companies are working jointly to:

- Derive shared information
- Plan, based on that shared information
- Execute with greater success than when acting independently
- Measure performance
- Reward success.

In such a collaborative environment where the internet is a communication medium, the potential for SCM (originally envisaged as 'the elimination of barriers between trading partners') is that multiple levels of trading partners can all benefit from information on consumer behaviour as it occurs. This is because it can be communicated through the collaborative network, thus immediately allowing all partners access to the same information and joint decision-making capabilities. The key communication requirements for this type of business model are:

- *Real-time.* If information is old, it is worthless, and you might as well wait for the re-order point replenishment and build-up the inventory.
- *Global.* It is for the first to stamp their leadership role in any given value chain.
- *Secure.* It must be secure if sufficient trust is to be established across multiple trading partners.
- *Simultaneous.* The information needs to be shared between numerous interested parties at the same time and there must be trust that this is so.

Collaborative commerce techniques directly applied to supply chain and electronically enabled business models, along with the entire range of trading partners and employees that are impacted by them, are collectively known as the Commerce Chain. The ability of organizations to employ collaborative commerce techniques and technologies to optimize their supply and value chains, bolster eCommerce activities and foster inter-enterprise communication and information exchange, will largely determine tomorrow's market leaders. Success will go to the companies and markets that provide real value-added services to the end-to-end supply chain (Gaudin, 2002).

Similarly, those companies that deliver more productive and efficient ways to connect with customers, suppliers and partners will dominate the new B2B eCommerce environment. Collaboration must also extend into all areas within the enterprise such as planning, procurement, engineering, manufacturing, product testing etc. (Ferreira et al 2002).

The eLandscape enables information to be shared simultaneously at multiple levels of the supply chain, with organizations and enterprises moving to non-linear information flows, so that all supply chain participants are 'working off the same page' with the most up-to-date and accurate information available. Passing point-of-sale (POS) data from the retailer to the distributor, who then passes this on to the manufacturer is an example of a linear information flow. In a nonlinear information flow, a retailer might simultaneously share the POS data with the distributor and the manufacturer (Woods and Jimenez, 2002).

ECollaboration then, is about optimizing business processes and business value in every corner of the extended enterprise - from the supplier's supplier to the customer's customer. eCollaboration uses the eLandscape and electronically enabled business technologies to manage beyond the organization, both upstream and downstream. It is the strategic approach that unites all steps in the business cycle, from initial product design and product data management, through production, shipping, distribution, and warehousing until a finished product is delivered to a customer.

A collaborative supply chain relationship must be based on trust between organizations that previously may not have considered themselves partners. This means that strong leadership and change management are essential ingredients of a successful collaboration effort. There are three main issues associated with supply chain performance that are relevant here. They are visibility, velocity and variability.

1. *Information visibility* is knowing what's happening in the supply chain. Essentially, the broader the visibility of information within the chain, the better the service that can be offered to customers.
2. *Supply chain velocity* is the speed of the supply chain. It is governed by the speed of access to good data, planning cycles and execution cycles.
3. *Supply chain management variability* is the ability to manage change. Average companies are still focused on managing/coping with variability through simply better forecasting.

The top companies understand that change is inevitable, and while continuing to forecast, they change their business and supply chain management to be more adaptive and agile in the face of change.

Companies with a network of suppliers, partners, and customers need a fast, efficient way to review designs, disseminate information and enable two-way communications. This is done over the internet using:

- Extranet web sites (e.g. 3Com)
- Web servers (e.g. IBM's webserver platform)
- Groupware (email-integrated collaborative software – eg Lotus Notes)

Web-enabling a company's systems provides visibility to selected partners, suppliers, and customers and allows it to integrate them into the company's other processes. In this way, companies can:

- Tie together all players in the extended enterprise, from design concept to production
- Provide real-time information thus allowing the participants to anticipate and adjust their operations in response to market conditions
- Improve productivity through better data integrity, fewer data entry errors, less rework, and faster communications
- Improve customer satisfaction by providing a more responsive and higher quality service
- Lower the costs, improve the speed, and increase the accuracy of data sharing within the extended enterprise.

Although the internet has been an important catalyst toward advancing data availability, tools such as email leave much to be desired for successful mass collaboration. To be truly effective, collaboration tools must take advantage of all levels of the organization and workflow process, affording ready data access to employees, whether they are next door or half way around the world.

The rapid growth seen in video and data conferencing markets and constant improvements in streaming technologies are enabling companies to leverage existing internet/intranet investments to enhance communication with remote sites bringing information to people, rather than bringing people to information. There are two main points to note however:

1. *There is a difference between communicating and collaborating.* Properly deployed, interactive collaboration tools can provide effective, competitive, and cost-saving advantages to an organization by streamlining the interactive processes of project management. As discussed in Chapter 3, internet and intranet solutions are providing good electronic collaboration options that can be extended to incorporate outside suppliers and vendors as well – for example through the development of *enterprise information portals* (EIPs).

2. No matter how good the technological improvements are in collaborative groupware there will remain the *socio-technical hurdle of user adoption*. Getting people to modify their behaviour is typically the greatest stumbling block, requiring a transformation of user habits (see Chapter 8).

Collaborative Supply and Demand Forecasting

Forecasting is a crucial method for managing inventory control and in a demand-driven environment where the focus is on meeting customer expectations, accurate demand forecasting is only achieved when a collaborative process integrates various forecasting systems. In addition, if you want your organization to succeed, you need to do more than create forecasts. You've got to use the results of the forecasts to improve your future. Having demand information from everyone in your supply chain lets you synchronize what you are doing with your customers and suppliers. It also helps you avoid some fatal mistakes, such as building too much manufacturing capacity – or too little.

For manufacturers, collaborative forecasting could be one of the greatest benefits of internet-aware process integration. By sharing demand forecast data in real time amongst the members of the supply chain, manufacturers can vastly improve the accuracy and usefulness of data used to drive business operations and decisions (Lewis, 1998). The problem: when there are multiple entities in the supply chain, forecasting becomes a very complex proposition.

To implement collaborative forecasting, manufacturers must form collaborative partnerships with their partners, develop compatible processes which will require agreement on how to negotiate and approve changes in the supply chain, and implement technology changes so the technology works with the processes and ERP systems in place in the supply chain. Change processes rather than deploy customized solutions, which generally prove costly to maintain.

The technology exists to enable seamless transfer of data between dissimilar ERP systems, so that instantaneous demand data can ripple down the chain. By leveraging the open environment of Web-enabled internet applications, electronically enabled businesses are well positioned to accomplish this transfer of data. For manufacturers in particular, better data transfer means better collaborative forecasting and ultimately, a more competitive business.

Collaborative Planning, Forecasting and Replenishment (CPFR).

CPFR is an integrated process of marketing, production and sales processes in consumer goods firms (retailers and suppliers) (Peterson, 2003; VICS, 2004). The aim of CPFR is to seamlessly link the manufacturer to the consumer, allowing trading partners to see information along the entire supply chain from one end to the other (iSource, 2003). CPFR is essentially a set of business processes that eliminate supply and demand uncertainty through improved communications between supply chain trading partners.

CPFR is part and parcel of collaborative commerce (cCommerce) in that it calls for complete collaboration and information sharing between trading partners, including the merchandising process, item/category selection and seasonal and promotional planning. Combined with real-time updates based on hourly activity, trading partners will be able to engage in total supply chain visibility and forecasting. CPFR differs from its predecessors such as electronic data interchange (EDI), and vendor-managed inventory (VMI), in that it is designed to link the supply and demand processes allowing for a more consumer driven supply chain.

Collaborative Data Modelling is a critical phase in the development of CPFR since this activity focuses on the relationship between two or more companies. Compatible data from all organizations involved in the relationship is necessary, as are defined business rules on how the data may be used and how that data is to be aligned and organized. Advocates of CFPR suggest that this approach can provide three main benefits (Peterson, 2003):

- *Inventory reductions.* Currently about 8% of customers are not able to purchase what they intend to when they enter a shop because it will be not in stock. If inventory were in the pipeline, including the stores, then consumers would never have to deal with products being out of stock. Unfortunately, inventory bears a high cost in terms of capital consumption and expense, something participants in the value chain are aggressively trying to manage. CPFR has a significant impact on the management of value chain uncertainty and process inefficiencies, which are the drivers for holding inventory.
- *Improved technology ROI.* Through the CPFR process, technology investments for internal integration can be leveraged by extending these enabling technologies to trading partners.
- *Improved overall ROI.* The return on investment from CPFR for most companies will be substantial because investments in the technology necessary for CPFR are relatively small compared to the technology that it leverages (e.g. decision support systems) and compared to the expense reductions from improved supply-process efficiency and marketing-expense productivity improvement as well as stimulating inventory reductions that free up working capital. The only area of potentially significant investment is the change management required to move to a corporate culture that supports collaboration.

Where CPFR has most impact is in supply chains where good inventory management can substantially reduce working capital outlay (generally by millions of dollars). For example in the food retail sector, the CEO of Woolworths Australia, recently stated that he wanted Woolworth's current inventory holding of 40 days' worth of supplies to drop to 'best practice' levels (which he identified as being that of Tesco Supermarkets in the UK) of 17 days. This type of inventory management and control in what is a highly competitive and consumer driven market, can only be achieved by having: good quality (i.e. very few errors) real-time information on the current demands from customers around the country, logistics (supply levels in warehouses and the whereabouts of trucks), and suppliers' capability to provide product as well good forecasting models.

Practically, collaborative supply chain management can only occur when issues of data quality, technological capability and integration between organizations, and the technical competencies of the people contributing to the supply chain, are aligned. The company i2 – see Byte Idea this chapter, is currently a leading player in providing software to electronically manage supply and demand forecasting and potentially, collaborative supply chain management.

Byte Idea – i2
www.i2.com

Integrated Supply and Value Chain Management

i2 was founded in 1988 by Sanjiv Sidhu and Ken Sharma to apply technology and best practices to eliminating inefficiencies in business. Today, i2 software solutions are targeted at within business and external to business information management to allow businesses to respond to volatile markets as quickly as possible. The company has grown to over US$1 billion in revenues in the last 25 years.

There are five core components of i2's software approach:

1. To improve the supplier relationship by coordinating processes across product development, sourcing, supply and demand planning
2. To move beyond the four walls of an enterprise by providing multi-enterprise visibility, intelligent decision support and execution capability using open, real-time collaboration with trading partners
3. To focus on the customer because core resources must be focused on customers to ensure profitability and loyalty
4. To speak the universal language of the supply/value chain from product design through planning, procurement, sales and service. This requires that everyone is fully integrated and that data and content are capable of being fully quality checked
5. Reliability of the technology infrastructure including networks for the supply/value chain which allows companies to run both their enterprise solutions as well as collaboration and planning solutions across their value chain. i2's Global Network is an internet collaboration space that enables buyers, suppliers and marketplaces to rapidly connect to each other for content, collaboration and commerce.

i2's customer list includes Boeing Aerospace, Dell, the European Space Agency, the US Navy and Army, Sanitarium Health Foods, Hewlett Packard, Incitec, BHP, Shell and Coca Cola to name a few.

Information sourced from CIO Magazine (www.cio.com) and the i2 company website

Supply Chain Electronic Information Management

As discussed throughout this chapter, successfully managing a supply chain requires accurate information to flow smoothly and in a timely fashion through a variety of business functions and organizations (Lummus and Vokurka, 1999). For example, in manufacturing, information provides the routings and instructions necessary to assemble products: no information often means no parts, no assemblies and no products moving through the manufacturing area. From there, if the information is not accurate in the warehouses or is not flowing smoothly, then the lists needed to deliver parts to the manufacturing operations won't be printed and the labels and routings needed to move in-process products won't be created when they need to be. Finally if the mission-critical documentation needed to distribute products to customers doesn't happen and the trucks aren't loaded and dispatched, then the business's own supply chain has created a business disaster.

This situation is amplified with the current globalization trend because it means that modern, successful businesses are looking to employ global resources to maximize their potential opportunities in the global community. Whilst it is recognized that a supply chain that makes decisions based on global information will clearly dominate one with disjointed and disparate decisions made by separate and independent entities in the supply chain, such coordination is difficult to achieve.

The technologies of the eLandscape have all contributed to providing physical pathways for coordinated global supply chains – the next hurdle is how to share and manage information (which includes inventory data, sales data, order status data (for tracking and tracing orders), sales forecast data, production delivery schedule data, performance metrics and capacity data) in a collaborative environment. The major challenges (Lee and Whang 1998) to information sharing across the supply chain are:

- Aligning the incentives of the different partners (i.e. minimizing the potential for exploitation by chain partners)
- Confidentiality of information shared
- Cross organizational information systems are expensive and risky
- Timeliness and accuracy of shared information.

To truly be collaborative, these issues need to be addressed early in the planning procedure. The First4Farming agribusiness 'hub' is a good example of how this can be achieved – see Byte Idea this chapter.

The Agile Supply Chain

The increasing volatility or rate of change in the business environment is creating the need for businesses to be highly agile in responding to change, ie they can respond in shorter timeframes to changes in volume requirements and variety change.

Martin (1999) put forward the idea of an *agile supply chain* which is a supply chain that is capable of reading and responding to real demand, using a 'pull' philosophy for supply chain management rather than 'push' whilst at the same time being able to recreate itself rapidly to meet the challenge of shorter product lifecycles and development times. There are four main distinguishing characteristics of an agile supply chain:

1. *Market sensitivity*. In contrast to a lean supply chain which aims to minimize inventory, the agile supply chain aims to improve market sensitivity with inventory minimization happening as a by-product. The agile supply chain model is demand driven (vs. forecast driven).
2. *Virtual*. The agile supply chain is information driven vs. inventory driven and the use of information technology to share the data between buyers and suppliers creates essentially a virtual supply chain.
3. *Process integration*. Shared information between supply partners can only be fully leveraged through process integration, collaborative buyer-supplier teams, and transparency of information, joint product development and common systems.
4. *Network based*. The agile supply chain is a confederation of partners linked together as a network. Trust is critical and agile supply chains tend to be tightly focused private exchanges.

The drivers of the agile supply chain approach are a combination of the need to address an increasing customer demand for variety in less predictable environments with shorter life cycles, the need and greater ability to forecast demand, the need to cut out supply chain inefficiencies, to get closer to the customer, an increasingly knowledgeable and thus powerful customer base, and a need to create value for all participants. The theory is that providing partners access to business-critical data will produce better informed 'real-time' performance, ordering and production decisions.

The Agile Supply Chain Versus the 'Traditional' B2B Marketplaces

The similarities between the agile supply chain and B2B marketplace models (European Union Commission, 2003) are that they involve the same players in the supply chain, i.e. suppliers, buyers and customers; and both focus on using technology to create a virtual meeting place for players. However, there are a number of differences. These include:

- That the B2B marketplace needs a critical and large mass of players, while the agile supply chain concept is essentially associated with private exchanges that link only key partners and suppliers.
- The relationship structure in the B2B model is adversarial while in the agile supply chain model, the hallmarks are trust and collaboration.
- The B2B marketplace presents more of a cultural barrier due to unfamiliarity and discomfort in dealing with a large number of faceless partners. Migrating a brick-and-mortar private exchange to a virtual one presents less of a culture shock.
- There is difficulty in assessing the quality of players in the B2B marketplace vis-à-vis the hand-picked players in the private exchange.
- The up-front cost and complexity may prove a drawback in both models, but more so for the B2B marketplace model. The loose B2B arrangement is less amenable to positive net benefit projections and hence is less attractive.

Essentially whichever model survives over time – the reality is that survival itself will only happen if there are cost savings and additional value that can be obtained by adopting the model. Fundamentally, if the cost savings or benefits aren't there then the markets don't exist. In order to succeed, a new technology or business model must be demonstrably better than the technology it replaces. The telephone is still a very quick way to place an order.

Byte Idea – First4Farming
www.first4farming.com

Go 'Agri-ehubs'!!

First4Farming (F4F) is an independent UK-based company that was born in 2000 as a result of a growing need to have an online exchange for agribusiness companies. Initially, seven major UK and international companies formed a consortium (Banks Cargill Agriculture, BOCM PAULS, Bank of Scotland Business Banking, Dalgety, DuPont, Kemira GrowHow, and Terra) to support the development of the First4Farming agri 'eHub'. The hub allows businesses the ability to exchange transactional documents such as: *Contracts* (contracts, offers and quotations), *Orders* (standalone orders and call-offs), *Goods movements* (e.g. notices of intent to move goods; instructions to logistics support to move goods; notices of goods issued; confirmation of goods received etc.), and *Invoices* (both invoices and credit notes); electronically between their back office systems regardless of system type.

The concept of electronic document exchange is not a new one, but what is different with F4F is the range of ways in which all businesses, big or small can participate in the 'electronic community'. First4Farming's philosophy is not to 'force' companies to take on board major back-office changes and upgrades to enable them to participate in the eHub. They have developed software 'adaptors' that enable even unsophisticated back office systems to link into the First4Farming exchange and services – therefore truly allowing all sizes of company to be participants in the eHub – within their own budgets. The developments made by F4F mean that any company wishing to 'e' enable their systems in order to trade electronically, can do so in the knowledge that there are ways in which they will be able to leverage that investment to deal with all of their customers and suppliers who are on line.

In addition F4F has a website which is essentially an online portal, meeting place and business exchange for everyone in the agricultural and rural business industries throughout any, and all agri-food supply chains.

Today, First4Farming has a direct membership of in excess of 120 companies representing 78% of UK agrochemical manufacturers by market share, 65% of ex-farm UK grain traders, 70% of UK fertilizer manufacturers (by market share), 65% of fertilizer retailers, 50% of UK feed manufacturers and 83,000 farmer members of UK farming co-operatives.

First4Farming have also set up a similar but private eHub in both mainland Europe and in Australia. The Australian business involves most of the major Australian agribusinesses and uses the same philosophy and technologies as the UK and European eHubs.

Information sourced from interviews with Mr Nick Evans, Managing Director,
First4Farming and F4F website.

Extended Supply/Value Chains

The extended supply/value chain platforms work almost entirely behind the scenes and are invisible to end consumers (Hoque, 2000). The ultimate aim of these applications is to share enterprise information with suppliers, buyers and business partners so as to enable supply planning, demand planning, production planning, and logistics to occur in real time. Eventually the traditional isolated supply chain model breaks down and new value grids are created where enterprises of all kinds share information to provide each of them important strategic advantages over their competitors.

Cox (1999) indicates that value chains improve with age. The longer they've been around, the more value they add to an organization and the older the chain within an organization, the more likely it will be used on a daily basis. Today, these supply chains are being integrated in a process or workflow sense as well as electronically so they can be managed as an integrated overall process with appropriate end-to-end measures. More important, a manufacturer can configure multiple supply chains to deal with the many supplier, distribution, and customer service channels needed to effectively reach a variety of customers around the globe. Value chains are being made into multiple-path, multiple-node *value webs*.

Customers searching for the highest product and service value, and their urgent desire to reduce the time they waste in their daily routines, are driving the value web/supply-chain revolution. At the same time, today's advanced value webs (for example see the Byte Idea on Cargill this chapter) are getting easier to create and operate effectively because:

- Standards continue to make it easier for disparate information systems to plug and play effectively.
- Data communications capability has undergone astounding advances – from 64-KB/sec telephone circuit speed to today's 2.5-billion-B/sec high-speed networks, thus allowing manufacturers to communicate throughout the value web and accommodate the constant changes that occur in today's volatile business environment.
- Computer processing power has gone up by a factor of 100,000 since 1971, and this rate of change shows little sign of abating. Data storage capability has shown equal performance gains.
- Middleware has emerged to knit together communications between disparate systems and databases.
- Object-based software programming in a component-based architecture lets software be more flexible and scalable.
- There is a growing awareness of the power of process thinking as a basis for software design or configuration. Sound, efficient business processes mirror the way customer-focused work is accomplished. These processes must be robust and repeatable to ensure quality, low cost, and customer satisfaction. Common business processes then allow common information systems, which greatly reduce operating, maintenance, and training costs.
- Enterprise-wide software application packages, such as those from PeopleSoft, and SAP that cover most of the traditional functional and process activities core to a business' operations, are pervasive in the majority of large and midsize manufacturing companies.
- The internet, with its browser-based technology as a global communication medium that can readily communicate with any computing platform, has provided a quick and inexpensive base for electronically enabled business models and applications – whether B2B or B2C – and for interpersonal communication.

Change Management

Many people still don't understand why things in the business have to change. Management has to vigorously seek and absorb information about the business and technical worlds, and to personally mix with customers and suppliers around the world to appreciate the intensity of global competition and rate of new knowledge discovery and application today. Even when people know why they have to change, there's still the universal human resistance to doing so. Moving from the familiar and comfortable to the new and uncomfortable is still a challenge for all of us.

To meet the challenges to business that are being laid down by the eLandscape, managers should maintain a constant quest for new ideas, customer feedback and suggestions, and best practices from external to the company – in the home industry and, more important, from other industries. In reality, process-focused, customer-focused, and time-based performance metrics will play an increasingly key role in pulling a company toward world-class performance. The recipe for success is:

1. *A Systems viewpoint*. That is, a closed-loop control and feedback mechanism (Chapter 2) is essential in understanding the interactions between activities and business processes within and external to a company.
2. *Process-based thinking*. Here, activities are initially viewed independently of who or what function performs the work. Where it gets done geographically is also critical as intricate, tightly coupled, and real-time business processes are developed across companies and around the world.
3. *An appreciation of the value of time*. Specifically, the need for real-time responses to customers which has become possible with real-time information that aids fact-based decision making and communication to accommodate today's increasingly volatile business environment.
4. *An appreciation of the internet's role*. In enabling a team-based business environment in which knowledge can be shared and leveraged to more effectively serve the customer.

Byte Idea – Cargill Incorporated
www.cargill.com

The Largest Privately Owned
Agribusiness in the World

Cargill Incorporated is a 140 year old multinational agribusiness specializing in: agricultural services (the company is a major global producer of phosphorous and nitrogen, and recently added potash to its supply capability with the formation of a partnership with IMC Global); food ingredients and applications (the company supplies food manufacturers, food service companies and retailers with food and beverage ingredients, and meat and poultry products); industrial applications (e.g. fertilizer, salt and steel products and services, as well as developing industrial applications for agricultural feedstocks); processing (the company connects producers and users of cotton, grain, oilseeds and other agricultural commodities via an international marketing and distribution service); and risk and financial management services in the global marketplace.

Cargill is privately owned, largely by the MacMillan and Cargill families (83%). The remaining 17% is in a company controlled Employee Share Ownership Plan. The company has over 97,000 employees in 59 countries and rightly sees itself as one of the largest (if not *the* largest) agribusinesses in the world. The following statistics give a flavour of the company's reach:

- 2003 revenues = US$60 billion
- 25% of all US grain exports
- 22% of US beef market
- 21% of US turkey market
- 9% of US pork market
- Largest exporter from Argentina
- Biggest poultry processor in Thailand
- Cargill has operations all over the world including Australia, Indonesia, Thailand, Malaysia, Canada, Singapore, Japan, Argentina.

As part of Cargill's extended enterprise approach, it has formed a joint venture with Radian Group, a supply chain management and information technology consulting firm which has provided innovative electronic enablement solutions to many of Fortune 1000's top companies. It also has a policy of actively supporting diversity in its supply chain believing that in so doing it creates added value for its customers.

Information sourced from Cargill webpage, the Financial Times
(www.ft.com), and The Rams Horn (www.ramshorn.ca).

Electronic Integration

Electronic integration within the business and across the supply/value chain is essential for true ESCs to develop. Integration is a mutual responsiveness and collaboration between distinct activities or processes. The extent of integration between two processes or activities is related to the speed with which one responds to another. The speed depends on the immediacy of communication and the degree to which the processes respond to the information communicated (Alter, 1999). Information systems play a role in both aspects of integration by:

• Supporting the communication
• Making it easier for each business process to use the information to respond effectively.

Integrating information, across the enterprise and within the industry sector, should be at the top of a business's corporate strategy (Davenport 1997; Ulrich, 1999a), particularly when considering migrating into the electronically enabled business environment. Integration needs to happen at multiple levels, across multiple organizational units, and needs to be carefully coordinated. The problem with IT is that the amount of assembly required to benefit from any technological advance increases as the size and complexity of the business grows, because each new application must be interfaced to all related applications already in use in the enterprise. Electronically enabled business makes this problem even worse as it requires near real-time interplay between the sell-side, customer-focused application and the operational systems that run the company. The reality is that most companies cannot get their eBusiness site to communicate with their procurement systems. Application integration takes place at two distinct levels:

1. *Data integration.* This is where a common data repository is shared between two or more systems. This is inexpensive and quick to implement and does not alter custom business logic. It does *not* do a complete process integration.
2. *API/method level integration.* This involves direct access to internal systems such as process/workflow modelling, application objects and application data.

Enterprise Application Integration (EAI)

With *enterprise application integration (EAI)* the focus is integrating a set of applications, whether built or bought, inside an enterprise in order to automate an overall business process for that enterprise (Ulrich 1999b; Yee, 2001, 2002). EAI identifies and links user workflow and application functions through sophisticated message queuing and web-based technologies. EAI tools identify, capture, integrate, and deliver data and system functionality to users under a series of cross-functional, multiplatform interfaces. EAI technologies have matured to the point where they can support the integration of these functions without major retooling of complex legacy environments.

For a company, technically, an EAI solution is needed that can (Yee, 2004):

• *Transform the data* – data names, formats, data types, character sets
• *Connect the applications* – supplying multiple and flexible communication models (synchronous/asynchronous), store and forward, publish and subscribe, request/reply, file/forget

- *Control transactional behavior* – defining and managing the scope of transactions, processing multiple updates, maintaining data integrity
- *Externalize process flow from application functions* – flexible, reliable, fault-tolerant and scalable process management. Only then can the real integration task – dealing with the specific semantics of the processes and data structures within the new context – be started.

Business-wise, information provided to and by the systems should:

- Be consistent with information in other business units
- Enable data architectures to eliminate or synchronize the redundancies typically found in legacy architectures
- Enable disparate business units within an enterprise to provide customers with the same answer to the same question
- Enable physical and electronic supply chains to share consistent, rationalized views of the same data across the spectrum of companies defined in that supply chain.

Internet Application Integration (IAI)

With *internet application integration (IAI)* the focus is integrating applications, whether built or bought, across multiple enterprises in order to automate multi-enterprise business processes where the internet provides the communications backbone. Specific examples include trading groups or associations, virtual companies (e.g. components and assembly of a final product are totally out-sourced, the virtual company handles distribution, marketing and finance) and integrated supply chains (Figure 4.8).

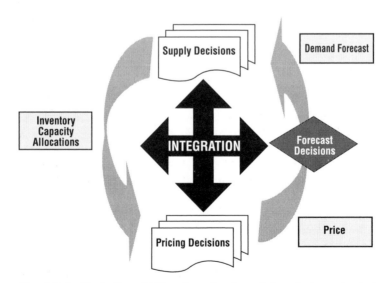

Fig. 4.8. An illustration of IAI as it applies to an integrated supply chain

The Eight Deadly Sins of Electronic Enablement and ESCM

The following are some classic practical problems that tend to occur over and over again with companies either developing as a stand-alone electronically enabled business, or those that are developing an online adjunct to their business, or even those who are remaining very much bricks and mortar but who are looking to the future to automate their supply chains. Unfortunately, all the 'sins' listed have the capacity to create a terminal failure of the project – thus the title of this section being 'The eight deadly sins of electronic enablement and ESCM'.

1. *Not having a good business plan.* Electronically enabling a business or a supply chain is no different from developing any other business component – a well thought out and planned business strategy that links to appropriate business processes and workflow is essential.
2. *Putting technology before business strategy.* Technology enables, it does *not* drive, the business. Putting technology before the value proposition of the business is a recipe for disaster and is a significant reason why non-IT trained managers should be at least aware of the issues discussed in this chapter associated with enterprise architecture.
3. *Ignoring the power of legacy.* Supply chains and legacy systems go together and, just because a system is old, it does *not* mean it should be thrown out unless the proposed new technology has a guaranteed value proposition. Ensuring that integration across the enterprise technology and between partners is a significant area of concern and must be addressed if the project is to succeed.
4. *Creating an IT spaghetti junction.* Integration is a mission-critical component in establishing a robust electronic enablement of the business. Short term solutions (patch-up jobs) lead to short term disasters in the IT world. The absence of a of a well-thought out strategy will give rise to an 'IT spaghetti junction' which will be a set of isolated applications joined by a spaghetti-like point-to-point tangled infrastructure which is expensive, fragile, difficult to maintain, non-extensible and a barrier to communication. In short: *Disaster!*
5. *Underestimating the complexity of the task.* Organizations must be aware of the changing business landscape and the increasingly complex nature of interactions that a 24 x 7 communication facility such as the internet provides. Supply chains are also morphing and becoming more complex for the same reason, with traditional businesses opening up business and technical vaults to new customers and partners. In addition, technology advances have created a shrinking shelf life for enabling technologies so there is a constant need to upgrade facilities to maintain any competitive advantages.
6. *Lulled by standards.* There are many 'standards' and like opinions, everyone has their favourite. It becomes particularly problematic when the enterprise needs to internally integrate across EDI, legacy systems, ERP, CRM and a myriad of other applications as well as integrating externally with partners and their systems.
7. *Biting off more than can be chewed.* Think *big* but think *smart* and *do* small until you know it works – then expand. Make sure that every step provides a firm foundation for the next.
8. *Ignoring the bottom line.* Last but by no means least – electronic enablement is about value creation which we already know from Chapters 1 and 2 relate to the bottom line by increasing revenue, decreasing operating costs, reducing working capital and reducing fixed costs. If the bottom line is ignored, the goal of creating value has failed and so has

the project. Strategies for maximizing value through integration and ESCs and ESCM include: reducing inventories, reducing time-to-market, improving asset utilization, reducing costs of people, process and technology, reducing cost of input, and improving customer satisfaction. In addition, the strategic key to a company's competitive success will be its ability to handle the demands customers, partners, suppliers, buyers make as well as those from within the company.

The Future – Demand-Based Management?

What will be the next competitive battleground in the 21st century? It appears more and more likely that it will be *demand-based management* linked to an agile supply chain. According to Lee (2001), demand-based management is critical to an enterprise in managing its supply chain, product development, technology strategy, service support, and organizational design for the total value maximization of the enterprise. While supply chain management deals with the buy-side of the enterprise, demand based management addresses the sell-side of the enterprise. Tackling one independently of the other, leads to less than optimal solutions.

There are two aspects to demand-based management. First, management of the many business issues used in influencing demand is necessary. These include pricing, promotions (discounts, rebates, and many others), assortment, shelf management, and deal structure (terms and conditions, price protection, return policies, etc.). In order to make the right decisions regarding these issues, the following must be understood:

1. The impact of changing the levels of these instruments on the demand for the product under consideration, as well as all related products (the effects of these instruments can be interactive)
2. The management objectives and constraints faced by the enterprise
3. True supply chain costs corresponding to the demand resulting from the use of these instruments must be incorporated
4. Demand-based management decisions must be linked with supply chain planning and execution decisions, so that demand can be anticipated and met with the right amount of inventory, otherwise, either large amounts of leftover inventory or stock-outs will result
5. Actual performance must be carefully monitored and measured.

Secondly, demand-based management deals with the coordination of the marketing efforts in the supply chain and working closely with customers so that the overall incoming demand for the enterprise and the supply chain will give rise to maximum values for all parties concerned. For example, pricing or promotion efforts at a manufacturer could be wasted if the pricing or promotion efforts at the retailer are not coordinated.

Summary

This chapter has looked at supply chains and supply chain management in some detail – and has addressed the automation of the supply chain using electronic technologies to better collaborate and integrate with partners both internal and external to the business. Very briefly, the chapter has touched upon some of the issues that managers should be aware of – and why, when moving down this path. Perhaps most importantly, the chapter dwells on the issue of optimizing the electronic supply chain through collaboration, what this means and entails, and why managing electronic information is so critical to a successful supply chain. The concept of the 'agile supply chain' is introduced towards the end of the chapter which essentially encapsulates where ESCM is currently, and why such an approach would sensibly lead into 'demand based management' which, it is suggested, is 'where it will be at' in the future.

Nick Evans – CEO of First4Farming is the Smart Thinker for this chapter, having managed the development from concept through to implementation of the 'ehub'.

The Review and Critical Thinking Questions below have been provided for you to revise and reflect on the content of this chapter.

Review Questions

1. What are the business trends leading to electronically enabled business and what are the components of an electronically enabled business framework?
2. Why is electronically enabling an agribusiness an attractive option and what benefits does it convey?
3. What is a supply chain and why would automating it be of benefit to a business?
4. What does supply chain management (SCM) involve and what are some of the most important issues for a manager when moving the supply chain to electronic status?
5. Why is electronic collaboration and electronic information management so important to the success of an agri-food supply chain?
6. Why are business and IT integration issues associated with ESCM so critical – but so problematic?

Critical Thinking Questions

1. Why is Nick Evans worthy of 'Smart Thinker' status – particularly in relation to a chapter on supply chains and SCM?
2. Research Lean Chains in Agribusiness and develop a training tool for an agribusiness industry chain of your choice.
3. Find and examine an agile supply chain of your choice and compare and contrast it to a large B2B marketplace of your choice.
4. Do you think the agile supply chain and demand based management approaches will provide the basis of a new competitive advantage for businesses in the future? – Why?
5. Why are the big Retail companies moving towards the eMarketplace concept and ESCM? What do they expect to get out of it? Give examples.

Smart Thinker – *Nick Evans*

Managing Director, First4Farming Limited

Nick Evans is the Managing Director of First4Farming Limited, an international IT business operating an eTrading community in the agri-food supply chain. Established in 2000, First4Farming now operates from offices in the UK, Belgium and Australia. In 2005 through a strategic merger with its principal IT supply company Adaptris (www.adaptris.com) it now offers customers a one stop shop for solution development, solution deployment, system integration and business process consultancy as well as transactional document exchange solutions and supply chain applications. The Group employs over 25 people across the world.

With a background and qualifications primarily in agriculture, Nick Evans previously worked in large corporations prior to forming the F4F business in conjunction with some of the largest industry stakeholders in the UK in 2000. The disciplines and rigid business processes that tend to be adopted by large corporations served him well in the early days of requirements gathering, process analysis and system specification. Using the standard business practices and procedures together with tight fiscal controls and focus on cash generation derived from 12 years' banking experience with the Bank of Scotland, Nick created a successful point of differentiation between F4F and the other early electronic businesses formed at the time.

In summarising what's been important from his experience to date of running an agribusiness related enterprise in the eLandscape: industry knowledge and an understanding of the needs of the customer, doing things the right way and not necessarily the quick way, and a strong team of key individuals supporting the growth who are able to operate to the agreed strategy without the need for overt supervision.

Where to next? The solutions developed for agriculture have a fit with many other industry vertical portals and, with the right industry knowledge, can be cross-sold. The applications developed by the IT side of the business enjoy a synergy with many leading integration technologies and so the opportunities for this businessman in the 'e' world are limitless.

Source: Author interview with Nick Evans, the First4Farming and Adaptris websites

Chapter 5

Managing Uncertainty in the 'E' World

It's a risky business

This chapter looks at the 'how' side of managing an electronically enabled agribusiness (or any business for that matter), in a world that is full of uncertainty. The chapter looks at the issues of *trust* and *privacy* (what is trust and how does it differ from privacy?), as well as their roles in both eCommerce and electronically enabled business. The issue of *risk* is discussed and defined from an operational, market, credit and business risk perspective, and the task of *managing risk* particularly in terms of the assessment and management of risks associated with electronic technologies, is examined.

As with earlier chapters, this chapter will only look at the *main* electronically enabled business issues a manager is likely to come across in the everyday workings of their job. Additionally, where agribusiness and agri-industry issues are particularly explicit, these will also be discussed in the appropriate context.

Chapter Objectives

After reading this chapter you should understand and be able to describe:

1. What trust and privacy mean and be able to distinguish between them
2. How trust is created – particularly in the agri-industry sector
3. The factors that facilitate trust over the web
4. The trust and relationship issues involved in business alliances and digital commerce
5. The concepts of risk and risk management
6. What the issues and activities associated with risk management are
7. How to manage the common risk scenarios in the electronic landscape
8. What disaster management planning is and how to develop a business continuity plan for an electronically enabled business.

Understanding Trust and Privacy

What is *trust*? According to Webster's Dictionary trust is a term that has many meanings and perceptions and may be defined as:

- Assured reliance on the character, ability, strength, or truth of someone or something (i.e. one in which confidence is placed)
- Dependence on something future or contingent (i.e. hope)
- Reliance on future payment for property (as merchandise) delivered (i.e. credit)
- A property interest held by one person for the benefit of another
- A combination of firms or corporations formed by a legal agreement, *especially* one that reduces or threatens to reduce competition (i.e. a charge or duty imposed in faith or confidence or as a condition of some relationship).

All of these definitions have some meaning in the context of 'doing business' and particularly so in the world of electronically enabled business which this chapter will explore.

Various reports and authors over the last few years have indicated that the internet is expected to have a significant effect on the economic growth of countries (Litan and Rivlin, 2000; US Government, 2001). However, Huang *et al.*, (2003) show that trust (or lack of it) has a statistically significant influence on levels of internet penetration across countries which is creating not only a digital divide among nations, but also an economic developmental divide. Essentially what these authors are saying is that depending on how 'trusting' the population of a country is generally, the level of internet adoption will be directly correlated as will economic growth. If this is the case, it has serious implications for many countries, particularly those that are predominantly low-middle income.

So, how do you increase trust in anything – let alone the use of electronic technologies and in particular electronic technologies in agribusiness?

The primary condition that creates a need for trust is a lack of full information or a lack of transparency of information which is generally associated with a person who is separated in time or space or a system whose workings are not fully known (Shore, 2001). When lacking complete information about a system, a user has to develop a trust-related attitude towards it which can range from complete trust through mistrust to angst – a situation known as the *Trust Continuum* (Fig. 5.1).

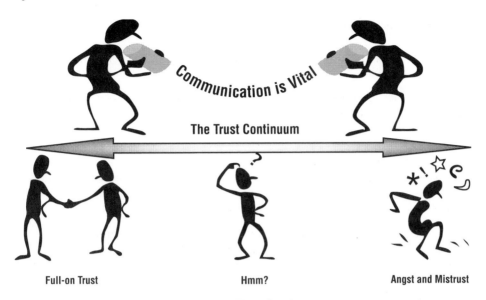

Fig. 5.1. The Trust Continuum

Dealing with a commercial organization necessarily requires a trust-related attitude towards it. When transactions with a commercial entity/system are mediated by a technological system such as the internet and associated technologies leading to the development of an electronically enabled or eCommerce platform, trust becomes of even greater importance. Trust does not come naturally to human beings – it is dynamically generated and has to be carefully structured, worked on and maintained. In interpersonal relationships, one party's actions, particularly self-disclosure (or lack of), can reinforce, diminish or destroy the other party's trust (Busch and Hantusch, 2000).

Conceptualizing Trust

There are many theories on how to conceptualize trust although most cluster within the 'rational' or 'social' perspective. The *rational perspective* centres on the view of self-interest which is based on calculations weighing the cost and benefits of a certain course of actions between participants. The *social perspective* centres on the view of moral duty and is based on individual shared common values.

Ishaya and Macaulay (1999) postulate a complementary view of trust in the eLandscaped world where to conceptualize trust one needs to link the rational and social perspectives. They define trust as:

> **...a characteristic for collaboration where members believe in the character, ability, integrity, familiarity and morality of each other...such characteristics of trust emerge after a series of successful collaborations... thus implying that an element of calculation may be present...**

Building Trust-Based Relationships

The basic principles of trust development include open and honest communications, the professional competence and integrity of the parties, and the willingness to adapt and implement changes for the betterment of the relationship (Lewicki and Bunker, 1996).

Ishaya and Macaulay (1999) hypothesize that trust is built on incremental step-by-step agreements between individuals and members of a team who start out with very imperfect knowledge and experience of each other. They suggest that there is a five stage process in developing trust:

1. *The Transparent Process.* In this stage there is an unclear, doubtful and swift kind of trust between members (e.g. amongst a film crew or a cockpit crew or similar), where group members are keen to address the task at hand and assume that, like themselves, there has been some filtering process applied to the other members of the group before they are grouped.
2. *The Calculative Process.* In this stage, trust is rooted in rewards and punishment associated with a particular collaborative task and is attained when individuals and groups believe that collaborating will provide expected benefits.
3. *The Predictive Process.* In this stage, trust is dependent on members knowing each other well (e.g. via shared experiences). A member uses information about others' past behaviour to predict their future behaviours.
4. *The Competence Process.* In this stage, the capability of accomplishing tasks is of paramount importance in trust development and, as with the predictive process, is dependent on earlier successes.
5. *The Intensive Process.* In this stage, trust follows on from parties identifying with each other's objectives and goals.

This five stage process is a similar one to that identified by Dove (2000) in his paper on trust and relationship management, where he defines three stages:

Stage 1. Economic/Calculus based (equates to the Transparent and the Calculative Processes)
Stage 2. Familiarity/Knowledge based (equates to the Predictive and the Competence Processes)
Stage 3. Empathy/Identification based (equates to the Intensive Process)

Figure 5.2 shows how these two theoretical approaches of how trust is developed can be combined. In practical terms, trust development is a slow process and involves getting to know a person or organization and the way he/she/it works and carries out the business. In the current business environment where the rate of change is fast, and the goal of many companies is to form successful alliances with organizations that are sometimes not well known to them, great risk is involved because Stage 3, shown in Figure 5.2, is rarely arrived at by the time major decisions are being made.

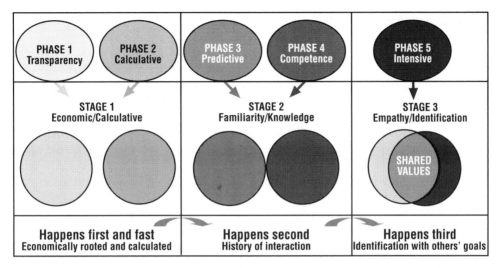

Fig. 5.2. A diagrammatic encapsulation of the combined theories of trust development outlined in the text showing how the five stage process of Ishaya and Macaulay (1999) aligns with the three stage process of Dove (2000)

Trust and Privacy

Privacy is a precondition of trust – and trust affects privacy. Privacy is defined in Webster's Dictionary as:

1. The quality or state of being apart from company or observation
2. Freedom from unauthorized intrusion
3. The capability to explicitly or implicitly negotiate the boundary conditions of social relations.

As with the definitions of trust, all of the above have application in an electronically enabled world, particularly in relation to the increased uses of consumer data on a global scale. Indeed, the updating of privacy laws associated with the use of such data is of paramount importance to ensuring the integrity of electronically enabled business operations.

A recent report (White and Case LLP, 2002) outlining a survey of 15 commercially prominent jurisdictions around the world, including France, Germany, Russia, Spain, Sweden, the United Kingdom, China, Hong Kong, Japan, Singapore, Brazil, Chile, Mexico, the United States and Australia, has indicated that all these countries apart from Hong Kong have updated their privacy laws in recent times, e.g. Commonwealth of Australia *Federal Privacy Act 1988,* updated 14 January 2003 incorporating amendments up to Act No. 125 of 2002 (www.privacy. gov.au/publications/privacy88_240103.doc), to take account of the proliferation of computer based data as well as the internet as a facilitator of information flow.

Having said that, there are many areas that will remain steadfastly offline because of the lack of mechanisms to ensure privacy. The following sections outline some of the major issues associated with trust and privacy in an eLandscaped world.

Trust in Agribusiness

The philosophy that trust must be dynamically generated and worked on can be extended to commercial organizations from any industry sector but, in the case of agribusiness, the issues associated with developing trust and trust-based relationships are magnified enormously by the products that are produced. To varying degrees – dependent on history, cultural background, and economic development – the very nature of agricultural commodities provides for mistrust to develop (Ickis and Brenes, 1999).

At the production end, seasonality associated with most products means that prices vary from day to day, week to week. In some circumstances price controlling in relation to seasonality by a country can ensure almost total dominance of the world market for that product – for example, in the guar industry where India holds sway (Bryceson and Cover, 2004). In other cases, particularly in relation to fruit and vegetables, perishability of most of the various products being produced, as well as individual product piece variations in taste and capability of withstanding postharvest stresses (e.g. between individual avocados or melons in a box), alters the price received dramatically – in a non-transparent way, according to the feelings of some producers.

From a supply chain perspective agri-industry chains or agri-food chains have highly unbalanced power scenarios with the multinational corporate food retailers holding the majority of power. As discussed later this power dynamic, which is generally perceived to be associated with coercive partnership relationships (Gaski, 1984) and an associated lack of transparency in dealings, creates significant mistrust.

At an international level, despite GATT and WTO agricultural trade negotiations over many years (Hausman, 1999; Zeeuw, 1999; WTO, 2005), market regulations, pricing controls (minimum and guaranteed prices), interventionist policies, import restrictions, export quotas and subsidies have all created and continue to create angst and mistrust across the board – domestically and internationally.

The situation in general in agribusiness, along with current globalization of the food system and consumer trust issues particularly in relation to food safety and quality (Noddle,1999), means that an understanding of the trust related issues and a knowledge as to the implications of a lack of trust in conducting successful agribusiness at all levels, is of paramount importance.

Trust in Digital Commerce

Trust and confidence are fundamental to successful business models, and privacy in relation to access, release and anonymity of personal information has become the main trust issue challenging consumer-oriented businesses operating in the global internet-linked economy (Huberman *et al.*, 1999; Hogg *et al.*, 2000; Kozlowski, 2001).

The promise of personalized and intimate consumer relationships is threatened as companies struggle to build the trust and confidence necessary for consumers to lend their personal information. Leading companies are having to build solutions into their business models to gain and maintain consumer confidence.

In the main, trust in digital commerce has been understood primarily as one of how to guarantee the security and privacy of transactions for customers, particularly in relation to the data generated by those transactions. The solution to date has been a combination of **data encryption** (the transformation of data into a form – a cipher – that is impossible to read without the appropriate decoder – a cryptographic key), and **legal controls** to ensure the integrity of data and prevent its misuse by unintended recipients (ArticSoft, 2003).

- *Encryption* ensures privacy by keeping information hidden from anyone for whom it is not intended, even those who have access to the encrypted data.
- *Decryption* is the reverse of encryption; it is the transformation of encrypted data back into an intelligible form. However, the problem still remains – can these protocols be trusted?

Access Control

The major concern in the electronically enabled environment is access control of which there are two forms:

- *Physical* – who can access a computer and how software and data are stored
- *Logical* – defines the software and data that can be accessed from the user through groups of users as defined by the network administrator.

Access control involves **identification, authentication, privacy** and **authorization.** These issues are explored in the section below:

1. *Identification* is the method that provides the means to recognize a user to a computer system – for example, it should enable every system user to be identified with a password, prevent passwords from being compromised and eliminate mechanisms by which identification can be bypassed.
2. *Authentication* is the act of positive identification, coupled with a level of certainty before granting specific rights or privileges to the individual or station that has been positively identified. Protocols include:

   - *Secure Electronic Transaction (SET).* SET was developed by MasterCard and Visa, working in conjunction with partners including IBM, Microsoft, Netscape and others. It is an open, multi-party protocol, transmitting bankcard payments via open networks like the internet. SET allows the parties to a transaction to confirm each other's identity. Employing *digital certificates*, SET allows a purchaser to confirm that the merchant is legitimate and conversely allows the merchant to verify that the credit card is being used by its owner. It also requires that each purchase request includes a *digital signature*, further identifying the cardholder to the retailer. The digital signature and the merchant's digital certificate provide a certain level of trust. SET is important because it offers protection from repudiation and unauthorized payments.
   - *Digital certificates.* Purchasers and retailers generate these certificates through the bilateral use of secret keys that authenticate the legitimacy of each party to the transaction. The majority of digital certificates conform to the CCITT (ITU) standard X.509v3. Many major companies that develop groupware products, such as Lotus, Novell and Microsoft, have decided that the X.509 standard is the best choice for the securing of information on the internet.
3. *Privacy* includes control of outflow of information that may be of strategic or aesthetic value to the person and the control of inflow of information, including initiation of contact.

Protocols to ensure privacy include:

- *Secure Sockets Layer (SSL)*. Netscape Communications Corporation developed this security protocol, designed to reduce the chances that information being sent through the internet would be intercepted. It does not offer a means to confirm the customer, merchant or financial institution involved in a given transaction.
- *Platform for Privacy Principles*. Known as P3, it is supported by the World Wide Web Consortium, and the Direct Marketing Association (http://www.w3.org/P3P/). This developing standard tries to define and describe limits on the use of users' private information garnered from websites.

4. *Authorization* is the act of granting a user or program the requested access. Protocols include:

- *Open Profiling Standard for Authorization and Single Sign-On (OPS)*. Supported by Netscape and VeriSign, this obviates the necessity for customers to re-enter information that identifies them more than once at a website.
- *Use a certificate authority*. This is an authority in a network that issues and manages security credentials and public keys for message encryption and decryption.
- *Basic audit requirement*. This ensures that the system audit trail is protected from unauthorized reading, writing or destruction, defines the minimum types and number of events recorded by the system, defines minimum information to be recorded in audit records and specifies how the audit trail is to be configured, managed and reviewed.
- *Trusted recovery*. This ensures the integrity of the security mechanisms following a system crash or unauthorized user shut down.

Trust and Branding on the Internet

Development of brand names to distinguish products and promote reputations for quality has been carried out since the second half of the 19th century. Brand names provide a certainty and familiarity that customers are happy with and, in addition, convey an assurance of quality and established trust between buyers and sellers who have no 'normal' means of communicating with each other.

Trust is required when there is '*a lack of full information, generally associated with a person who is separated in time or space or a system whose workings are not fully known*'. Fukuyama (1995, 2001) states that what is needed to establish trust in commerce that is conducted globally over the internet is:

- To be able to assess the competence, reliability and reputation of the product or service and to then extend the branding process for the whole range of services and products that could 'conceivably be exchanged over digital networks'
- To create new intermediaries into the process of branding such that firms do not have to establish their own reputations for quality and reliability but rather rely on the intermediary to have the reputation (the virtual handshake).

In other words, it is important that a forum is available where information about a product/company/person online can be gained or ratified by a **reliability** or **privacy seal program** such as that provided by the Better Business Bureau Program (BBBOnLine – www.bbbonline.com/) and by Truste (www.truste.org/about/truste/index.html).

- A *reliability program* can help web users find reliable, trustworthy businesses online, and can also provide the means for reliable businesses to identify themselves as such through a voluntary self-regulatory program that promotes consumer trust and confidence.
- A *privacy seal program* helps web users to identify companies that stand behind their privacy policies and have met the program requirements of notice, choice, access and security in the use of personally identifiable information.

Trust, Business Alliances and the B2B Marketplace

An alliance is commonly defined as any voluntary initiated cooperative agreement between firms that involves exchange, sharing or co-development, and it can include contributions by partners of capital, technology or firm-specific assets (Gulati, 1998a and b). The governance structure of the alliance is the formal contractual structure that the participants use to formalize the alliance. An example of such a contractual structure is a joint venture, where partners share equity and the hierarchical controls associated with the organization are quite strong.

Strategic business alliances involve coordinating two or more partners to pursue shared objectives and thus satisfactory cooperation is essential to their success; however, strategic alliances have been shown to have a high failure rate – *Why*?

> '*Those you trust the most can steal the most*'
>
> (Lawrence Lief).

Das and Teng (1998 and 2001) suggest that confidence in a partner firm pursuing mutually compatible interests in the alliance, rather than acting opportunistically, is a major factor in a successful versus unsuccessful alliance. He suggests that uncertainty and/or level of risk in the undertaking is a major issue and confidence in this sense comes from two sources:

- *Trust* – the high probability of a positive outcome from a partner in a risky situation
- *Control* – used by firms to make the attainment of organizational goals more predictable, thus ensuring certain outcomes (i.e. minimize uncertainty/risk).

Trust and control have a supplementary relationship – that is, the trust level and the control level jointly and independently contribute to the level of confidence in partner cooperation. Confidence and trust building techniques in strategic alliances include:

- *Gaining trust from risk taking.* Trust leads to risk taking, and risk taking bolsters a sense of trust given that the expected behaviour happens (e.g. a significant level of non-recoverable investment in an alliance signals a commitment and trust which boosts the trust level amongst partners).
- *Gaining trust from equity preservation.* The firm contributing the most resources should get most of the equity. Potential inequities in profit distribution create angst and a significant drop in the level of trust.
- *Gaining trust from communication.* Open and prompt communication and proactive information exchange is essential in creating and maintaining trust.
- *Gaining trust from inter-firm adaptation.* Trust can be earned if firms adapt to the needs of the cooperation in the partnership over and above their own needs.
- *Goal setting as a control mechanism.* Alliance goal setting is regarded as critical because of the potential for goal incongruence amongst partners.

- *Establishing structural specifications as control mechanisms.* This includes rules and regulations such as reporting, checking devices, cost control, accounting examination and quality control.
- *Managing cultural blending.* Organizational culture provides a sense of control – thus managing the blending of two or more organizational cultures is critical. The challenge is to make cultural blending work while preserving the separate cultures. It can be very difficult when very disparate and inherently discordant cultures, such as a hierarchical large company and a more flexible small company, come together.
- *Relationship commitment.* This can take several forms including power, cooperative or socially-based approaches. Of these, power-based approaches involving a high use of coercive power by one partner tend to create tight structural bonding but low performance satisfaction and trust generation. Cooperative and socially-based relationships tend to create better performance satisfaction and high levels of trust.

Trust and the B2B Marketplace Continuum

B2B exchanges (Sculley and Woods, 2001) bring multiple buyers and sellers together (in a virtual sense) in one central market space and enable them to buy and sell from each other at a dynamic price which is determined at a moment in time in accordance with the rules of the exchange. Examples include Cyberlynx which is an exchange focused on non-core business materials – see Byte Idea this chapter, and First4Farming – see Byte Idea Chapter 4.

To be successful these B2B marketplaces need to address trust and control issues discussed above (European Union Commission, 2002, 2003). To date this has not happened, with many B2B marketplaces not delivering the expected benefit to the partners involved (an example is that of Corprocure which was originally formed from 14 of Australia's largest companies, but which ended up being sold off to Australia Post in late 2001 to be used by that company as an internal exchange).

Why is this so? The phrase *always look on the bright side of things, but if you're buying something, look on both sides* is salutary in that it reflects the uncertainties not only associated with buying anything at all, but particularly so if your business partners are at arms length or you are buying online. The reality is that most B2B marketplaces have been set up as a way to reduce transaction costs for individual members of the market (Fig. 5.3 – 'Win at Expense of Partner') rather than to benefit each partner (White, 2000). This approach inevitably leads to a breakdown in trust between partners and a resulting critical business failure somewhere in the business life cycle. However, there are a number of alternative models of the B2B situation which are illustrated in Figure 5.3.

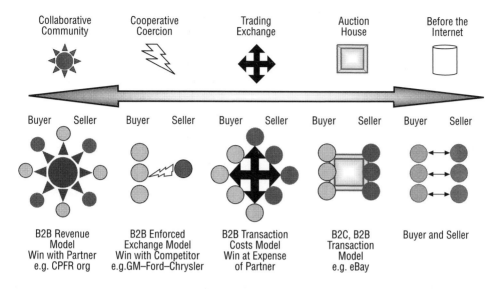

Fig. 5.3. The B2B marketplace continuum (adapted from White, 2000)

Model 1 – Pre-Internet. The business model here focused on internal optimization – B2B relationships were handled on a one-on-one buyer/seller basis with no synergy or benefits being derived from combining processes or transactions across the value chain.

Model 2 – Auction House. The ultimate in B2C purchasing where all buyers and sellers share technology to reduce the transaction costs associated with doing business with each other.

Model 3 – Trading Exchanges. Electronic data interchange (EDI) made possible the many-to-many environment powered by digital exchange of information and where the focus was/is the physical transaction in which, once executed, buyer and seller need never meet again. This model generally focuses on vertical industries and in this type of setup, a company plans to win at the expense of its partner/s.

Model 4 – Cooperative Coercion. This is a very powerful hybrid model that attempts to marry the transaction cost benefits of the Trading Exchange with the additional benefits of Collaborative Communities. As a long-term business model this is unlikely to last since a company will attempt to win with its competitor/s at the expense of their supplier/s.

Model 5 – Collaborative Community. The model here is that a company will win with its partner – that is, a buyer will strategically align with their supplier so that in effect, it is value chain against value chain rather than company against company. Trust in this instance is a major factor that needs to be addressed and cultivated.

Byte Idea –
CyberLynx Procurement Service
www.cyberlynx.com.au/

B2B Marketplace

Cyberlynx Procurement Services is an Australian company that provides a solution to reducing indirect or non-core goods and services to a wide variety of clients via the internet.

The business model is pure B2B and delivers tangible benefits for both suppliers and buyers through the application of focused resources and economies of scale to enable major organizations to share the cost of strategic sourcing in indirect procurement areas. Cyberlynx focuses on:

- Sourcing goods and services which are non-revenue critical, at a cost that generates a competitive advantage for clients. A process that otherwise would require a significant investment of time, money, resources and technology.
- Harnessing the combined buying power of a network of major client companies to effectively create improved supply chain solutions. These include the Commonwealth Bank Group, Woolworths Limited, Lion Nathan, Telecom New Zealand, Telstra Australia, EDSA, Carter Holt Harvey, Royal and SunAlliance and Nestle.
- Increasing its purchasing categories which include office supplies, travel and travel management, IT products, fleet management, contract labour, facilities maintenance and print services.
- Fleet management which allows organizations to take immediate advantage of significant cost and process savings.
- Major spend categories reducing the total cost of indirect products and services. In so doing, Cyberlynx allows clients to improve their financial performance and focus on their core business.

Creating value through a disciplined sourcing process plus ongoing process improvement as well as through pooling resources and strategic sourcing. The return comes from aggregation fees on volume spent through the Cyberlynx network.

Information sourced from Cyberlynx website

Collaborative Supply Chain Management

As indicated in Chapter 4, successful collaboration in the business sense means that two or more groups or companies are working jointly to create shared information and can plan their strategies based on that shared information and are then able to execute their plans with greater success than when acting independently.

The reality however is that these arrangements can degenerate into '*coordination*' rather than '*collaboration*' (A. Dunne Brisbane, 2005, personal communication) if the participants are not mindful of what drives true collaboration. That is: a *collaborative supply chain* relationship must be based on trust and this in fact may end up being between organizations that previously may not have considered themselves partners. Strong leadership and change management skills are thus essential ingredients of a successful collaboration effort (Borck, 2001).

Risk and Risk Management

The next section deals with the concept of risk and managing the risk associated with doing business on the internet. So what is risk? Risk is defined by Websters Dictionary as:

- The possibility of loss or injury
- Someone or something that creates or suggests a hazard
- The chance of loss or the perils to the subject matter of an insurance contract and/or the degree of probability of such loss
- A person or thing that is a specified hazard to an insurer.

In general terms, risk is generally thought of as the chance of something 'bad' happening – thus 'bad' and 'chance' are two key elements of risk:

- *Bad* – refers to an event or outcome that is adverse: it is also relative – losing more money is worse than losing less money
- *Chance* – risk involves *uncertainty* that an adverse event will occur.

If something 'bad' is absolutely guaranteed to happen, there is no risk because uncertainty isn't present. For example, there is no risk associated with jumping out of an aeroplane without a parachute. You will die, guaranteed. It is stupid, but not 'risky.' However, jump out of the plane with a parachute, and you'll probably live, but there's a chance you won't. Thus, most people consider sky-diving risky, but jumping out of a plane without a parachute, suicide.

There are elements of risk in every aspect of running a business – from the operational side of things through to the financial, from the availability of a market through to the whole of business scenario and partner relationship management (Chapman and Ward, 2002; Hillson, 2002 a and b). What makes the situation so much more difficult in the eLandscaped world is the rate of change and the uncertainty of what will happen next, thus making conventional forecasting mechanisms difficult to operate with any accuracy.

Agribusiness Risk

At whatever level in an agri-industry supply chain, the success of an agribusiness depends on being able to identify, assess, understand, minimize and manage risk. As discussed previously in this book, the agri-industry sector is in the midst of significant change, stemming from shifts in regulations and the impact of globalization. It is also an industry sector which has a number of external to the industry risks associated with it that must be recognized:

- Effects of global warming (e.g. extreme weather events)
- Food safety and security
- Commodity price risk
- Genetically engineered food
- Loss of revenue due to volumetric risk (i.e. inability to supply constant volume and quality)
- Environmental impairment (long term sustainability issues – e.g. salinity)
- Governmental and regulatory reforms
- Accidental and malicious contamination.

At each level in an agri-industry chain aspects of some, or all, of these risks will need to be managed. Take for example commodity price risk management: the rationale for a producer in using this strategy to minimize on-farm risks associated with price volatility in the markets is that market-based tools such as derivatives or hedging instruments (Varangis and Larson, 1996, Varangis *et al.*, 2003) can effectively insulate them from short-term price volatility if used appropriately – see Byte Idea this Chapter on ARMS. The same applies to a trader further downstream in the chain. There are two basic types of risk management tools in common use (Varangis and Larson, 1996):

1. *Futures contracts.* These involve the buyer (or seller) of a futures contract agreeing to purchase (or sell) a specified amount of a commodity at a specified price on a specified date. Contract terms (for example, amounts, grades, delivery dates) are standardized, and transactions are handled only by organized exchanges. Profits and losses in trades are settled daily through margin funds deposited in the exchange as collateral. Futures contracts are usually settled before or at maturity, and do not generally involve physical delivery of the product.

2. *Options contracts.* These offer the right but not the *obligation* to purchase or sell a specified quantity of an underlying futures contract at a predetermined price on or before a given date. Like futures contracts, exchange-traded options are standardized, over-the-counter options offered by banks and commodity brokers. Purchase of an option is equivalent to price insurance and therefore requires that a price (premium) be paid. Options include: *calls* (which give the buyer the right to buy the underlying futures contract during a given period and are purchased as insurance against price increases) and *puts* (which give the buyer the right to sell the underlying futures contract during a given period and are purchased as an insurance against price declines).

Throw in the risks associated with electronic enablement of the business and/or industry specific supply chain risks and risk management has become a very large chunk of an agribusiness manager's portfolio.

In the following sections, risk and risk management in general within a business are briefly described before the chapter moves into risk management associated with electronic enablement.

Byte Idea – ARMS
www.arms.graincorp.com.au/

Agricultural Commodity Price Risk Management

As price and production variability increases, commodity price risk management as a component of a producer's overall risk management strategies becomes an option. A reduction in risk may be facilitated by the use of the commodity futures exchange markets to hedge the potential costs of commodity price volatility in a similar fashion to taking out an insurance policy on a car against having an accident.

ARMS (Agricultural Risk Management Services Pty Ltd) is a subsidiary of GrainCorp (www.graincorp.com.au) one of the largest grain handling and storage agribusinesses in Australia. ARMS core business is about providing producers and end-users of grains and oilseeds with commodity price risk management services. ARMS services include:

- Market intelligence and reporting on significant events so that significant commodity price risks can be identified.

  ARMS Advantage is a subscription based service offered to clients which provides market information, commentary and daily prices and includes a daily SMS text service detailing the daily futures close and resulting change for Chicago Board of Trade (CBOT) for wheat and corn and the Winnipeg Commodities Exchange (WCE) for canola. Spot FX quotes for the USD and CAD against the AUD are also provided.
- A facility to help manage and report against contract details and finalization including providing Grower Optional Pricing Contracts (GOP) and Consumer Optional Pricing Contracts (COP).
- Financial services and advice – ARMS is fully licensed with the Australian Securities and Investment Commission and holds an Australian Financial Service (AFS) Licence to provide financial product advice in derivatives and foreign exchange contracts.

Information sourced from ARMS and the ARMS/Graincorp website

The Risk Jigsaw

The Risk Jigsaw is a conceptual model of the different types of risk involved in managing risk in an organization (Fig. 5.4).

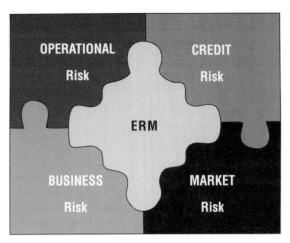

Fig. 5.4. The Risk Jigsaw (adapted from www.erisk.com/Learning/RiskJigsaw.asp)

The model (Lam, 1999), which was developed based around the financial industry, is nevertheless appropriate for all industries and is particularly relevant to electronically enabling a business. The model defines five main areas of risk for a business:

1. *ERM or enterprise risk management.* This is an approach to fully integrating risk management into the way a company conducts its business. It involves assessing, measuring and managing the different types of risk to which an organization is exposed. Depending on which component of an agri-industry supply chain is involved, the relative importance of the various risks associated with agricultural production and consequent agri-industry supply chain management will be important.
2. *Operational risk.* This is the risk of loss due to physical catastrophe (e.g. drought), technical failure and/or human error in the operations of a firm. It also includes the risk of loss due to fraud, failure of management and process error (see Chapter 9 in relation to eGovernance)
3. *Market risk.* This is the risk to the company's financial position from changes in market factors such as interest rates, foreign currency value and/or prices of commodities and equities.
4. *Credit risk.* This is the risk that a party might become unable, or less likely, to fulfill its contractual obligations. This is a constant risk for agribusinesses – particularly those involved in export.
5. *Business risk.* This is the risk that a change in the variables of a business plan will destroy the plan's viability. It includes quantifiable risks such as the demand forecast risk and the unquantifiable risks such as innovation by competitors and technology developments.

Within these five main areas the risk drivers can be further split into: strategic, operational, stakeholder, financial and intangible (see Table 5.1). Associated with these five main drivers are the in-house capabilities of managing the risks that are listed – and it is essential that all types of risk are assessed, planned for and managed when running a business.

Table 5.1. Main risk drivers within a business

	Risk Driver				
	Strategic	**Operational**	**Stakeholder**	**Financial**	**Intangible**
Risks	Competitors	Processes	Partners	Accounting	Knowledge
	Resource allocation	Physical assets	Suppliers	Credit	Intellectual property
	Programme/ project management	Technology and infrastructure	Customers	Cash management	Information systems
	Governance	Legal issues	Management	Marketing	Databases
	Brand/ Reputation	Human resources	Employees	Taxes	Information for decision making
	Partnerships/ JVs	Environmental hazards	Government	Regulatory compliance	
	Macro trends		Community		
	Strategic planning				
	Organizational structure				

Risk Management

Risk management, in a business context, is about reducing the cost of risk, which includes the cost of managing risk. Risk management strategies ask the question, *Is the risk appropriate for the return?* It is important to note that risk management would not help the individual intent on jumping without a parachute in the example above! (i.e. *recognition of the risk* is necessary for effective risk management).

Risk management is an intricate process that combines market and external influences, asset performance capability and acceptable levels of risk to determine optimum direction (Fig. 5.5). The inputs to the process are highly dynamic and business processes and models need to be flexible enough to respond to changing conditions.

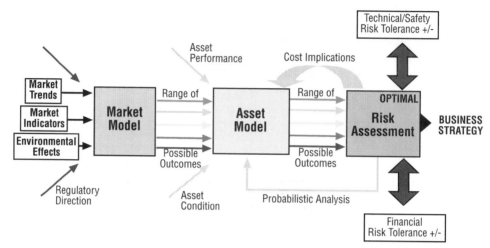

Fig. 5.5. A diagrammatic illustration of the inputs to a risk management scenario

No business is risk-free, and risk management will not eliminate all risks. Risk management is about the **3Rs**: **Returns, Risks** and **Ruin**.

- *Returns.* People want a higher return on their investment in exchange for taking a greater risk. The best risk management strategy to implement is therefore to understand and know as well as is possible the chances of the possible outcomes.
- *Risks.* Here, the best risk management strategy is not to risk more than can be afforded.
- *Ruin.* By the time you arrive at this point, which is the result of losing more than can be afforded, risk management is essentially a thing of the past. The lessons learned will hopefully enable you to take greater account of risks and risk management in the future.

The 3Rs in practical terms can be reduced to understanding what the risks are, why they are risks and how they can be avoided or minimized. Table 5.2 outlines these basic points in a One-Minute Risk Management Strategy.

Table 5.2. The One-Minute Risk Management Strategy (Source: Mehr and Hedges, 1963, *Risk Management in the Business Enterprise*, Irwin Press)

One-Minute Risk Management Strategy	
WHAT	Assuring financial solvency against possible risk at lowest cost
WHY	To ensure the continuing profitability and viability of the organization
HOW	Understanding financial statements provides the foundation on which a sound risk management plan can be devised and appropriate risk management tools employed
GUIDING RULES	• Don't risk more than you can afford to lose • Don't risk a lot for a little • Understand the likelihood and severity of possible losses

Functional Activities of Risk Management

There are a number of functional activities associated with risk management – identification of risks, analysis of risks, planning associated with management of risks identified, tracking of risks, controlling risks and communication of risks and management strategies. These are defined in more detail in Table 5.3, and should occur continuously and simultaneously as depicted in Figure 5.6.

Table 5.3. A description of the functional activities of risk management

Function	Description
Identify	Search for and locate risks before they become problems
Analyse	Transform risk data into decision-making information. Evaluate impact, probability, and timeframe, classify risks and prioritize risks
Plan	Translate risk information into decisions and mitigating actions (both present and future) and implement those actions
Track	Monitor risk indicators and mitigation actions
Control	Correct for deviations from the risk mitigation plans
Communicate	Provide information and feedback internal and external to the project on the risk activities, current risks and emerging risks. *Note*: Communication happens throughout all the functions of risk management

Fig. 5.6. The functional activities associated with risk management

Risk Evaluation

A number of distinct factors help firms to evaluate a particular risk. These factors, which are defined below, include: **exposure, volatility, probability, severity** and the **time horizon** involved (Davis, 2004). It's the interaction of these factors with two other notions – capital and correlation – that determines the effect of a specific risk on a specific company (Lam, 2000).

- *Exposure* is the maximum amount of damage that will be suffered if some event occurs. All other things being equal, *the risk associated with that event increases as the exposure increases.* For example, a lender is exposed to the risk that a borrower will default. Exposure can be controlled in a number of ways: for example, it might be reduced by transferring the risk to another company (such as an insurer), financed (cushioned by capital) or simply retained.
- *Volatility* loosely means the variability of potential outcomes and is a good alternative to the word 'risk' in many of its applications. This is particularly true for risks that are predominantly dependent on market factors, such as options pricing. Generally, *the greater the volatility, the higher the risk.* For example, the number of loans that turn bad is proportionately higher, on average, in the credit card business than in commercial real estate. Nonetheless, it is real estate lending that is widely considered to be riskier, because the loss rate is much more volatile – and therefore harder to cost and manage.
- *Probability* – how likely is it that some risky event will actually occur? The more likely the event is to occur, *the higher the probability and thus the greater the risk.* The assignment of probabilities to potential outcomes has been a major contribution to the science of risk management. Certain events, such as interest rate movements or credit card defaults, are so likely that they need to be planned for as a matter of course and mitigation strategies should be an integral part of the business's regular operations. Others, such as a fire at a computer centre, are highly improbable, but can have a devastating impact.
- *Severity* – How bad might it get? Whereas exposure is typically defined in terms of the worst that could possibly happen, severity is the amount of damage that is, in some defined sense, likely to be suffered. Essentially, *the greater the severity, the higher the risk.* Severity is the partner to probability: if we know how likely an event is to happen, and how much we are likely to suffer as a consequence, we have a pretty good idea of the risk we are running. But severity is often a function of our other risk factors, such as volatility – the higher a price might go, the more a company might lose.
- *Time horizon* – a measure of how long you are exposed to a risk and how long it takes to reverse the effects of a decision or event. In other words *the longer the duration of an exposure, the higher the risk.* For example, extending a ten-year loan to the same borrower has a much greater probability of default than a one-year loan. Hiring the same technology company for a five-year outsourcing contract is much riskier than a six-month consulting project – though not necessarily ten times as risky. The key issue for financial risk exposures is the liquidity of the positions affected by the decision or event. Positions in highly liquid instruments, such as government bonds, can usually be eliminated in a short period of time, while positions in, say, real estate, are illiquid and take much longer to sell down.

Risk Management Strategy

A proactive risk management approach from a business perspective should, in essence, be similar to an internal 'due diligence' audit which refers to an audit of the operations, solvency, and trustworthiness of a particular firm (Concise Finance Encyclopedia, 2004). The term due diligence itself refers to the care a reasonable person should take before entering into an agreement or transaction with another party (Investopedia, 2005).

A risk management approach based around due diligence is thus the process of examining the financial underpinnings of a business with the goal of understanding the risks associated with a deal such as a pending merger, equity investment or large-scale IT purchase. In a due diligence situation, issues that could be reviewed include corporate capitalization, material agreements, litigation history, public filings, intellectual property and IT systems. A risk management approach should therefore address the following areas:

- A comprehensive and regularly updated assessment of the drivers of risk impacting the business strategy, for example:
  - *Financials* – ensure their accuracy
  - *Assets* – confirm their value and condition
  - *Employees* – evaluate each of them
  - *Sales* – what's working, what's not and how to improve
  - *Market and marketing* – what is driving the business? How can you improve it?
  - *Industry in which it operates* – understand trends and new technologies
  - *Competition* – what are they doing? Will you be able to compete and how?
  - *Systems* – are they effective and secure?
  - *Legal and corporate issues* – minimize potential for costly surprises
  - *Company contracts and leases* – are they in order? Confirm obligations
  - *Suppliers* – will they continue to do business? Are there any hidden agendas or exposure?

- The creation of policies and procedures for managing each driver of risk in the business
- A risk assessment platform for creating transparency throughout the organization's investment portfolio, especially in export activities
- An integrated and focused due diligence and internal audit program designed to treat risk as an opportunity
- The establishment of an early warning reporting framework to identify issues before they become problems.

Figure 5.7 illustrates the risk management approach discussed in the text as a framework linking leadership, people, risk drivers, management and outcomes.

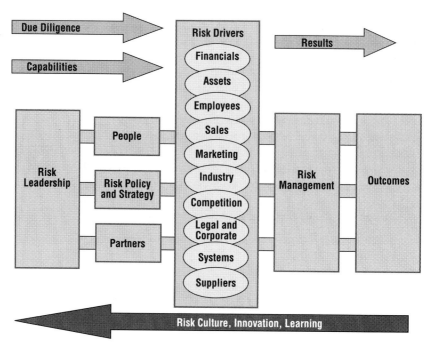

Fig. 5.7. A risk management approach based on a due diligence philosophy (adapted from the European Framework of Quality Management – www.efqm.org)

Most organizations can handle identification and analysis of their risks – the problems arise in planning how to respond to them. There are seven principles which provide a good planning framework for effective risk management (Hillson, 1999):

1. *A global perspective.* It is necessary to recognize both the potential value of opportunity and the potential impact of adverse effects.
2. *A forward-looking view.* This enables the identification of uncertainties, allows anticipation of potential outcomes and management of resources and activities while anticipating uncertainties.
3. *Open communications.* It is important to encourage free-flowing information at and between all business levels; to enable formal, informal, and impromptu communication; to use processes that value the individual voice, thus bringing unique knowledge and insight to identifying and managing risk.
4. *Integrated management.* This approach makes risk management an integral and vital part of the business's management strategy and involves adapting risk management methods and tools to the business's infrastructure and culture.
5. *Continuous process.* This means identifying and managing risks routinely.
6. *Shared product vision.* There should be a mutual business vision based on common purpose, shared ownership, and collective communication while focusing on results.

7. *Teamwork.* Pooling talents, skills, and knowledge while working cooperatively to achieve common goals is the best way to achieve a solid risk assessment and management situation

If the risk response is to be effective Hillson (1999) indicates that the responses must be:

- *Appropriate.* The correct level of response must be determined, based on the 'size' of the risk. This ranges from a crisis response where the project cannot proceed without the risk being addressed, through to a 'do nothing' response for minor risks.
- *Affordable.* The cost-effectiveness of responses must be determined, so that the amount of time, effort and money spent on addressing the risk does not exceed the available budget or the degree of risk exposure. Each risk response should have an agreed budget.
- *Timely.* An action window should be determined within which responses need to be completed in order to address the risk. Some risks require immediate action, while others can be safely left until later.
- *Achievable.* There is no point in describing responses which are not realistically achievable or feasible, either technically or within the scope of the respondent's capability and responsibility.
- *Effective.* All proposed responses must work! This is best determined by making a 'post-response risk assessment' of the size of the risk assuming effective implementation of the response.
- *Agreed.* The consensus and commitment of stakeholders should be obtained before agreeing on responses.
- *Owned.* Each response should be owned and accepted to ensure a single point of responsibility and accountability.

A good example of a company in the agribusiness world that has addressed these issues is Withcott Seedlings – see Byte Idea this chapter.

Byte Idea – Withcott Seedlings
www.wseedlings.com.au

A Business Built on Trust and Sustainability

Withcott Seedlings is located in Queensland, Australia and is the major supplier of vegetable seedlings to the entire east coast of Australia. The company is one of a number in the Erhart Group of companies all owned and managed by Wendy and Graham Erhart who are both innovators in terms of business processes and practices – while being managers of risk par excellence!

From a business perspective, Withcott Seedlings was established in 1983 and has grown at a rate of 25% per annum every year since then – a major growth pattern that has needed to be managed with care. The risks of overextending a company's abilities in such a situation are significant. Thus to stay on top of the business risks and to address general production risk which is ever-present in the Australian production environment, the way forward has always been seen to be through innovation, collaboration and sound business practices.

Innovative developments to ensure quality of produce have been a watchword of this company. They include implementing stringent quality management systems (to ensure that seedlings meet individual customer demands exactly and consistently); designing and implementing environmentally sustainable water recycling techniques (which have virtually made the company immune to the variable water availability and rapidly declining water quality issues in their local region); and implementing in-house electronic enablement for most business processes – including in November 2005, establishing an online ordering system for buyers of baby leaf salads from Smart Salads (www.smartsalads.com) which is another of the Erhart Group of companies. In fact, the Smart Salads process itself is highly innovative and all about managing risk – food safety risk. It involves a purified watering system with all seedlings being grown above ground on aluminum benching – and it also involves being able to harvest and chill the baby salad leaves within 20 seconds. This process eliminates the potential for food safety hazards such as foreign bodies or microbial contamination. The added value? An extended shelf life which has enabled the company to extend the company's reach into a growing export market.

However, as essential as the innovative use of technology is, it has been complemented by the implementation of innovative business processes focusing on facilitating network-based opportunities based around trust development, cooperation, communication and management across their supply chain including employees, suppliers and customers.

With business awards – including being the Ernst and Young Entrepreneur of the Year (2005) Northern Division Finalists, and for Wendy, the Veuve Clicquot Business Woman of the Year Award (Australia) in 2005 – Withcott Seedlings and the Erhart Group look set to continue their skyrocket journey of well managed success for years to come.

Information sourced from Wendy Erhart, Stroudgate (Innovation Australia) (www.stroudgate.net)
and the Withcott and Smart Salads websites

ETechnology and Electronic Enablement Risk

Electronically enabled business risk management defines organizational strategies encompassing all reasonable efforts to preserve the integrity of information assets and corporate tangible and intangible assets. The emergence of the internet as a critical business tool has created a host of new exposures for companies establishing a presence on the web, as well as for those providing the products and services needed to operate on the internet. If the organization has a website, intranet or extranet, accepts credit card details online or even simply sends email to customers, then the business is potentially wide open to a number of legal and financial risks.

As the complexity and breadth of business alliances increase, and the reliance on cross-organizational supply chains increases, overall risk management needs to address these trends. Because outsourcing is also on the rise, so is vulnerability to corporate asset theft. Any damage or security breaches to networked connections have ramifications for financial loss throughout the supply chain.

Computers streamline internal and external communication, track information, and help companies retain their competitive edge. Electronic enablement trends are towards real time enterprises with automated information flow and greater transparency throughout the extended enterprise. Such technology also raises significant legal issues in the workplace that every company must face – some old risks in new forms, some entirely new risks that are now emerging such as:

- Electronic security issues of identification, authorization and authentication
- Infrastructure is vulnerable to unsophisticated attacks, particularly interconnected infrastructure such as telecommunications and power generation and distribution which has ripple effects through the economy if the situation goes on long enough
- Careless employee email messages that result in multi-million dollar liability when produced in the discovery stage of litigation
- Offensive jokes on the company email system may trigger employer liability and adverse publicity
- Postings in chat rooms about the company that within a matter of hours could drastically reduce the company's share value or jeopardize a public offering
- A fired employee using the company email system to distribute confidential salary information to all remaining employees.

EBusiness risks are not all physical and neither are the solutions. Protecting the company against the financial risks of copyright infringement, credit card fraud and transmission of computer viruses can be difficult if not impossible to achieve in a physical sense. Companies can help manage these risks by establishing effective policies and procedures such as:

- Electronic communications policies
- Electronic record keeping policies
- Confidential information policies and security procedures
- Intranet policies and oversight of intranet content to ensure a no harassment/discrimination/ hostile work environment risk
- Electronic monitoring guidelines
- Employee departure procedures to minimize risk to IT systems
- Procedures to bullet-proof the company's confidential information from electronic security breaches

- Insurance policies against credit card fraud and internet liabilities associated with ownership of a website and the day-to-day use of email by employees and others. This should include indemnity for legal liability arising from claims for infringement of copyright, defamation, unauthorized use of trade names or logos, and the unintentional transmission of a computer virus.

All businesses require a life cycle approach to risk management. This approach considers that a business, particularly one that is undergoing electronic enablement, is in a constant state of development, reengineering and refinement. As a company's electronic development evolves from concept to initial development, production and ongoing maintenance, enterprise-wide risk management solutions need to apply accordingly at each stage.

The integrated process addresses the evolution of critical electronic business elements including: evolving business goals, corporate structure changes, rapidly changing technologies, changes in outsourcing relationships, physical location changes and expansion, and changing international considerations.

Disaster Management Planning

Part of a good overall risk management strategy is the development of disaster management and business continuity plans. These are particularly pertinent in a business world in which both natural (e.g. floods and fire) and human disasters (e.g. terrorism) occur regularly around the world – see Byte Idea this chapter on ESRI.

Immediately after a major disaster, there is the possibility of loss of life resulting in the afflicted company being deprived of key strategy and direction management at a critical juncture. In addition to loss of life, loss of data and infrastructure may occur. This latter issue has serious implications for financial loss, loss of customer confidence, loss of trust and for the disruption of the operations of business partners. Thus, every company must, when considering disaster management strategies, also consider the impact of strategic business allies being inflicted with a disaster.

The shocking events of 11 September 2001, when the World Trade Center in New York City was destroyed by terrorist attacks, caused a significant toll on human life and disrupted communications and business operations.

Apart from the difficult task of ensuring their employees were safe or being helped, companies also had to ensure the continuity and viability of their businesses. In the aftermath of the disaster, it was obvious that companies with a disaster management strategy that included having a good *business continuity plan (BCP)* in place were relatively less damaged in terms of business discontinuity than those without. The following section deals with developing business continuity plans which comprise the essential ingredients of a disaster management plan.

The Business Continuity Plan (BCP)

A business continuity plan (BCP) defines what needs to be done and the resources required to run operations in the event of an interrupting incident or disaster. Its purpose is to minimize the impact of financial and human loss and provide an acceptable level of business capability until normal business operations are continued to its customers, business partners, investors and suppliers. Having a BCP ensures an organized and orderly response in the event of a major incident or disaster, and enables management to provide direction and advice to affected business departments or units. A BCP comprises three major components: **risk assessment**, **contingency planning** (which is the process of creating, testing and maintaining a plan to recover from any form of disaster) and **disaster recovery** (the purpose of a disaster recovery plan (DRP) is to recover mission-critical technology and applications at an alternative site. A DRP is a critical part of the BCP process and should be prepared to address each level of risk.

Byte Idea – ESRI (Environmental Systems Research Institute)
www.esri.com

GIS for Disaster Management

ESRI was founded in 1969 by Jack and Laura Dangermond, who still own the company. ESRI initially focused on developing software to help organize and analyse geographic information and is the developer and leading provider of geographical information system (GIS) software packages such as ArcInfo, ArcView and ArcGIS which are currently in common use throughout the world.

GISs are computer-based systems that are designed to allow spatial data (information that is geographically located in space by a coordinate system such as latitude and longitude), to be captured, stored, manipulated and integrated with other data sources and/or used in modelling scenarios. With a GIS it is possible to create a detailed picture of a given point on the earth as well as locating that point within the surrounding landscape.

GIS technologies can be used for many purposes but land suitability, risk/hazard mapping, environmental management and precision farming needs are the most common uses. In businesses, GIS technology can be used for market and network locational analyses and it is often used for disaster management. Disasters are usually spatial events (floods, fires, plagues, famine, political unrest, etc.) and mapping and gathering information about the disaster is vital for disaster management. The ESRI suite of software packages (including ArcSDE, a business to business product which allows the storage of spatial and tabular data in commercial DBMS products) support all aspects of disaster management including disaster planning, response, mitigation and recovery.

Information sourced from the ESRI website

1. *Risk assessment.* Each organization needs to assess the major risks relevant to its own operating environment and determine the probable disaster scenarios that could affect its operations. As a minimum requirement, an organization should include the following disaster scenarios in its risk assessment where applicable:

   • *Physical and environmental disasters* that cause the workplace to be inaccessible or inoperable. This could be due to fire, explosion, flood, broken water pipes, accidental activation of water sprinklers, etc.
   • *Internal infrastructure failures* that cause a cessation of internal information flow such as power failures, data centres or mainframes not being available, voice and data telecommunication failures and network failure (LAN, WAN), etc.
   • *External infrastructure failures* which causes a cessation of inter-business information flow such as an interface failure between business partners, a failure by service providers, a widespread telecommunication failure, widespread power failures.
   • *Situational disasters* such as terrorist attacks, a run on the stockmarket, national instability (e.g. political upheaval, military coup), or localized instability (e.g. a riot).

2. *Contingency planning.* The BCP process should be centrally coordinated and managed with input from all the departments/units that comprise the organization. In order to ensure that the plan can be fully implemented, all staff should be made aware of their roles and responsibilities, and regular drills held to monitor the effectiveness of the plan and to reveal any shortcomings which should be addressed (Fig. 5.8).

   The practice of ensuring a disciplined documentation process and monitoring of a defined BCP process, the transparent provision of information to the public with timely press releases, the accountability of the various teams mobilized to make decisions and act on issues, sound management direction in a crisis and showing responsibility for social issues such as the safety and recovery of staff and tenants, all point towards good corporate governance that will promote trust and confidence in the organization from its suppliers and customers.

3. *Disaster recovery plans (DRP).* In a DRP, which should focus on the most probable set of risks for a particular location, business functions must be identified and prioritized for recovery at an alternative site. Each plan should include manual processing procedures in the event that business systems cannot be fully automated on resumption. In the *worst case scenario*, in which access to the corporate site is unavailable, the alternative site should be activated, key personnel relocated to the site and the DRP activated. The following key procedures, each with the sequence of operations, tasks to be performed, individual action plans (who, what, where and when), and communication and contact procedures associated with it, should be documented in the DRP:

   • Notification and activation of key staff
   • Retrieval of vital records
   • Re-location to alternate site
   • Reconciliation of vital records for work-in-progress at the time of the disaster
   • Resumption of business operations in 'recovery mode'
   • Acquisition of internal services and vendor support
   • Expense monitoring and control
   • Manual processing instructions
   • Restoration of data when systems become available.

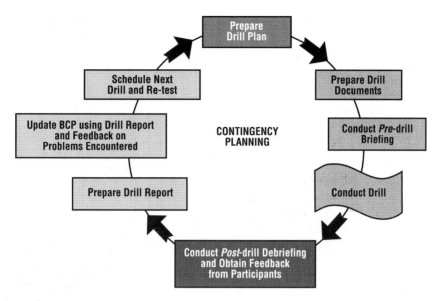

Fig. 5.8. The business contingency planning cycle

Business Continuity Planning

Business continuity planning should be ongoing throughout a business life cycle and has four phases: **crisis prevention, emergency response, recovery of operations** and **restoration of facilities**. ISO17799 (BS7799) (ISO, 2005) is an international security standard which covers disaster recovery and crisis management planning. The standard is very clear and requires a formal approach and the creation of a quality plan. Compliance with ISO17799 represents a statement on your company's disaster recovery arrangements and gives assurance that the plan is sound and that the disaster recovery practices are likely to be adequate.

BCP Phase 1. Crisis Prevention and Survival

In this phase, various measures and activities can be taken by organizations to lessen the possibility or the impact of an adverse incident occurring to the business. Such activities include protecting corporate assets, which itself involves safeguarding the human assets, facilities and contents and vital records of the corporate entity, undertaking an analysis of the risks involved and creating appropriate risk minimization strategies, and conducting a business impact assessment.

• Safeguarding human assets involves protection from harm as well as the identification of key employees required in a crisis situation and how to contact them at all times (seven days a week, 24 hours a day).
• Facilities and contents represent substantial capital investment and include buildings, furniture, equipment and other assets. Chief among these are data centres, branches and overseas offices if they exist. Each asset must be evaluated in terms of the threats most

likely to materialize impacting the company's business or that of a strategic business partner, and the capacity of the company to cope with and minimize effects of threats which may include:

- ◆ Natural disaster
- ◆ Terrorist attack
- ◆ Cyber-terrorism
- ◆ Market manipulation
- ◆ Accidental damage.

Risk minimization for facilities also includes the assessment of vulnerability to factors such as the use of a hazardous process or materials, storage of combustible materials, floor layout and arrangements which may concentrate or crowd control equipment, inadequate building exits, lack of shelter areas and limited evacuation routes. Analysis of facility vulnerability can provide the basis for developing practical and workable response plans.

- A business impact analysis (BIA) should be conducted to determine the financial and operational impact and/or exposure of a business outage for different periods of time and should include an analysis of the:

  - ◆ Maximum tolerable down-time for each business process
  - ◆ Critical resources and systems needed to support business recovery
  - ◆ Alternative site requirements
  - ◆ Dependencies associated with each business process.

Other issues that must be addressed in this first phase of BCP should include:

- The establishment and maintenance of a contact list of all vendors and suppliers for non-computer recovery resources required by the organization at an alternative site if that is required
- An estimate of the cost of resources required for recovery activities and a check that existing insurance coverage is adequate
- The establishment of guidelines and procedures for insurance claims in the event of a disaster
- The maintenance and updating of an inventory list of resources kept offsite
- The establishment of procedures for processing requests for recovery expenses
- The arrangement for purchase or lease of IT-related resources and telecommunications
- The arrangement of back-up servers for systems on servers/workstations to be established and maintained at the alternative site.

BCP Phase 2. Emergency Response

Following staff evacuation from the office building under the survival phase, the emergency response phase involves immediate reaction to assess the damage or impact. The BCP standards should cover procedures such as incident identification, damage assessment, activation of recovery plans and management/staff notification.

Remote-access-aware disaster management teams should be activated in order to coordinate staff and restart business processes. These are cross-functional teams with representatives from human resources, facilities, relevant departments and IT. Teams should have direct access to registry lists of employees per site and the lists should be organized in chained levels of priority for easy access to home addresses, contact numbers and remote-access capabilities.

These disaster management teams should coordinate activities to create virtual business units that leverage remote access to minimize points of failure and risk. They should use pre-assigned toll-free numbers for emergency contacts, create remote centres for pre-prioritized staff to work out of on a separate power grid, activate website/automated call centres for check-in by staff who may have been affected by the disaster, with confirmation entered into a database which can be queried via the internet or telephone.

BCP Phase 3. Recovery of Operations

This phase involves initial resumption of time-sensitive and essential business operations and related computer and application systems, followed by less time-sensitive operations. How much of the less time-sensitive operations to be resumed at an alternative site would depend on their criticality to the organization.

Disaster or business recovery plans (BRPs) for each department or business unit in the organization should be prepared in advance and centrally managed by a designated business unit – most likely the unit that handles corporate security. In an eBusiness scenario, the following techniques may be employed if an alternative site is being deployed:

- Remote disk mirroring – which involves rolling-over a live system to backup facilities with little or no interruption
- A hot network node
- Remote journaling
- Deploy the standby operating systems
- Replicate the main server/s
- A tape library – to support the backup and recovery needs of data centres with mission-critical data backups.

BCP Phase 4. Restoration of Facilities

During this phase, all business and computer processing operations that are transferred to the alternative site will be restored at the primary site or relocated to a new structure. Normal business operations and services will be re-established at the primary/new site and post-recovery operations will be completed.

Business continuity planning and the resultant business continuity plan are imperative business processes in the 21st century. Without such planning, and in the face of a major disaster, chaos can rapidly result and the business can go to the wall. BCP is particularly important in a company that uses electronically enabling mechanisms to do business – losing data and corporate information is critical and has a long lasting effect which some companies can never recover from.

Summary

Developing and maintaining trust-based business relationships – the cornerstone of the electronically enabled business – is a difficult task requiring good communication and negotiation skills. This chapter has looked at some of the theory behind the development of trust, and how these theories and associated issues are related to the practical requirements of doing business in the eLandscape. Of particular importance is the need to recognize that the 'unique selling point' of the internet is that it opens up and extends communication channels that may not have been present previously. In so doing, it facilitates what electronically enabled business is all about – that is, the increased flow of good quality information and the creation of 'gateways to deals' between parties separated by time and distance. As such, the issue of privacy in terms of access to that information and the systems involved is a major concern and the physical and logical issues necessary to control access must be understood and acted upon to ensure that buyers, sellers, employees and customers are all sufficiently comfortable to enable them to conduct business with the organization.

Managing the risks associated with doing business in an uncertain environment where trust related issues can make or break a business, is also a key management task addressed in this chapter. Risk management and risk evaluation strategies are put forward with the key areas necessitating serious contemplation for a manager – including disaster management and business continuity planning. Wendy Erhart, co-owner of Withcott Seedlings has proven time and again to be a good forseer and manager of risk. She is the Smart Thinker for this chapter.

Smart Thinker – *Wendy Erhart*

Owner / Director Withcott Seedlings
and Smart Salads

Wendy Erhart is an interesting woman – a former nurse, she won the 2005 Australian Veuve Clicquot Businesswoman of the Year Award for her work with husband Graham in growing a small nursery called Withcott Seedlings (www.wseedlings.com.au) in the Lockyer Valley in Queensland, Australia, into the largest supplier of vegetable, fruit and plant seedlings in the southern hemisphere. She confesses to being 'full-on' in terms of thinking up innovative developments for the business to meet any potential problems head on – and believes that in an industry such as horticulture, innovation is necessary for successful operation. She believes that a product that will make money needs to have gone from concept to implementation in two years – and it is this approach that led to the development of sustainable water management solutions for her water hungry business, and the Smart Salads brand.

Smart Salads (www.smartsalads.com) consists of baby leaf salads for elite food markets in Australia and overseas. The process of harvesting and packing the salad leaves came about somewhat by accident a couple of years ago – but it is now patented in Australia, the EU and the Asia Pacific and has enabled the Erhart Group to expand into the overseas Asian markets.

Not only do the innovations in terms of technology and process keep on coming, thus helping to manage general business risks and uncertainties, as well the specific risks associated with a water-intensive agricultural industry in a very dry continent, Wendy has also made it a clear business goal to involve her staff in the management of the businesses involved. In so doing she has developed a loyal and trusting employee base which delivers as a result, excellent quality based customer service which she calls 'QA on a Tree' because dotted around the huge acreage involved, QA crib boards are nailed to trees strategically located so that suppliers, employees and customers are left in no doubt that quality is the watchword of this business.

Source: Author interview, Veuve Clicquot Award (www.veuveclicquotaward.com.au) and
Withcott Seedlings websites

The Review and Critical Thinking Questions below have been provided for you to revise and reflect on the content of this chapter.

Review Questions

1. What is trust, how can it be conceptualized and how is it different from privacy?
2. What are the major issues associated with trust in agribusiness
3. What are the major issues associated with trust in the digital environment?
4. Why is trust so important in a B2B marketplace and how is it created?
5. Define risk and give three examples of risky practices in the digital environment.
6. Describe the 3Rs and discuss the functional activities associated with managing them.
7. Define and discuss the factors used in evaluating risk.
8. Outline some risks that are particularly relevant to electronically enabled business and describe how these might be minimized in an organization.
9. Why have a business continuity plan?

Critical Thinking Questions

1. What are the major drivers of risk? Describe, using a specific example, a risk management strategy that could be put together to address them.
2. Compare and contrast the concepts of collaborative supply chain management and ordinary supply chain management – where do electronic technologies fit in, what are the biggest risks and how can these be managed?
3. What do you think Withcott Seedlings' biggest risk factors will be in the future and discuss, using the what, why, how framework, Wendy Erhart's biggest challenge in the coming years.
4. Discuss the potential use of GIS technology in disaster management planning by devising a scenario to: a) look at production risk in a 'normal' and in a 'drought' situation; b) describe how spatial data and a GIS could be used to manage, from a business angle, a disaster situation such as the World Trade Center attack.

Chapter 6

Food Tracking and Traceability in the 'E' World

We are what we eat

As consumers become more and more sophisticated they are demanding more information about where the food they are buying comes from and what has been done to it. In addition, the recent major animal health scares such as foot and mouth disease (FMD) and BSE (bovine spongiform encephalitis or Mad Cow disease) outbreaks in the UK, Europe, USA and Canada – along with the potential for bioterrorism to cripple the marketplace, have highlighted the need for systems to be in place along the food production chain to enable full tracking and traceability of a product from paddock to plate. As a result, both governments and the commercial sector have focused, and continue to focus attention on the development of processes, standards and regulations to ensure that food safety within production systems is a key issue in ensuring economically, environmentally, ethically and socially sustainable food production.

This chapter looks at the issues of *biosecurity, food safety, product tracking* and *traceability* (including *identity preservation*) as they relate to the agri-food industry. The chapter also outlines 'What, Why, and How', electronic enablement within and across agri-food chains facilitates and enables the processes involved. Precision farming technologies, electronic identification systems, and product packaging and labelling systems will be discussed as illustrations of major points.

Chapter Objectives

After reading this chapter the reader should understand and be able to describe:

1. What biosecurity means in relation to agri-industry food chains.
2. What food safety entails and what HACCP is.
3. Food traceability systems and what is involved in product tracking.
4. Identity Preservation and where it fits into the traceability debate.
5. How 'E' technologies are being employed for traceability systems – particularly in relation to agri-food products.
6. What, Why and How precision farming and livestock electronic identification schemes aid in food traceability.
7. The part that food packaging and labelling play in today's quality controlled world.

Biosecurity

Biosecurity refers to the policies and measures taken for protecting a nation's natural resources including its food supply, from biological invasion and threats both from accidental contamination and/or deliberate attacks of *bioterrorism* (i.e. ensuring a nation's bio-safety). Bioterrorism in this case covers such deliberate acts as introducing pests intended to kill food crops; spreading a virulent disease among animal production facilities, and poisoning water and food supplies or purposively contaminating the environment. (Wikipedia, 2005).

Biological weapons that cause disease, such as the anthrax that was used through the US postal system in 2001, are distinct from chemical weapons such as ricin, which was used on the Japanese subway in 1995, and nuclear weapons. Because of the technical factors unique to the development and deployment of biological weapons, historically it has been Nation States, rather than non-state groups, that have been likely to be the deliberate perpetrators of bioterrorist attack (CRC, 2003). This chapter will not pursue this aspect of biosecurity further but readers can go to the Institute of Biosecurity at St Louis University in the USA (www.bioterrorism.slu.edu/) for further readings.

Bio-safety in relation to the agri-food sector is about reducing the risk of viral or transgenic genes, or prions such as BSE/Mad Cow and reducing the risk of food bacterial contamination. By international agreement, bio-safety involves the application of the *precautionary principle* which is the idea that if the consequences of an action are unknown, but are judged to have some potential for major or irreversible negative consequences, then it is better to avoid that action. Some of the major issues that need to be thought about under the umbrella of biosecurity and biosafety are:

1. *Emerging infectious diseases* – for example, new, previously unrecognized diseases such as Severe Acute Respiratory Syndrome (SARS), or known diseases which have increased in incidence, virulence or geographic range over the past two decades (e.g. foot and mouth disease), as well as diseases which threaten to increase in the near future (e.g. avian influenza).
2. *Pathogen pollution* – which is human-mediated introduction of alien pathogens such as pests.
3. *Livestock disease surveillance* – which is a systematic series of investigations of a given population of animals to detect the occurrence of disease for control purposes.

In an effort to counteract the general public's concerns, a number of pieces of legislation have been enacted around the world that address these issues – for example:

* *Regulation No 178/2002 of the European Parliament and of the Council* which specifically addresses the general principles and requirement of Food Law, establishes the European Food Safety Authority and lays down the procedural requirements in relation to food safety (Euro-Lex Portal, 2005)
* The USA's *Public Health Security and Bioterrorism Preparedness and Response Act of 2002* (the Bioterrorism Act, Section 36) (FDA Portal, 2005)
* The UK's *Food Safety Act 1990 (Amendment) Regulations 2004 and General Food Regulations 2004.* (UK Food Standards Agency, 2005)
* *The Australia New Zealand Food Standards Code* (up to Amendment 81, 2005) which has force in Law under the Australia New Zealand Food Authority Act 1991 (incorporating amendments made for the inclusion of New Zealand in the National Food Authority Amendment Act 1995). (Australian Food Standards Agency, 2005).

It is not the intention of this chapter to go into detail on the legislative issues associated with biosecurity – rather this chapter will look at food safety focusing on those aspects associated with product tracking and traceability throughout agri-food chains. It is however suggested that readers familiarise themselves with the appropriate legislation of the country in which they are interested in doing business in, or with, to ensure that their processes and documentation are appropriate and up-to-date.

Quarantine

One aspect of biosecurity that should be mentioned here as it relates to the agri-industry sector is the issue of quarantine and import/export practices. Generically, a country's quarantine service (for example in Australia, the Australian Quarantine and Inspection Service (AQIS), in the USA the Division of Global Migration and Quarantine (CDC), and in the UK the Department of Animal Health and Welfare), are government organizations that exist to protect a country's unique environment against exotic pests and diseases. In order to do this they provide inspection services at official entry ports to a country and they provide inspection and certification for import and export purposes.

Legislation as to what may or may not be brought into a country, or may be exported from a country to another one, is strictly country-unique – details of which can generally be found at the website of the country of interest's quarantine service (quarantine services around the world have quickly adapted their information services to internet delivery). However, in terms of agricultural produce these differences can have a dramatic impact on trade. A good example is that of Australian tropical fruit being exported to China (UQ, 2005).

Currently, Australian horticultural companies are endeavouring to expand their export interests particularly into Asian markets. China is seen as a potentially lucrative market to access because of Australia's geographical proximity, counter seasonal production advantage and potential to service mass markets. At present, only Tasmanian apples and Queensland mangoes (as of 2005 harvest) have legal access to China and require stringent quarantine post harvest treatments before they are allowed into the country via official trade channels. China imposes stringent quarantine processes on tropical fruit because of the many possible pests and diseases that can contaminate the products. Australia is a very large island country and comprises a Federation of States; however, it is examined under a single protocol system by Chinese authorities rather than by protocols that govern each state. This means that individual state trade potential is inhibited because of pests and diseases that may be prevalent in one state but not in another. Thus, Queensland produce is considered to harbour the same pests and diseases from a Chinese perspective as that of the 3000km-distant State of Western Australia – which in reality, it does not. This particular issue has already impacted the export of mangoes from Queensland to China, in that they can only be exported after expensive and quality reducing Vapour Heat Treatment (VHT) – a treatment that is not required by China from other mango importing countries such as South Africa (UQ, 2005). Such quarantine requirements present two barriers for Australian exporters:

- Increased costs of conducting the necessary treatment
- Loss of quality as the process can change the texture and taste of fruit.

The effects of this are again two-fold. The first being that it deters exporters because they can simply target markets where these costs are not incurred (e.g. New Zealand) and secondly, importers would prefer to source foreign fruit that has retained its superior standard from other countries such as South Africa.

With SARS, FMD, Avian Influenza and BSE all becoming real threats to world health in recent years, quarantine services are a significant infrastructural component in a country's biosecurity armour.

Food Safety

There is no doubt that '*we are what we eat*' and there are many things that we eat! Food safety is defined by Wikipedia (2005) as:

> **Protecting the food supply from microbial, chemical (ie rancidity, browning) and physical (ie drying out, infestation) hazards or contamination that may occur during all stages of food production and handling – growing, harvesting, processing, transporting, preparing, distributing and storing.**

Recent trends in global food production, processing, distribution and preparation are creating an increasing demand for a safer global food supply. How can this be achieved? One avenue is obviously research into what creates a problem and what can be done to solve it – the main areas of interest being:

- Microbiological risks
- Chemical risks
- Biotechnology (GMO) issues
- Food Standards
- Food borne diseases

From the perspective of food tracking and traceability, what stands out is the issue of food standards because these by their very nature, incorporate what research is currently available into a set of measurable indices that can be used to assess food quality and safety as well as minimize risk. *The Codex Alimentarius* is the main set of reference tools globally on food standards.

The Codex Alimentarius

The Codex Alimentarius is the most important international reference point in matters concerning food quality – in fact it is *the* global reference point for consumers, food producers and processors, national food control agencies and the international food trade. It is a collection of international food standards (the Codex Standards) that have been adopted by the Codex Alimentarius Commission and which cover all the main foods, whether processed, semi-processed or raw. The main objective of the code is to protect the health of consumers and facilitate fair practices in the food trade (FAO, 2005). Codex Standards cover the hygienic and nutritional quality of food, including microbiological norms, food additives, pesticide and veterinary drug residues, contaminants, labelling and presentation, as well as methods of sampling and risk analysis. Specifically these standards address:

- Animal Feeding
- Cereal, Pulses and Legumes
- Cocoa Products and Chocolate
- Fats and Oils
- Fish and Fishery Products
- Food Additives and Contaminants
- Food Hygiene
- Food Import/Export and Certification Systems
- Food Labelling
- Food derived from Biotechnology
- Fruit and Vegetable Juices
- Fruit and Vegetables
- Meat Hygiene
- Methods of Analysis and Sampling
- Milk and Milk Products
- Natural Mineral Waters
- Nutrition and Foods for Special Dietary Uses
- Pesticide Residues
- Residues of Veterinary Drugs in Foods
- Sugars
- Vegetable Proteins

To ensure that food safety standards are maintained and both domestic and international legislation is adhered to, a business needs to develop quality control and management systems (Malcolm, 2005). There are a number of internationally recognized systems that businesses can employ including **HACCP (Hazard Analysis Critical Control Point)** and **ISO 9001:2000.** In Australia, the **AQIS Certification Assurance** and **SQF 2000** systems are also employed.

1. *HACCP (Hazard Analysis Critical Control Point system)*. This is a process control system that identifies where hazards might occur in the food production process and puts into place stringent actions to prevent the hazards from occurring. HACCP has been adopted by the Codex Alimentarius Commission set up by the Food and Agricultural Association of the United Nations (FAO) and the World Health Organization (WHO), as the international standard for food safety. HACCP focuses on identifying and preventing hazards from contaminating food and involves seven principles (CFSAN (2005) quoting Codex Alimentarius Commission, (1997)).

   i. *Analyse hazards*. Potential hazards associated with a food and measures to control those hazards should be identified. The hazard could be biological, such as a microbe; chemical, such as a toxin; or physical, such as ground glass or metal fragments.
   ii. *Identify critical control points*. These are points in a food's production – from its raw state through processing and shipping to consumption by the consumer – at which the potential hazard can be controlled or eliminated. Examples are cooking, cooling, packaging, and metal detection.
   iii. *Establish preventive measures with critical limits for each control point*. For a cooked food, this might include setting the minimum cooking temperature and time required to ensure the elimination of any harmful microbes.

iv. *Establish procedures to monitor the critical control points.* Such procedures might include determining how and by whom cooking time and temperature should be monitored.

v. *Establish corrective actions to be taken* when monitoring shows that a critical limit has not been met, e.g. reprocessing or disposing of food if the minimum cooking temperature is not met.

vi. *Establish procedures to verify that the system is working properly* – for example, testing time-and-temperature recording devices to verify that a cooking unit is working properly.

vii. *Establish effective record keeping to document the HACCP system.* This should include records of hazards and their control methods, the monitoring of safety requirements and action taken to correct potential problems. Each of these principles must be backed by sound scientific knowledge: for example, published microbiological studies on time and temperature factors for controlling food-borne pathogens.

2. *ISO 9001:2000.* ISO (International Organization for Standardization) is a global network that identifies what international standards are required by business, government and society, develops them in partnership with the sectors that will put them to use, adopts them by transparent procedures based on national input, and delivers them to be implemented worldwide (ISO, 2001, 2004). ISO 9001:2000 is one of ISO's internationally recognized quality assurance systems upon which many other similar systems are based (http://praxiom.com/).

3. *AQIS Certification Assurance.* This is the quality assurance system monitored by the Australian Quarantine Inspection Service (AQIS) who provide inspection and certification for a range of exports.

4. *SQF 2000.* This is based on HACCP and recognized by various horticultural businesses as a food based quality system alternative to ISO 9001:2000.

Food quality systems have evolved substantially in the last few years, and they have become an integral part of the food industry's operational processes in most countries. HACCP and ISO standards, particularly the ISO 9000:2000 family of standards, form the basis of good tracking and traceability systems which are discussed in the next section.

Food Tracking and Traceability

Much has been written about product tracking and traceability generally and more specifically, in the food industry, in recent times. As indicated, this has been as a result of recent well documented outbreaks of foot and mouth and Mad Cow disease, food poisoning, genetically modified food introduction, concern over the environment and the fear of bioterrorism. The question is: 'What is traceability?' The International Standards Organization (ISO) defines traceability as:

> **The ability to trace the history, application or location of an entity by means of recorded identifications.**

In other words it is a tool that allows something to be tracked. It is about collecting information on a particular attribute of a product and recording it in a systematic fashion from its creation through to being sold to the consumer.

Despite ISO 9001:2000 indicating that traceability is one of the aspects that should be considered in a quality management system, the ISO definition of traceability is too broad for current food safety legislation as it gives no specifications as to what needs to be measured or the processes involved. According to Golan *et al.* (2003) this means that in the USA, food and other agribusinesses have developed their own versions of traceability systems varying in the *breadth* (i.e. the number and type of attribute of the food item collected), *depth* (how far back and forward the system tracks relevant information) and *precision* (the degree of assurance that the system has in detailing a particular food product's movement) of the information they collect about their product – which is not ideal.

The European Union definition is more precise and has enshrined in law in Article 3 (15) of EU regulation 178/2002, that traceability is defined as:

> **The ability to trace and follow a food, feed, food producing animal or substance intended to be, or expected to be incorporated into a food or feed through all stages of production, processing and distribution.**

In Europe, traceability is – as of 1 January 2005 – mandatory, and has the principal aim of protecting public health and consumers' interests in relation to food. Many other nations are rapidly following suit – so what does this actually mean in operational terms compared to the ISO standard requirement? Basically, it means that food businesses will be, or are being, confronted with the legal requirement to build traceability systems that collect, and make easily accessible, far more detailed, precise and accurate information about food products throughout the food chain than is currently required (CIES, 2004).

The following section deals with generic issues of traceability systems as they relate to record keeping, before moving onto more specific food traceability system criteria.

Traceability and Tracking Systems

All organizations, irrespective of their unique business activities, require systems that can capture full and accurate records and perform processes for managing those records over time. Traceability systems are record-keeping systems that act as a tool for making information available either within an organization or between organizations.

Records

Records are traditionally regarded as documents in paper files or bound volumes although nowadays, they can exist in any physical format, eg photographic prints, video cassettes, microfilm and/or any of the many current electronic formats. Records have three distinctive characteristics:

1. *Content* (the information associated with the attribute being recorded)
2. *Structure* (the format and relationship between the elements comprising the record)
3. *Context* (why the record was created, received or used, which includes when and by whom and under which circumstances, and what links there are to other documents making up the total record).

In short, a record must reflect accurately the message and the decision, or what was done.

The International Standard on records management (ISO, 2001) defines records as being:

Information created, received, and maintained as evidence and information by an organization or person, in pursuance of legal obligations or in the transaction of business.

Thus it is very important that records be **authentic, reliable, integral** (or **whole**), **useful** (and **useable**) (Gunnlaugsdottir, 2002; quoting ISO 15489-1:2001):

- *Authentic.* Records must not be forged and should be truly created or sent by the person purported to have created or sent them – and at the given time. Organizations need to introduce records management policies and procedures to provide the necessary control on the creation, receipt, transmission, maintenance and disposition of records to ensure that records, creators are identified and have the needed authorization. Additionally, organizations need to protect records against additions, deletions, alterations, uses and concealment which is unauthorized.
- *Reliable.* A record's content should be capable of being trusted as a full and accurate account of the transaction, activity or facts to which it attests and must be able to be depended upon. To fulfil this, records must be created at the time of the event or soon afterwards and by individuals who have direct knowledge of the transaction or incident.
- *Integral or whole.* A record must be complete and unaltered. It is necessary to protect records against unauthorized alterations, and those authorized must be explicitly indicated and traceable.
- *Useable or useful.* Records must be capable of being used and in order to be used, the record needs to be located, retrieved, presented and interpreted. The connection to the business activity or transaction which produced it should be clear.

Traceability Systems

Traceability systems are manual or automated systems in which records are collected, organized, and categorized to facilitate their preservation, retrieval, use, and dissemination. Such systems (National Archives of Australia, 2000) enable organizations to:

- Conduct business in an orderly, efficient and accountable manner
- Deliver services in a consistent and equitable manner
- Support and document policy formation and managerial decision-making
- Provide consistency, continuity and productivity in management and administration
- Facilitate the effective performance of activities through an organization
- Provide continuity in the event of a disaster
- Meet legislative and regulatory requirements including archival, audit and oversight activities
- Provide protection and support in litigation, including the management of risks associated with the existence of or lack of evidence of organizational activity
- Protect the interests of the organization and the rights of employees, clients, and present and future stakeholders
- Support and document current and future research and development activities, developments and achievements, as well as historical research, and provide evidence of business, personal and cultural activity
- Establish business, personal and cultural identity
- Maintain corporate, personal or collective memory.

In order to capture, maintain and provide evidence over time, systems must also be capable of performing various fundamental recordkeeping processes (National Archives of Australia 2000). These processes involve:

1. *Record capture* – formally determine that a record should be made and kept
2. *Record registration* – formalize the capture of a record into a designated system by assigning a unique identifier and brief descriptive information about it (such as date, time and title)
3. *Classification and indexing* – identification of the business activities to which a record relates and linkage of that record to other records to facilitate description, control, retrieval, disposal and access
4. *Access and security* – assigning rights or restrictions to use or for management of particular records
5. *Appraisal* – the identification and linkage of the retention period of a record to a functions-based disposal authority at the point of capture and registration
6. *Storage* – maintaining, handling and storing records in accordance with their form, use and value for as long as they are legally required
7. *Use and tracking* – ensuring that only those employees with appropriate permissions are able to use or manage records and that such access can be tracked as a security measure
8. *Disposal* – identification of records with similar disposal dates and triggering actions, review capability of any history of use to confirm or amend the disposal status, and maintaining a record of disposal action that can be audited.

All of these processes generate metadata that plays an essential role in the accountable management of records. In addition *system integrity* is imperative – that is, it must be possible within the traceability system to implement control measures, such as access monitoring, user verification, authorized destruction, security and disaster mitigation to prevent unauthorized access, destruction, alteration or removal of records and to protect them from accidental damage or loss.

Product Traceability Systems

Product traceability systems are systems that manage the information related to individual products over the entire product life cycle, beginning with the planning stage and including design, manufacture, maintenance and disposal (Kojima *et al.*, 2004). Product traceability systems can serve many purposes but some of the benefits include (CIES 2004):

1. They ensure a fast product withdrawal or recall, thus protecting the consumer.
2. They minimize the impact of such a product recall by limiting the scope of the product implicated (e.g. the financial impact of recalling an entire commodity or brand versus a single batch, is massive).
3. They enable companies to show that their product is not implicated in a given product recall, by ensuring proper segregation and clear identification of product.
4. They strengthen consumer confidence, through the industry's ability to promptly identify and recall a problematic product.
5. They provide internal logistical and quality related information, improving efficiency.
6. They create a feedback loop to improve product quality, condition and delivery.

7. They provide transparency in distribution routes, improving supply chain efficiencies and trading partner collaboration.
8. They provide reliable information
    - to business to business
    - to consumers
    - to government inspectors
    - to financial or technical auditors.
9. They enable the establishment of responsibility and liability for a certain problem.
10. They facilitate the protection of the company and/or brand name.

Whatever the perceived benefits of these systems, implementation requires upfront investment, so the usual return on investment and risk analyses should be undertaken before companies go forward and purchase a system. However it must be kept in mind that the benefits and savings are not immediately obvious. The expenditure involved should be considered as a long-term strategic investment because it is linked to consumer's perception, the image of the company and the trust that consumers display when buying a product.

Food Traceability Systems

Food traceability systems are no different from any other product traceability system in terms of the technologies employed – they are all about maintaining good records. However, there are some specific issues associated with biosecurity and food safety that do need to be included in food traceability systems that are not needed in standard product tracking systems. While not all of the following list of issues apply to every business in every agri-food chain, each area will need to be included at some point in every agri-food traceability system. Areas to be aware of include:

- Government/Regulatory
- Import/Export
- Animal Identification
- Agriculture and Identity Preservation
- Food Processing
- Food Storage
- Food Transportation
- Food Security
- Food Services (catering and restaurants)
- Retail/Wholesale
- Food Hygiene
- Traceability and the Consumer

Food traceability systems are not only rapidly gaining ground in response to consumer demand for identified product, but are also being included in public policies associated with Corporate Social Responsibility (CSR) (European Commission (EC), 2004). The EC regard both sustainable development and CSR as:

> **innovative progress combining the social, environmental and economic dimension in an integrated approach .....in developing better social and economic governance and improving living and working conditions as well as enhancing companies' competitiveness.**

At this stage nothing is mandated – thus the reality is that many organizations, particularly the smaller businesses, are less likely to be developing systems. However, there is no doubt that ever more stringent requirements in terms of health and safety reporting and even labelling (see section later in chapter) will have a major impact on the future requirement and development of these systems.

Full traceability, of course, can be problematic in non-collaborative supply chain environments. A transparent traceability system means good collaboration between trading partners to ensure data and information flow freely. As mentioned in Chapter 5, this is in fact quite rare in practice, even with the best of intentions from all concerned. Issues of trust, privacy, competition, corporate information hoarding and unequal power domination in agri-food chains can, and do, all play a role in blocking the information flows needed to set up good traceability systems. Add to this, the unpleasant reality that on-farm regulation in terms of agricultural production and associated recording and reporting, varies dramatically between countries, and that even in the more regulated agricultural production environments, information flow across the farmgate is poor (Bryceson and Kandampully, 2004), and the issue of whether food product traceability is actually feasible, is open to question.

An example of a relatively inexpensive software system for record keeping on-farm that extends beyond the farmgate into the supply chain is found at Muddy Boots (see Byte Idea this chapter).

Golan, (2004); Golan *et al.* (2003, 2004) make the point that companies have for some time been involved in developing food traceability systems despite the fact that they have not been a requirement under law. They suggest that food suppliers have three motives for establishing traceability systems: for **managing supply and inventories,** for **food safety and quality control** and for **differentiating micro-markets in relation to subtle quality (credence) attributes.**

Traceability Systems for Managing Supply and Inventories.

Managing production and tracking retail activity is essential for most retail firms in order to manage inventory levels and coordinate ordering systems. In addition to the standard accounting software packages that almost all businesses use, there are a number of industry standard coding systems. Two systems worth describing here are the **barcode** system and the **RFID** system.

1. A *barcode* is a machine-readable representation of information in visual format on a surface (Brighan, 1995). Almost all packaged food products in Australia, North America and Europe have a barcode on their labelling. Barcodes originally stored data in the different widths and spacings of printed parallel lines (Fig 6.1), but today they also come in patterns of dots, and images that allow more information to be encoded into a small space.

   Barcodes need to be read by optical scanners called barcode readers (Table 6.1) which generally consist of a light source, a lens and a photo conductor translating optical impulses into electrical ones. Nearly all current barcode readers contain decoder circuitry to enable automatic data capture into a computer based information management system which improves the speed and accuracy of computer data entry.

Byte Idea – Muddy Boots
www.muddyboots.com/home.asp

Traceability Is the Name of the Game!

Muddy Boots is a UK-based company that operates in the UK, Australia, New Zealand and South Africa specializing in software development for managing traceability issues from paddock to plate in agri-food chains. Products include:

- Cropwalker is a software package that allows the capture and recording of all aspects of crop management from seed procurement to produce dispatch. A new development has been Pocket Cropwalker which enables the software to be used with a personal digital assistant (PDA) while the user is on the move. The software incorporates Produce Manager which provides capability of managing the data including converting what is needed of the data into industry standard barcode technology.
- Quickfire is software that enables users to capture data in the field using the latest hand held computer technology, and then to communicate the data to a central database. It provides audit, inspection and quality assurance capability to the user. Quickfire can be set up to deliver the information recorded in the database into almost any inspection or verification protocol thus providing the capability to enable conformity to standards management and risk assessment.
- Procheck is an interactive electronic pesticide database containing up-to-the-minute product information. The database is maintained in conjunction with agrochemical manufacturers and is maintained daily by Muddy Boots, and can be accessed via the internet with information then being downloadable if required.

With alliances across three continents (including Vitacress (www.vitacress.com), Europe's leading grower and packer of watercress, rocket, baby leaf salads and specialty vegetables; The Sekem Group (www.sekem.com), which is a vertically integrated Egyptian biodynamic company involved in fresh produce production and processing; and Homegrown Kenya, which is part of the Flamingo Holdings Group (www.flamingoholdings.com) and grows flowers and fresh produce for distribution throughout Kenya and for export to the UK), Muddy Boots is out there with innovative supply chain traceability software solutions for the fresh product market.

Information sourced from Mr Julian Shelbourne, CEO Muddy Boots Australia, and the company website

Fig. 6.1. Examples of bar codes taken from a packet of pork sausages obtained at a speciality butcher and a container of Roundup obtained from a hardware store. Note the start, middle and end bars are longer than the rest, which are where the product details reside (for further details on the bar code numbering system *per se*, it is suggested that readers go to http://en.wikipedia.org/wiki/Barcode)

Table 6.1. Types of bar code scanners/readers and how they work

Scanner Type	Mechanism
Handheld scanner	These have a handle and typically a trigger button for switching on the light source
Pen scanners or wand scanners	A pen-shaped scanner that is swiped across a barcode
Stationary (wall- or table-mounted) scanners	The barcode is passed under or beside these, which are commonly found at the checkout counters of supermarkets and other retailers
PDA scanners	A personal digital assistant (PDA) with a built-in barcode reader

The UPC (Universal Product Code) was the original barcode widely used for items in stores, and only used numbers with no other characters involved (Brown, 1997). The UPC is now officially the EAN.UCC-12 (European Article Numbering-Uniform Code Council (EAN-UCC*).*

In 2005 the EAN-UCC changed its name to GS1 (www.ean-int.org) and developed the GTIN (Global Trade Item Number) which is an umbrella term used to describe the entire family of EAN/UCC data structures for trade items (products and services) identification. The GTIN is usually represented by the existing bar code symbology: UPC, EAN, Code 128, or Interleaved 2 of 5. By January 2005, all trading partners with the EU were expected to support GTINs.

The *Global EAN Party Information Register* (GEPIR) (www.gepir.org) is a distributed database that contains basic information on over 1,000,000 companies in over 100 countries linked via the GTIN. GEPIR allows searching for information on a company by its product, brand, name, address and identification (GLN- Global Location Number) and shipment identification (SSCC – Serial Shipping Container Code). It can also be used to synchronize trade items (any product or service upon which there is a need to retrieve pre-defined information) ordering, procurement and dispatch operations in an EDI exchange with trading partners at any point in the supply chain. This includes individual items as well as all of their different packaging configurations.

2. *RFID (Radio Frequency Identification)* is another technology being used more frequently in traceability systems (AIM, 1999). RFID is a method of receiving and storing data from a remote object using devices called RFID tags. RFID tags are small transponders that can be attached or incorporated into a product such as a box of razor blades, an animal, fruit and vegetables and even people. The tags contain antennas to enable them to receive and respond to radio-frequency queries from an RFID transceiver. Passive tags require no internal power source, whereas active tags require a power source. RFID technology is part of a suite of technologies that includes bar codes, biometrics, magnetic stripes, optical character recognition, smart cards, and voice recognition, that come under the umbrella term of 'Automatic Identification' or 'Auto-ID' which involves identifying objects, collecting data about them, and entering that data directly into computer systems (i.e. without human involvement).

RFID technology is based on international standards. Transponders and readers are approved as meeting ISO standards by the International Committee on Animal Recording (ICAR). ISO 11784 and ISO 11785 were published in 1996 and although there are some recognized flaws in these standards (RFID News, 2005a, b), they still remain the benchmark for the use of RFID technology in animal identification.

- *ISO 11784* – specifies the structure of the radio-frequency identification code for animals, but not the characteristics of the transmission protocols between transponder and transceiver.
- *ISO 11785* – specifies the technical concepts of electronic identification of animals such as how a transponder is activated and how the stored information is transferred to a transceiver.

The most well known uses of RFID technology in primary production are vehicle identification when loading and unloading bulk commodities such as wheat, and in the various livestock identification schemes currently being implemented around the world. This is discussed in more detail later in the chapter.

Both barcoding and RFID technology have their place in a product traceability system: different characteristics make them suitable for different tasks. Barcodes are cheap, standardized, can be used where RFID could cause problems, and they are easy to apply. On the negative side, they require line of sight scanning, they can only be used once and if the barcode becomes dirty – e.g. muddy or wrinkled, the scanners very often do not work.

RFID on the other hand, allows unattended scanning, does not need line of sight for downloading purposes, is reuseable, small and capable of storing a great deal of information. Multiple objects can be scanned at once and the tags are able to be incorporated into the product. It also allows an integrated approach to a whole of supply chain management scenario – if everyone in the chain has implemeted RFID technology. On the negative side, tags can be expensive, conductivity is negatively affected by metal and water, RFID standards are not ratified, legislation regarding power rating is still under review, and if a collaborative trading network is involved, all partners in the supply chain need to be electronically enabled before RFID can work for all.

None the less, the development of integrated management systems that code, track and manage transactions means that these technologies combined with electronic enablement within businesses and across the supply chain are here to stay. In some cases buyers manage these systems to monitor supply flow while other firms establish systems to link suppliers and buyers – all the while aiming to reduce inefficiencies and improve quality.

Traceability Systems for Food Safety and Quality Control

Traceability for food product safety and quality is a necessary part of doing business in the agri-industry sector today. The increasing number of new food pathogens such as *Escherichia coli* (found in contaminated ground beef, and unpasteurized milk), and *Salmonella enteritidis* (found in eggs), that are now recognized as important causes of food-borne illness; an increasing public health concern about chemical contamination of food, such as the effects of lead in food on the nervous system; plus the size of the food industry and the diversity of products and processes now available, are all contributing to a major push to develop better food traceability systems.

Food traceability systems should enable products to be traced from source (i.e. on farm in most instances in the food industry) through to the customer – and they therefore must have a transparent chain of custody to maintain credibility and to complete information transfer functions (McKean, 2001). A good food traceability system helps to minimize the production and distribution of unsafe or poor quality products, which in turn, minimizes the potential for recalls, liability and poor publicity. Such systems help minimize damages to the whole industry both up and downstream. In addition, if the business's quality control system fails and results in sales of unsafe or defective products, a traceability system can help to track product distribution and reduce the size and cost of recall.

The better and more precise the tracing system, the faster a producer can identify and resolve food safety or quality problems. By themselves, traceability systems do not produce safer or higher quality products. These systems and the electronic infrastructure that enables them to handle large volumes of data at speed, provide information about whether or not control points in the production or supply chain are operating correctly – for example Superquinn the Irish retailer is a leading exponent – see Byte Idea this chapter.

Many buyers now require their suppliers to establish safety/quality traceability systems and to verify, often through third-party certification, that such systems function as required. Most, if not all, third-party food-safety/quality certifiers such as Société Générale de Surveillance (SGS) recognize traceability as the centerpiece of a firm's safety management system. The growth of third-party standards and certifying agencies is helping to push the whole food industry toward documented, verifiable traceability systems (Golan, 2004).

Byte Idea – Superquinn
www.superquinn.ie

Quality Fresh Foods That Are Safe and Traceable

Superquinn is a leading retailer of fresh foods in Eire with a nationwide distribution of shops. The company strives to be innovative in its business practices and apart from its many shops has two online shopping avenues (www.Superquinn4Food.ie – selling over 12,000 products, and www.Superquinn4Wine.ie – which has 500 plus wines on its catalogues). It has won awards such as the Best in eCommerce Innovation Award from the Global Retail Technology Forum in 2002, an International Accenture Global Electronic Marketing Award (GEMA) in 2003, and the O2 Digital Media Award in 2003.

In addition, quality fresh food that can be traced from 'pasture to plate' is a major goal of the company. With a project called 'Nutrition at Superquinn', food safety and traceability have been targeted as key points of differentiation between Superquinn and its competitors. With €2.5 million/year being spent in this area alone, the company's food traceability initiatives in beef, chicken, pork and fresh salmon are leading edge and its safety initiative on beef traceability has recently been recognized with a major international award – the IGD Unilever Award.

Beef traceability uses a system developed at Trinity College, Dublin by IdentiGen (www.identigen.com) called TraceBackTM, which is based on DNA technology, enabling each animal's DNA code to be 'read' thus making it easy to identify not only the farm the animal has come from, but the genetics or animal of origin as well. An auditing facility in the TraceBackTM system also allows Superquinn to audit and inspect the quality systems the company already has in place. The whole process is part of the Beef Quality Assurance Scheme which ensures that all Superquinn beef meets the company's specifications for animal welfare and freedom from disease, growth hormones and harmful bacteria.

Information sourced from IdentiGen and Superquinn websites

Traceability Systems for Differentiating Micro-Markets in Relation to Subtle Quality (Credence) Attributes

Credence attributes are characteristics of a food product that consumers cannot discern even after consuming the product – e.g. the reality is that despite all the noise and worry, very few if anyone, can tell the difference between a product made from genetically modified material (GMO) or non GMO materials. There are two types of credence attribute – **Content** and **Process**.

1. *Content attributes* affect the physical properties of a product but are difficult to perceive. For example calcium enhanced milk or bread basically does not taste any different from non enhanced milk or bread.
2. *Process attributes* refer to characteristics of the production process and include: country of origin, organic, sustainable etc.

The only way to monitor these types of attributes is via a traceability system where stringent records are kept throughout, thereby enshrining the process or content attribute involved. Carrefour, the leading French retailer, is a case in point – monitoring whether their suppliers manage their production in an environmentally sustainable fashion is now a component of their own 'Filiere Quality Chain' traceability system – see Byte Idea this chapter.

Producers themselves are also developing sophisticated systems for tracking and establishing value for credence attributes including the use of precision farming technologies (see the section on precision farming later in this chapter).

Byte Idea – Carrefour
www.carrefour.com.fr

Food Safety and Environmental Sustainability

Carrefour provides one of the largest retail outlets in the world for food through its supermarket, hypermarket and hard discount stores. In fact, Carrefour is the largest retailer in Europe and the second largest in the world (after WalMart) with over 400,000 employees and 11,000 stores in 31 countries.

Carrefour, which has roots in France, has always had a strong food safety and environmental sustainability ethos. The Carrefour food safety seal is now regarded as having four essential and measurable components:

1. *Product traceability and product recall.* Carrefour first introduced the Filiere Quality Chain concept which looked at product traceability from paddock to plate with their beef suppliers in France in 1991. This concept has now been extended to over 300 supply chains that Carrefour has established covering meat, fish, fruit and vegetables. Electronic capability in monitoring and managing the data associated with traceability has played a big role in facilitating product traceability throughout Carrefour – in fact, any product under any brand can be recalled in under two hours from anywhere in the world. Environmental sustainability is also part of the Filiere Quality Chain approach which is taken to mean environmental preservation through the control of environmental impact using 'sustainable' fertilizing and spraying, appropriate crop rotations, maintaining the well-being of animals, non-intensive production and the preservation of natural resources.
2. *Product bacteriological quality.* This involves third party testing of product suppliers and distributors and is undertaken to ensure stringent bacteriological standards are met.
3. *Safety first principle.* Food safety is of paramount importance.
4. *Non-GMO alternatives.* Carrefour offer a non-GMO alternative to any GMO product that they retail which is in direct response to a recognition that many of their customers have no wish to buy GMO products.

From a general electronic enablement perspective, Carrefour is a leading exponent within its business. It is also an equity partner in GlobalNetXchange (www.gnx.com) a major online retail exchange and service provider for the global retail industry. Other partners in GNX include: Metro Ag (third largest retailer in Europe), Coles Myer Ltd (largest retailer in Australia), The Kroger Co (one of the largest retail grocery chains in the USA), Sainsburys (one of the largest retail chains in the UK) and Sears (a founding partner in GNX and a leading retailer of clothing and homewares in the USA). Of particular interest is a subsidiary exchange – the GNX Perishables Exchange – which is advertised as the retail grocery industry's first fully-integrated exchange for managing the purchasing and replenishment process for perishable goods with full electronic traceability of products from source to destination.

Information sourced from Just-Food.com (www.just-food.com), The Financial Times (www.ft.com), GlobalNetXchange (www.GNX.com) and the Carrefour website

Breadth, Depth and Precision of Traceability Systems

As with any system development, when looking to develop and implement a food traceability system, the business has to determine the issues that it needs to address to add value while making sure it addresses what legal requirements exist. Golan *et al.* (2004) point out that there are three components to a traceability system which need to be determined and balanced to ensure that the system is adequate (i.e. it can efficiently and effectively identify and remove contaminated food).

- *Breadth.* This describes the amount of information the traceability system records. A record keeping system cataloguing all of a food's attributes would be very large, very expensive, and generally unnecessary. The example often used is that of a cup of coffee – the beans could come from multiple countries with different pesticide regimes and management practices the detail of which would be difficult to pick up and not required by consumers. On the other hand, whether the coffee is decaffeinated or not and how it was decaffeinated is probably something that consumers would like to know about and thus worth recording.
- *Depth.* The depth of a traceability system describes how far back or forwards it can track – i.e. once the attributes to track have been decided upon, the depth of the system is basically determined. For example, a traceability system for the above mentioned decaffeinated coffee, would only extend back to the processing stage. For food safety issues, the depth of the traceability system depends on where hazards and remedies can enter the food chain. For some health hazards, such as BSE or aflatoxin (see precision farming section later in this chapter) ensuring food safety requires establishing safety measures at the farm. For other health hazards, such as bacterial pathogens, firms may need to establish a number of critical control points along the entire production and distribution chain.
- *Precision.* This reflects the degree of assurance with which the tracing system can pinpoint a particular food product's movement or characteristics. Precision is determined by the unit of analysis for the attribute used in the system and the acceptable error rate. The unit of analysis, whether container, truck, crate, day of production, shift, feedlot, etc. is the tracking unit for the traceability system.

Identity Preservation

Identity Preservation (IP) refers to a system of production, handling, and marketing practices that maintain the integrity and purity of agricultural commodities in order to enhance the value of the final product (Sundstrom *et al.*, 2002). As crops and production systems have diversified to meet market demands, the need for segregation and identity preservation of agricultural commodities has increased.

Seed certification programs play a major role in maintaining seed purity standards at levels established by the industry for national and international trade. Similarly, commodity traders, marketing organizations, and food processors have established purity and quality tolerances for specific end-product uses.

Crop varieties with unique product quality traits, such as high oleic peanuts and low linolenic canola require IP programs to channel these commodities to specific markets to capture the added value. The introduction of crops developed using biotechnology (genetically modified crops (GMO)) also requires new IP programs, as markets differ in their acceptance of these commodities. For example, the United States, Canada and China, have adopted

GMO crops readily but particularly in the European Union (EU), Japan and Australia, this has not been the case. In addition, some countries are instituting labelling laws that require the segregation and identification of seed and food products developed using biotechnology.

Sundstrom *et al.* (2002) emphasize that *an IP program is a system of standards, records and auditing* throughout the supply chain (Figure 6.2), and that it is a process that results in a certification that a product meets specific quality standards. As such the IP fits in well with most traceability systems since these too are based around record keeping, auditing and authenticating a product through certification.

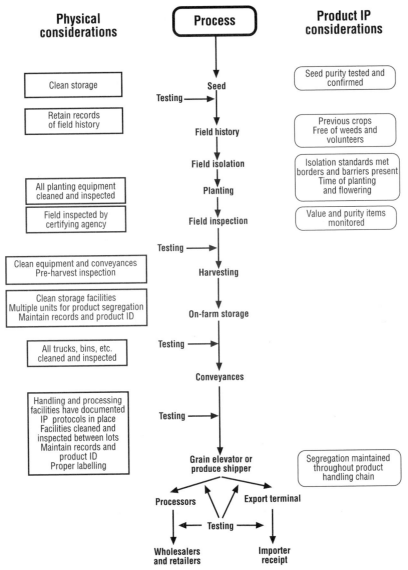

Fig. 6.2. The IP process including physical factors and product considerations (adapted from Sundstrom, F. J., van Deynze, A. and Bradford, K. (2002) Identity Preservation of Agricultural Commodities. Agricultural Biotechnology in California Series: Publication No: 8077)

'E' Enabling Biosecurity, Food Traceability and Food Safety

It almost goes without saying in a book based around 'E' issues in Agribusiness that the focus of this chapter would be on the electronic enablement of the systems associated with biosecurity, food traceability and food safety, particularly those technologies that are more innovative than the ususal business computing systems which have been discussed in some depth in earlier chapters. The three areas of technology chosen for discussion are **Precision Farming** – for reasons that such an approach and the technologies involved provide both traceability and food safety risk management mechanisms; **Electronic Identification Schemes**, and **Packaging and Labelling.**

Precision Farming (PF)

PF is a tactical approach to handling sub-paddock variability. It is the tailoring of soil and crop management to match conditions at every location in a field from year to year. The size of each 'location' depends on the particular application and more commonly, on both time and dollar resources available (Robert *et al.*, 1994; Bryceson and Marvenek, 1998). PF incorporates the use of technologies not normally associated with agricultural management; these include:

- *Global Positioning System (GPS)* is a satellite navigation system which when accessed via a ground receiver enables every point on earth to have a unique spatial 'address'; depicted by a set of co-ordinates e.g. (latitude, longitude) or (easting, northing).
- *Geographic Information Systems (GIS)* which are computer systems designed to allow spatial data storage, manipulation and integration, particularly those data that are geographically referenced.
- *Remote Sensing (RS)* which is the acquisition of information about an object (usually the earth) from some distance away. The classic data sets acquired in this manner are satellite data, aerial photos and airborne video data.
- *Yield Monitoring and Mapping* which involves accurate measurements on a grid basis of production within a paddock. Monitors are on-board the headers and they enable high and low-yield locations to be pinpointed within paddocks.
- *Variable Rate Technologies (VRT)* allow variable rate application of seed, fertilizer, herbicide and pesticides, based around the GPS recorded locations within paddock.

The core issue in PF is knowing exactly where each location is in the paddock (obtained using the GPS), and having detailed knowledge about the physical and chemical characteristics of that location and how these have been managed. All data for every variable collected at each location (including yield) is thus also geographically located within the surrounding physical landscape of farm and region. PF provides information at the sub-paddock level in greater detail than has traditionally been available.

The use of PF technologies to date has mainly been associated with property management planning (Robert *et al.,* 1994) where the tactical management value of detailed information at the sub-paddock level has been to adjust fertilizer and herbicide application rates to address location specific soil properties and conditions, to adjust seeding rates for similar reasons, for yield data to be similarly monitored, or more recently, to develop environmental management systems (see Byte Idea on LEAF – this chapter). The strategic management value of collecting this type of detailed information has always been that it contributes to building a good database of information about the farm. Now, it would seem that such a database and such technologies when used in conjunction with other farm management and/or analysis softwares, could provide the clear, concise and recorded picture of what has happened to a product on farm at source, information that is needed to document such product from a traceability perspective and which in the long term will contribute to better deals for the producer.

Byte Idea – LEAF
www.leafuk.org/leaf/

Linking Environment and Farming

LEAF (Linking Environment And Farming) was established in 1991 to develop and promote integrated farm management (IFM) which is an approach that integrates beneficial natural processes into modern farming practices using advanced technology. This approach is seen as a way in which to address the concern that the gap between consumers and farmers is becoming 'the great divide' – a situation which does not bode well for the future of sustainable agricultural production.

Recently, as part of LEAF's ongoing goal (currently LEAF is one of the six national associations under the European Initiative for Sustainable Development in Agriculture) of fostering IFM, it created the 'LEAF Marque' certification standard for farmers and growers which assures that the food produced by certified scheme members meets strict environmental standards. In other words the LEAF Marque system provides the environmental equivalent of crop and livestock assurance schemes and is the guarantee that producers have been inspected and are operating to the required standards and provides recognition for producing quality food alongside environmental protection and enhancement.

Associated with LEAF Marque is 'LEAF Tracks' which is an online service which lets consumers find out who produced their certified food. By entering the producer's number, the search displays the producer's public details and links to a map of the farm with the ability to view an aerial photograph of the farm by clicking on the camera icon. CMi Certification (which is a leading UK certification body for the food sector) is the biggest certifier of LEAF. The LEAF Tracks label is currently appearing on fruit and vegetables in Waitrose stores and it is hoped will be expanded into other retail shops in the near future.

Information sourced from LEAF management and from the company website

Precision Farming and Food Safety

A very good example of how PF technologies are at the cutting edge of the food safety arena is in aflatoxin monitoring in peanuts.

The peanut is a summer legume which is a native of South America and which requires warm conditions in which to grow. The fruit we know of as the peanut develops on 'pegs' extending down into the soil from the vegetative part of the plant which slowly matures above ground (Figure 6.3).

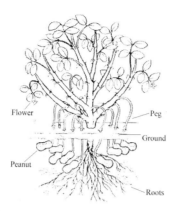

Fig. 6.3. Peanut plant showing vegetative and fruit development

Major peanut growing countries around the world include the United States, Israel, Egypt, Argentina, South Africa, China and India with Australia being a small player. The main food products created from peanuts include: peanut butter; nut-in-shell, snack peanuts and confectionery. The major processors and manufacturers involved worldwide include companies such as Unilever, Kraft, Nestle, Cadbury Schweppes, General Foods and Mars (Bryceson, 2003).

Aflatoxin is a toxin that is produced by two soil-borne fungi *Aspergillus flavus* and *Aspergillus parasiticus*. These toxins occur naturally in the soil and are found in a range of commodities including peanuts used for animal and human consumption (Mackson *et al.*, 2000). The issue is that aflatoxin, even in small amounts, is a carcinogen and in many developing countries is a major health risk to humans due to high levels of contamination.

As is usual at the production end of an agri-industry chain, financial returns from peanut growing are dependent on yield (tonne/hectare) and quality, with best results obtained with reliable rainfall or irrigation and intensive crop management. The main production management issue for peanuts is in ensuring that aflatoxin levels are kept to a minimum (Rachaputi *et al.*, 2001). The dilemma for farmers is that the longer they leave the peanuts in the ground, the higher the yield but at the same time the risk of contamination from aflatoxin increases as the soil becomes hotter and dryer towards the end of the season. The skill is in determining when to pull (harvest) the peanuts to maximize yield and minimize aflatoxin risk. Until recently this has been a gamble and has cost producers a great deal of money in penalties since aflatoxin status has a very tight specification criteria around the world (<8 parts per billion).

However, recently Robson *et al.* (2004) have developed a method for detecting variations in crop maturity in field-grown peanuts (as measured against internal chemical constituents of peanut leaves, kernels and shells that correlate with harvest maturity), by combining and analysing remotely sensed near infrared spectroscopy and infrared aerial imagery for detectable differences between maturity levels. The idea being that the detection method developed could be used to enable growers to formulate harvest regimes that can maximize yield and quality and reduce aflatoxin risk at the same time (ENVI Application Challenge, 2005).

In this instance therefore, the technologies themselves are being used to combat a food safety issue directly at source and the value added product is then being used to improve production and management practices.

Europe is a major importer of peanuts and the amended Articles 11, 14, 16, 17, 18, and 19 of Regulation (EC) No 178/2002 imposed in January 2005 will see more administrative responsibilities imposed on manufacturers involved in peanuts to ensure product integrity (and to supply evidentiary materials to support this). The technology development described in this section is likely to be a contributing component to this process on farm.

Livestock Electronic Identification Schemes

Livestock identification schemes are the first step in a traceability system for meat and meat products. Many countries are progressing down similar paths in relation to implementing such schemes, including the USA with the United States National Animal Identification Plan (USAIP) (www.usaip.info); Canada with their National Livestock Identification for Dairy (NLID) (www.nlid.org/English) and the UK with the Cattle Tracing System (CTS) or 'cattle passport' traceability scheme which was implemented in 1998 in compliance with the European Union requirements (Article 6.2 of Regulation [EC] No. 2629/97) (3) (www.defra.gov.uk/animalh/tracing/cattle/cts.htm), (Smith, 2004). In Australia, the National Livestock Identification Scheme (NLIS) (see Byte Idea this chapter) is run through Meat and Livestock Australia working with the various state departments of agriculture/primary industries (e.g. www.dpi.qld.gov.au/nlis/ and www.agric.nsw.gov.au/reader/nlis).

Essentially electronic identification of animals involves ear tagging (see Figure 6.4) each beast with an electronic ear tag at birth so that information such as genetics, weight, husbandry, movements onto and off properties etc. can be recorded against a number linked to each animal until it is slaughtered.

Fig. 6.4. Australia's National Livestock Identification Scheme (NLIS) RFID tags including ear tags and rumen bolus

Information from each ear tag is read by a RFID reader and downloaded to a computer. In Australia's case this information is then uploaded to the national NLIS database run by the MLA. Saleyards are required to notify the NLIS database of all cattle being sold. Abattoirs are required to notify the database of all cattle slaughtered. By July 2007 the whole Australian herd will be tagged for full traceability from birth to death. Further, the scheme will be rolled out for sheep starting in 2006.

The NLIS number is transferred from the live animal to a piece of meat in the abattoir when the carcase is halved and put on the chain. At this point, each half is given a number that corresponds to the NLIS number and note is taken of the chain colour. As the carcase moves along the chain the boner takes the cuts off and they are packed into boxes that are marked with the time and colour of the chain. While each cut can not be linked directly back to an individual animal, it can be traced back to a small group of beasts.

Additionally, at the time of slaughter, a small sample of tissue can be taken from each carcase and linked to the NLIS number. This sample provides a unique DNA profile of the animal which is recorded using computer-based technology and the meat can then be tracked as it moves through the processing chain to the consumer. Any problems that might arise at the consumer end can then be traced back to the individual animal with ease. An example of one company currently employing this advanced traceback technique is OBE Beef (www. obebeef.com.au) – an Australian beef producer specializing in organic beef. OBE Beef use an integrated production, handling and processing system to deliver quality certified organic beef that complies with the Australian Standards for Organic and Bio-Dynamic Produce and all AQIS standards for the export of organic products. Market research has shown the company that it is worth their while to employ this precise traceability system in addressing their niche market requirements.

In both the cattle and dairy industry, electronic identification tags as used in the NLIS system have proven additionally useful for herd management purposes – particularly when linked to appropriate herd management software such as Cattleworks, DairyManagementSystem 21 and CattleLink.

However, it should be remembered that electronic identification systems are not a remedy for managing disease outbreak, they are simply a tool to provide more timely traceback information. The main issues these systems are attempting to address are biosecurity and surveillance, quarantine and customs, reporting processes and research and development into disease and pest outbreaks.

Byte Idea – NLIS
www.mla.com.au

From Birth to Death Traceability
of Cattle in Australia

The Meat and Livestock Association of Australia is a producer-owned company with some 34,000 members that provides services to livestock producers, processors, exporters, foodservice operators and retailers. The MLA's core business is about building demand and improving market access for Australian red meat and conducting research and development (R&D) to provide competitive advantages for the industry. To do so, the MLA conducts a number of industry programs associated with the beef, sheep and lamb, and goat industries. Not least of these industry programs has been the introduction of the NLIS or National Livestock Identification Scheme.

The NLIS is a scheme to trace livestock from birth to death in Australia (other countries such as Canada, USA, the UK and Europe are all implementaing similar systems). In so doing it is hoped that the system will provide an improved capability to manage disease and chemical residue issues in the livestock (cattle) industry as well as improving access to export markets – which in Australia's case, are primarily in Japan, USA, Korea and Europe.

The NLIS system uses electronic ear tags or rumen boluses (an electronic tag deposited in the stomach) to individually identify cattle and their movements off property and through the saleyards which are recorded on a central database. This then allows fast and accurate tracking of cattle movements which is particularly useful for disease or residue checking. The system also provides documentary records of movements, such as waybills and National Vendor Declarations.

As of July 2005 it is mandatory that all cattle movements are to be recorded in the NLIS database and as of July 2007 it will be manadatory that the entire Australian herd – wherever the cattle are located – will be individually eartagged and NLIS compliant. This in turn means that no animal can be sold without an eartag and its movements in consequence are to be registered in the NLIS database. Because NLIS is both electronic and permanent, individual animals can be traced faster and more accurately than with the present tail tag and waybill system. In the event of a disease outbreak, quarantine measures can be deployed faster and with limited costs to industry and government.

The need for such a system has been long discussed by the major beef producing nations of the world but the global trends and attitudes towards food safety – particularly after recent BSE and foot and mouth problems – mean that such systems are now a requirement.

Information sourced from the MLA website

Product Packaging and Labelling

It is somewhat fitting that this section is the last in this chapter as what it will be talking about is the last stage in the agri-food supply/value chain, which is of course also what the average consumer sees and remembers of the food production process – a packaged food product.

Packaging of Food

A large variety of materials is currently used for food packaging including paper, fibreboard, glass, tinplate, aluminium and various types of plastics. The prevention of contamination of food by the packaging intended to protect it is the object of constant research and regulation. Any substance which migrates from the packaging into the food is a problem if it could be harmful to the consumer. Plastics are of particular concern since some contain a large number of additives including antioxidants, stabilizers, antistatics and plasticizers which are included to improve the functional properties of the plastics.

Both the United States of America and the European Community either have, or are preparing, complex regulations to control migration from food packaging materials. In some instances these regulations set maximum limits for potentially harmful migrating substances. In other cases where migration presents a minimal hazard, permitted additives may be listed without specific limits being set. Australian Standards, which are separate from food regulations, have been based on what is permitted in some overseas legislation.

Companies involved in developing packaging materials need to be dynamic, agile and innovative technologically to ensure that they can keep up with both legislation which changes frequently and with marketing requirements. Convenience Food Systems (www.cfs.com) is one such company. CFS is a global supplier of processing and packaging solutions primarily for the meat, poultry, fish, seafood and cheese industries and offers an extensive range of solutions from a single machine or packaging material to a complete production line. They have gone one step further in terms of TrackMax®, which is a software solution that provides completely automated management of production and maintenance parameters offering end-to-end traceability and complete food safety documentation and other operational management information.

Food Labelling

Labelling is extremely important in relation to food products – labels provide information about the nutritional content of the food you buy, the percentage of the characterizing ingredient of the food, and declarations of the presence of potential allergens in foods, however small the amount. Depending on the country in which the food is being marketed other requirements will also be in place – for example in Australia, all foods containing genetically modified material must be labelled and any irradiated foods must be labelled as irradiated.

As well as this basic information, food labels contain a wide range of other material including the 'use by' date, information on ingredients, storage requirements such as if the food must be refrigerated or kept frozen, and last but not least there should be information on food additives, which are normally represented by numbers as some of the additive names can be long.

In Australia, under new laws, nearly all manufactured foods will carry a nutrition information panel. The information must be presented in a standard format which shows the amount per serving and per 100g (or 100ml if liquid) of the food (Fig 6.5). There are a few exceptions to requiring a nutrition information panel such as very small packages; foods like herbs and spices, tea, coffee; foods sold unpackaged (if a nutrition claim is not made – see below) or foods made and packaged at the point of sale. Nutrition information panels provide information on the amount of energy (kilojoules), protein, total fat, saturated fat, carbohydrate, sugars and sodium (salt), as well as any other nutrient about which a claim is made (for example: fibre, iron, calcium).

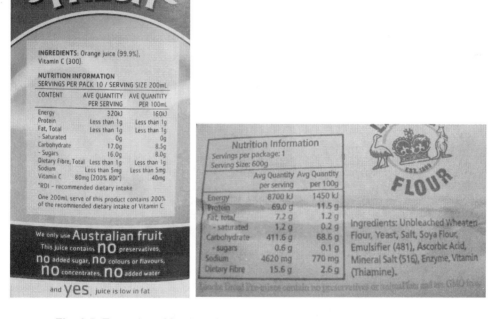

Fig. 6.5. Examples of food product labels used currently (2005) in Australia

As with packaging, speciality labelling companies exist for food labelling systems and products and there are many types of labels and labelling systems. The design and delivery of *smart labels* is ongoing. For example, in 2003 Zebra Technologies – see Byte Idea this chapter, developed a print engine that was capable of printing labels with bar codes and an embedded RFID transponder. The new device enables suppliers to integrate RFID into labels and print them at the same time (Collins, 2003).

Similarly, MPI Label Systems (www.mpilabels.com) is a company that has been in the business for 35 years and has many different products that are used worldwide in the food and beverage industry including shrink sleeve, digital labels and RFID.

There can be no doubt that a significant increase in RFID labelling usage can be attributed to the 1 January 2005 Wal-Mart RFID mandate for the company's top 100 suppliers – and the 1 January 2006 deadline for the next 200 largest suppliers. When a buyer of this power stipulates that their suppliers need to update their technology, it tends to be a case of 'do it or be damned'!

Byte Idea – Zebra Technologies
www.zebra.com

Smart Labelling

Zebra Technologies is a company that deals in 'Smart Labels'. They are a global provider of specialty printing solutions, including on-demand thermal bar code label and receipt printers and supplies, plastic card printers, RFID smart label printer/encoders, certified smart media, and digital photo printers. The Illinois based company commenced in 1969 but only became an on-demand labelling and ticketing systems specialist in 1986. Zebra has an international distribution and sales network that covers 100 countries in Europe, the UK, Latin America, and the Asia Pacific with more than 90 percent of Fortune 500 companies using Zebra-brand printers

Zebra solutions are innovative and cutting edge – printing solutions include bar code printers, mobile printers and RFID labelling engines and labels which include in-stock labels supporting leading transponder types as well as custom label options (www.rfid.zebra.com). In October 2005 the company released a document called 'Top 10 tips for optimal RFID smart label success' which can be obtained from their website. Additionally, Zebra provide networking products such as ethernet, WAP enabled connectivity and wireless connectivity for mobile phones to enable easy integration of their technologies with existing desktop applications.

Zebra is regularly named in Deloitte and Touche's prestigious 'Fast 50' list, a regional ranking of the fastest growing technology companies in the United States and Canada, and Forbes' list of the 200 Best Small Companies in America. In addition, Zebra was recently named 'Supply Chain Innovator' by Supply Chain.

Information sourced from RFID Journal (www.rfidjournal.com) and Zebra Technologies website

Summary

Food safety and traceability, plus the technologies currently used and those being developed, are of major importance for the future viability of the food industry. Physiologically *'we are what we eat'* and as consumers become better informed (and there are a wealth of education programs on the subject of food ranging from how to grow your own vegetables to the technological detail of RFID-based identification), their demands for products that both fulfill their nutritional requirements and their belief and value systems associated with safety and quality, get louder and more strident.

Disease scares such as foot and mouth and BSE have highlighted issues to legislators and have pushed food safety and traceability into the public eye – but the reality is that with the extended reach of the internet and electronically enabled information flows, it would only have been a matter of time before consumer demands for better documented systems had to be taken note of. No retailer can afford to ignore their customers and in order to satisfy this information 'pull' from the clients – while maintaining a strategy of reducing costs and maximizing profits – the only way to progress is to be innovative. Using both old and new technologies in a business smart way is not a bad start!

Someone who has had to deal with a major food safety issue that affects his product and who has upgraded his company with the most innovative of solutions to address it, is Bob Hansen, the CEO of The Peanut Company of Australia – the Smart Thinker for this chapter.

Smart Thinker – *Bob Hansen*

CEO and Board Member Peanut Company of Australia (PCA)

Bob Hansen has been the Executive Managing Director of The Peanut Company of Australia (www.pca.com.au) since 1993. Prior to joining PCA, Bob had already had a stellar career in the poultry industry culminating in five years as Victorian General Manager of Inghams Enterprises – one of Australia's largest most innovative poultry enterprises – so he was a seasoned campaigner. And he had to be – he has steered PCA through the growing pains of corporatization (PCA was the old Peanut Marketing Board – a government institution), increasingly tight food safety requirements worldwide for his product, and electronic enablement, with attendant demands from his major customers to comply with their systems.

What has emerged is a modern, streamlined highly profitable company that handles around 85% of Australia's peanut crop (about 55,000 tonne/annum). PCA not only supplies almost all of the Australian domestic market requirements but in the last few years has taken on the highly competitive Japanese and European markets as well, winning ground due to their capability of delivering high quality peanuts that have a solid QA system to back up marketing claims.

The QA system actually starts way back in the paddock – where Bob has invested in research associated with varietal development (high oleic peanuts are the go in today's shelf-life war), agronomic developments (from the standard 'how to grow a peanut' to supporting research into remotely sensed monitoring of peanut crops as a harvest maturity detecting tool), and both pre- and post-harvest technology development. In the laboratory, aflatoxin testing developments have been given significant support (in 2003, PCA opened a $1.76m Innovation and Technical Centre, which includes a pesticide laboratory, a microbiological laboratory and of course, the Aflatoxin and Chemical Laboratory which is NATA accredited for aflatoxin and chemical sampling and testing). Finally, in the factory, state-of-the-art electronic shape and colour sorters speed up the weeding out of foreign materials and the duds by using x-ray machines and five inline digital line scanning cameras that can detect 16 million colours down to a size of 0.5sq mm and enable 20,000 decisions a second to be made! Clean peanuts at speed!!

So – if you thought that a peanut was just a peanut... think again and check out the PCA website to find out about what a good peanut really is!

Source: Author interview with Bob Hansen, and the PCA Website

The Review and Critical Thinking Questions below have been provided for you to revise and reflect on the content of this chapter.

Review Questions

1. What are the major issues associated with biosecurity?
2. List the key points associated with the HACCP approach and discuss why it has become a worldwide standard.
3. What are food traceability systems? – describe the technologies being employed currently.
4. What is identity preservation? – why should it be included in a food traceability system?
5. Electronic identification schemes are becoming widely employed around the world – what are they, why are they regarded as important and what do they entail?
6. Why is product packaging and labelling so important – source and describe a system that you are familiar with.

Critical Thinking Questions

1. How can environmental sustainability be incorporated into a traceability system? Discuss what implications this would have on an agri-food chain if this was mandated?
2. Carrefour have their own QA system which they want their suppliers to follow, and Walmart is requiring its suppliers to adopt RFID technologies in their labelling. Discuss the issues associated with an industry chain so dominated by the retail sector.
3. Food safety and traceability are sops to consumer demand – discuss.
4. Take three examples of differing food products from your pantry, check the label and the packaging and list every ingredient and where it comes from. Then determine what attributes of those individual ingredients would be worth recording to enable a cusumer to know exactly what they are putting into their mouth.
5. Discuss the generic food-based issues and the customized solutions employed by Bob Hansen in running the Peanut Company of Australia.

Chapter 7

Is the Agribusiness 'E' Ready?

Think Big, Start Small, Scale Fast

The chapter takes a brief look at the concept of a business strategy and then discusses the issues associated with creating and implementing an *'E' strategy* paying particular attention to identifying the objectives of electronically enabling the business. The positional and bonding factors, including the very real issue of *leadership* and the effect of leadership on establishing clear guidelines for electronic enablement, are discussed in relation to developing an 'E' strategy, as are the *'E' readiness* of the business and the alignment of the business strategy and processes. Electronic enablement of key internal functional components such as human resource management, asset and facilities management and occupational health and safety management is also considered.

Chapter Objectives

After reading this chapter you should understand and be able to describe:

1. What a business strategy is and the associated development issues
2. What is needed to create and implement an 'E' strategy for the company
3. What 'E' readiness means and what it entails
4. What an economic web is and where it fits in terms of 'E' strategy
5. How electronic enablement is being used in human resource sections of more and more companies
6. How electronic technologies are changing the face of asset management in businesses.

Business Strategy Development

Strategy is the art of devising and employing plans that lead towards a goal. Strategy development is an integrated set of actions designed to create a sustainable advantage over competitors. Thus traditionally a business strategy is a plan that is meant to be the blueprint for a company to create or protect a defensible competitive advantage in the marketplace, the underlying assumption being that by applying a set of powerful analytic tools the future of any business can be predicted accurately enough to choose a clear strategic direction for a long period of time (e.g. five or ten year timeline). Today, this approach is proving a little unreliable because of the rate of change in the business world. This shift in thinking is discussed in later sections.

The underlying theory associated with business strategy development comes from Michael Porter (1980, 1985, 1996) where he considered the dynamic relationship between enterprise strategy and industry structure to be at the centre of his concept of 'competitive strategy'. In his Five Forces Model (Fig. 7.1), he makes three assumptions:

1. An industry consists of unrelated buyers, sellers, substitutes and competitors all operating individually
2. Wealth accrues to companies that can erect barriers against competitors and potential entrants
3. Uncertainty is low enough to permit predictions to be made about how participants in the market will behave.

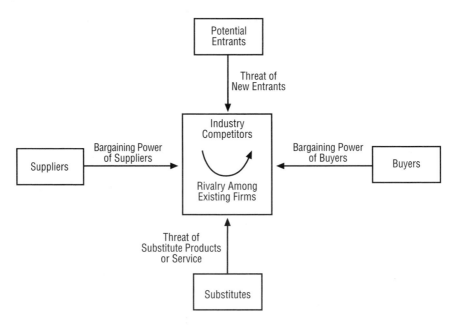

Fig. 7.1. Michael Porter's Five Forces Model (sourced from Porter, 1980, *Competitive Strategy: Techniques for Analyzing Industries and Competitors*, The Free Press, New York)

Porter's work, while identifying the main issues associated with developing competitive advantage, is actually less clear on the overall 'how' of creating a strategy. With much work going on in the area of strategy development in the following 20 years (Gluck, Kaufman and Walleck, 1980, 1982; Mintzberg, 1990, 1994), Gluck *et al.*, (2000) came to the conclusion that, traditionally, companies formulate and plan strategy in three main ways – that is, via strategic thinking, formal strategic planning and opportunistic decision making (Fig. 7.2).

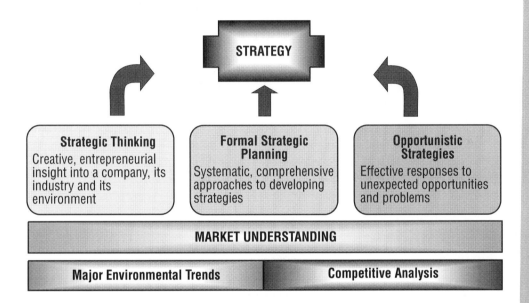

Fig. 7.2. Company strategy generation according to Gluck *et al.* (2000)

1. *Strategic thinking.* This activity seeks hard, facts-based, logical information. It questions everyone's unquestioned assumptions, is characterized by an unwillingness to expend resources, and is usually indirect and unexpected rather than head-on and predictable.
2. *Strategic planning.* This activity evolves through four discrete phases:

    • *Phase 1 – Financial planning.* This is the most basic component of strategic planning and is present in all companies. It involves setting annual budgets and using them to monitor progress and benchmark performance.
    • *Phase 2 – Forecast-based (portfolio) planning.* This is similar to Phase 1 but over a longer time frame. It evolves towards the development of sophisticated forecasting tools such as trend analysis, regression models and simulation models (see Armstrong (2001) for a comprehensive rundown of forecasting tools).

- *Phase 3 – Externally orientated planning.* This involves in-depth analysis of internal issues *plus* an external analysis of customers, potential customers, competitors, suppliers and others. Plans in this phase are dynamic and adaptive rather than static and deterministic as is the case in Phases 1 and 2.
- *Phase 4 – Strategic management.* This is the melding of strategic planning and everyday management into a single, seamless process. Five attributes of companies exhibiting strategic management include:

  i. A well understood conceptual framework of the strategic issues facing the company both currently and in the future
  ii. Strategic thinking throughout the business
  iii. A process of negotiating tradeoffs between competing objectives which involves iterative feedback loops
  iv. A performance review system that focuses the attention on key problem and opportunity areas
  v. A motivational system that rewards and promotes the exercise of strategic thinking.

3. *Opportunistic decision making.* Decision making of any sort is a process involving formulating the problem to be solved, and using information, knowledge and intuition to solve that problem. Opportunistic decision making is relatively unstructured and allows organizations to be flexible and open to making changes to the strategic planning process if it becomes necessary in the face of unexpected events and changes in the initial assumptions.

The development of a business strategy plan using the approaches outlined above still holds true for some of the large organizations around the world and many companies still use techniques such as discounted cash flow models, core competency diagnostics and Porter's Five Forces Model to plan their approach to doing business. However, the traditional model assumes that uncertainty in the industries involved is low enough to enable senior managers to make reasonably accurate predictions on which to base a strategy. The reality is that the speed of change in the business environment in the 21st century, which is driven principally by globalization and worldwide information flow via the internet, means that massive uncertainty is a fact of life and the long timelines associated with traditional business strategy planning and development are no longer viable. Businesses need to be more agile and flexible in addressing changing marketplaces and the changing nature of the competition - see Byte Idea on e-BAT this chapter.

Byte Idea – e-BAT
www.iiceb.org

EBusiness Assessment Toolkit

The e-BAT suite of eBusiness assessment tools was developed by and for the UK Council for Electronic Business (UKCeB) originally as a tool for the defence and aerospace sector. Essentially, e-BAT is a software and tool methodology that creates an eBusiness strategy. e-Bat has three main components: e-BAT Strategy, e-BAT Focus and e-BAT Benchmark.

1. *e-BAT Strategy* was the first of the e-BAT suite of tools to be developed and is intended to be used by a company or organization as an aid to development of an eBusiness strategy unique to the business, driven by business requirements rather than the availability of technology solutions.
2. *e-BAT Focus* consists of a high level questionnaire and is intended to be used as a mechanism for generating interest in eBusiness amongst Board members and senior executives. From answers to the questionnaire, a score is generated which assesses the company as belonging to one of five categories – e-novice, e-starter, e-advocate, e-exploiter or e-leader.
3. *e-BAT Benchmark* is designed to compare a company's eBusiness performance and capability with that of other organizations based on profile and industry sector.

The UKCeB is an independent not-for-profit organization funded by its members (the major UK defence prime contractors) and supported by its founding trade associations. Its mission is to accelerate secure information sharing to support collaboration – mainly in the defence sector. UKCeB distributes e-BAT itself and has four overseas distributors in Singapore, Japan and Korea.

Information sourced from the International Industrial Commission for Electronic Business (IICeB) website (www.iiceb.org) and the UKCeB website (www.ukceb.org.uk)

Strategy Development under Uncertainty

The traditional approach to strategy assumes that using powerful analytic tools can predict the future of any business accurately enough to choose a clear strategic direction for it. However, this approach to strategy development often underestimates uncertainty because while there is no doubt that standard tools can provide deep insight into strategic opportunities, they rarely generate any foresight into opportunities that arise suddenly. Thus companies are very often left in a position whereby strategies are developed that do not allow the take-up of opportunities as they happen, nor the flexibility to protect the company should a major problem occur. Figure 7.3 illustrates the four basic steps in dealing with uncertainty which are: 1) recognize the level of uncertainty, 2) tailor the approach based on the level of uncertainty, 3) make appropriate strategic decisions and 4) actively manage the chosen strategy to minimize the risks associated with the level of uncertainty, as well as their integration.

Fig. 7.3. Uncertainty and strategy development

Courtney *et al.* (2000) have presented a framework for determining the level of uncertainty surrounding strategic decisions and for tailoring strategy to that uncertainty so that a clear scenario can be developed. They believe that strategically relevant information tends to fall into two categories:

1. Information from which it is possible to identify clear trends, such as market demographics, which can help define potential demand for a company's future products or services
2. Information from which, if the right analyses are performed, many factors that are currently unknown to a company's management are in fact knowable – for instance, performance attributes for current technologies or a competitor's strategy.

The uncertainty that remains after the best possible analysis has been undertaken is called *residual uncertainty* – and in most instances some information or knowledge will be available about even this level of uncertainty. An example of a residual uncertainty would be in waiting for the outcome of some future competitive process where the competitors are well known. Courtney *et al.* (2000) believe there are four levels of uncertainty:

1. *A clear enough future.* This is a standard precise prediction of the future using market research, competitor analyses and Porter's Five Forces.
2. *Alternative futures.* This happens when analysis cannot identify which outcome will actually occur – for example, when a strategy depends mainly on competitors' strategies, which cannot yet be observed or predicted. Practically, some, if not all, elements in the strategy would change if the outcomes were predictable – for example, businesses facing major regulatory or legislative change such as telecommunication companies.

3. *A range of futures.* This is when there are a limited number of key variables that define the range of outcomes, but the actual outcome may lie anywhere within the range – for example, companies in emerging industries or those entering a new geographic market.
4. *True ambiguity.* This is when a number of dimensions of uncertainty interact to create an environment where it is impossible to predict outcomes or scenarios. This is a rare situation and is generally transitory – for example, a startup dot.com company.

Situation analysis at level 4 is highly qualitative but managers need to systematically isolate what they know and what it is possible to know. Even if it is impossible to develop a meaningful set of probable, or even possible, outcomes, managers can gain a valuable strategic perspective by addressing what information is available to them in a logical manner.

Strategies and Actions under Uncertainty

A company can assume one of three strategic postures to deal with uncertainty: these are, according to Coyne and Subramaniam (1996): **shaping, adapting** and **reserving the right to play**.

- *Shapers* aim to drive their industries towards a new structure of their own devising. Their strategies are about creating new opportunities in a market, either by shaking up a relatively stable industry or by trying to control the direction of the market in industries with higher levels of uncertainty.
- *Adapters* take the current industry structure and its future evolution as givens, and then react to the opportunities that the market offers.
- *Reserving the right to play* is a special form of adaptation. It involves making immediate incremental investments putting a company in a privileged position – through superior information, cost structures, or relations between customers and suppliers – that allows the company to wait until the environment becomes less uncertain before formulating a strategy.

Table 7.1 outlines the types of actions that are relevant to implementing strategy under conditions of uncertainty.

Table 7.1. Portfolio of actions associated with implementing strategy under uncertainty

Portfolio of actions Three types of moves are especially relevant to implementing strategy under conditions of uncertainty	• Big bets	*Big bets* are large commitments, such as major capital investments or acquisitions, that will produce large payoffs in some scenarios and large losses in others. Not surprisingly, 'shaping' strategies usually involve big bets; 'adapting' and 'reserving the right to play' do not.
	• Options	*Options* are designed to secure the big payoffs of the best-case scenarios while minimizing losses in the worst-case ones (e.g. conducting pilot trials before the full-scale introduction of a new product). Companies that 'reserve the right to play' rely heavily on options, although 'shapers' use them as well.
	• No-regrets moves	*No-regrets moves* are moves that will pay off no matter what happens (e.g. reducing costs, gathering competitive intelligence, or building skills). However, even in highly uncertain environments, strategic decisions such as investing in capacity and entering certain markets can be no-regrets moves.

In the current business world where globalization, the development of the internet as a major information and communications facilitator, and aggressive markets have created high levels of uncertainty, the decision whether to 'shape' or 'adapt' has become even more difficult. Courtney (2001) makes the point that the strategy posture chosen must depend on the uncertainties faced by the company and their approach to risk management. For example, with very high levels of uncertainty about variables the company can influence, shaping strategies can deliver higher returns with lower risk than adapting strategies. Adapting strategies on the other hand are more favourable when key sources of value creation are outside of the company's control.

However, there is no single solution and since the nature of the uncertainties a company faces changes with time, it must consider ways of varying how it thinks about adopting shaping or adapting strategies to enable it to meet new challenges.

Agribusinesses have always existed in highly uncertain business environments with climate in particular playing a major contributing role to that uncertainty in agricultural production (White *et al.*, 1999). However, the impacts of climate variability are not restricted to production but flow through to all parts of an agri-industry chain. One of the focuses of most climate variability R&D programs around the world is the development of tools such as decision support systems (Bryceson and White, 1993) that use climate science-based information to evaluate alternative courses of management before locking-in a decision.

Decision support systems (Appendix 1) are a well developed managerial tool that can be developed and implemented at all levels within an organization to address any functional or managerial issue that arises. They are generally interactive computer-based systems that use data, information, knowledge and models of processes and are designed to *support* rather than *replace* managerial judgement with their objective being to improve the *effectiveness* of decisions (Holsapple and Whinston, 2002; Turban *et al.,* 2005).

The Seven Dimensions of an 'E' Strategy for Agribusiness

As with most businesses, the strategy employed in agribusinesses thinking of undertaking electronic enablement development will vary depending upon the strategic positioning of the company, which will be influenced by the nature of the organization (whether the company was born on the net or moved to the net), the nature of the products (are they production, processing or service based – or mixed?) and the online model the organization wishes to adopt (e.g. B2C, B2B, etc.). Porter (1996, 2001) outlines six key principles that should be adhered to when developing a business's strategy – whether it be an 'E' strategy or not. These are listed in Table 7.2 with an accompanying explanation.

Table 7.2. The Principles of Strategic Positioning (Porter, 1996, 2001)

	Principle	Explanation
1	The Right Goal	Sustained profitability can only be attained by a good strategy grounded in real economic value.
2	A Deliverable Value Proposition	A set of benefits that must be different from a competitor's. The strategy defines a way of competing that delivers a unique value.
3	A Distinctive Value Chain	A company must perform different activities or perform similar activities in different ways. NB. Adopting best practice strategies will result in performing similarly to competitors.
4	Trade Offs Are Necessary	A company must give up some products or features, services or activities in order to be unique in others. Being all things to all people almost certainly guarantees that the company will not have a competitive advantage.
5	Tight Fit	A company's activities throughout the value chain must be mutually reinforcing – a good strategy defines how all elements of the company fit together.
6	Continuity of Direction	Defining a distinctive and/or unique value proposition allows a company to develop unique skills and assets.

For agribusinesses, managing the supply/value chain and ensuring a two-way information flow (Bryceson and Kandampully, 2004) is crucial to the success of a business and the industry chain in which it participates. The First4Farming hub concept (Figure 7.4) is a new 'E' agribusiness strategy that is beginning to provide both a deliverable value proposition and a distinctive value chain to those companies involved.

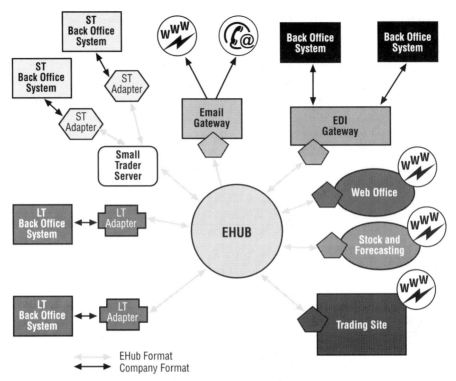

Fig. 7.4. First4Farming 'E" agribusiness hub concept – ST = Small trader; LT = Large trader

'E' Strategy Development – 'Think Big, Start Small, Scale Fast'

The essential question that comes before any strategy is developed is 'why are we electronically enabling the business, and what is our goal for the company by doing so?' – that is, what are the objectives of electronically enabling the business?

For example, is the aim of electronic enablement to improve internal systems such as financial account keeping/tracking, internal communications, supply and/or value chain management and inventory management – or is aim of electronic enablement to improve external systems such as extending the reach of the market business, improving access to and/ or speed of data/information flow, conducting financial transactions with other stakeholders? On the other hand, perhaps the aim of electronic enablement is to improve both internal and external systems?

In defining these objectives the phrase '*Think Big, Start Small, Scale Fast*' captures one of the most crucial concepts in electronically enabling a business. 'Think Big' is about having a grand vision and daring to dream of a better way of doing business unconstrained by current paradigms. 'Start Small' is all about recognizing the need to build a solid foundation and learning by doing. 'Scale Fast' is about capitalizing on the foundation you have created as soon as you know what you are doing. We will return to this concept in Chapter 8 when we discuss the practicalities of making electronic enablement happen, but from a strategic planning perspective it is *the* framework within which to work.

From the point of view of developing the appropriate strategy, Plant (2000) indicated that the differentiation between those companies that have a successful 'E' strategy and those that do not – once the objectives of electronically enabling the business are clearly determined – is a function of achieving a balance among seven major factors that contribute to the 'Think Big, Start Small, Scale Fast' mantra. These factors can be divided into **positional factors**, including technology, service, market and brand, and **bonding factors**, which include leadership, infrastructure and organizational learning (Fig. 7.5).

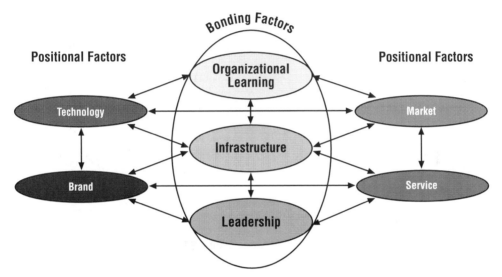

Fig. 7.5. The seven dimensions of an 'E' strategy and their linkages to each other

Positional Factors in 'E' Strategy Development

Positional factors are essentially about functionality – for example, how can *technology* be leveraged to define a unique service and create barriers to entry? Is it appropriate that the company should aspire to become a technology leader? How is it possible to gain a competitive advantage from existing IT systems? From a *branding* perspective, is the company an internet brand creator? Should the internet be used to reinforce the company's other branding channels or to reposition the brand? *Service*-wise, what is the company's internet service value chain? How do brand and technology decisions impact on the company's ability to deliver world-class service? How does the company deliver electronically enabled services that support the value proposition? Finally, the *market*: how is the company's online marketplace defined? How can 'bricks-and-mortar' markets be transferred onto the net? What partners are necessary to add flexibility and to develop a convergent branding strategy?

Note, the strategic decisions a company makes are always likely to involve branding, service levels, marketplace and technology, but the balance, focus and integration of their interactions change depending on organizational type (e.g. branding may be less of an issue in a business to government (B2G) environment than in a B2B eMarketplace environment), and the objective of electronically enabling the business – which could include being a technology leader, brand leader, service leader or market leader (Fig. 7.6).

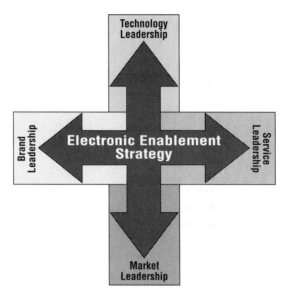

Fig. 7.6. Electronic enablement leadership propositions

1. *Technology leadership.* With this proposition, the strategy is to adopt an emerging technology early to achieve a pre-emptive position e.g. SUN Microsystems – see Byte Idea Chapter 2, Motorola Inc (www.motorola.com), AgLeader (www.agleader.com), John Deere (www.deere.com).
2. *Brand leadership.* Here, the internet is regarded as a new conduit to the customer, and as such it has an extensive ability to create a new corporate branding position, to reinforce the existing brand, or to enable the existing brand to be repositioned. Brand reinforcement comes through reflecting the values of the physical product through the medium of the internet. A brand reinforcement strategy does not necessarily imply the internet is used to transact, merely to interact e.g. Monsanto Co (www.monsanto.com) with their product RoundUp (www.roundup.com).
3. *Service leadership.* With this proposition, the value of service often simply consists of building relationships with, and gathering information about, potential customers and maintaining relationships with existing ones e.g. Elders Ltd (www.elders.com.au). The internet allows organizations to offer innovative types of service variations to more and more customers and it makes it possible for international companies to offer a level of service to all markets that was previously restricted to their home countries and major markets e.g. The Seam – see Byte Idea, Chapter 3.
4. *Market leadership.* With this type of proposition, nimble, creative, and agile corporations have achieved disproportionate market growth via the internet through responding to changing market conditions with product offerings as well as through their approach to understanding the market within which they operate. One successful approach has been to combine marketing, service and information systems groups to focus on issues as a cross-functional team e.g. Syngenta (www.syngenta.com).

Bonding Factors in 'E' Strategy Development

Bonding factors provide the foundations of a strong 'E' strategy. The three issues to address are leadership, infrastructure and organizational learning, and there is a strong interaction between these three components. For example, when eBay had problems in 1999/2000 the leadership learned from the experience, upgraded the systems infrastructure, and moved on. Other organizations fail to learn from their experiences and consequently die or, like Levi's, are forced to leave the internet space completely while they rethink their overall strategy.

1. *Leadership* – The primary drivers of change and the creators of strategic vision in an organization are the CEO and senior executives. An example of such a leadership-technology meld can be found within organizations such as Motorola Corporation, where Bob Clinton, the Director of the Internet Business Group in 1994, saw the opportunity for web-based communication between channel partners and created a concept that was called MOCA (Motorola Online Channel Access), or IBM, where Louis Gerstner, the then CEO, repositioned and transformed that organization based upon the eBusiness concept – or eBay, the online auction site, where Meg Whitman has explicitly taken a community-based leadership approach reflecting eBay's online community (see Smart Thinker this chapter).

   A critical aspect that must be addressed in any effective 'E' strategy is the impact of the strategy on the people in the organization and the impact of the people on the strategy. If the organization does not have the flexibility and capacity to deal with significant changes then initiatives such as electronic enablement have little chance of success. Business leadership has a critical role to play and the term leadership, rather than management, is used very deliberately in this context. The traditional models of management based around planning, measurement and control are necessary but not sufficient when considering the challenges that electronic enablement can create.

   Hay McBer (1996) identify four key factors that affect organizational performance. These are **individual competencies, job requirements, leadership style** and **organizational climate**, which they represent in 'The Four Circle Model' shown in Figure 7.7.

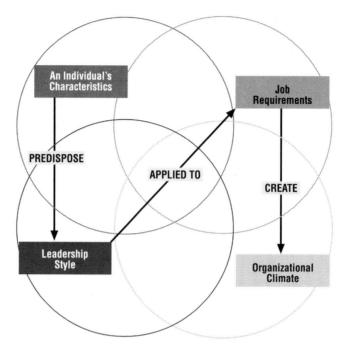

Fig. 7.7. The Four Circle Model of Organizational Leadership
(adapted from Hay/McBer Inc, 1996)

Essentially, an individual's characteristics and competencies will influence their type of leadership style which, when they apply this to the job requirement, will create the organizational climate or atmosphere of the workplace (Fig. 7.8). Where, and by how much, the four areas overlap is where a leader's effectiveness can be judged in relation to organizational performance: a large overlap tends to indicate a strong effective leader and a well performing organization and a small overlap is indicative of a less effective leader and a less well performing organization.

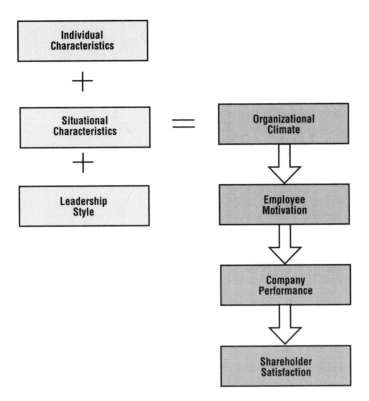

Fig. 7.8. A causal flow chart demonstrating how the elements of the Four Circle Model come together to impact on the performance of the business

The management tasks embodied in the model are around employing people with the individual competencies required by the organization and defining the specific job requirements to achieve the organization's goals. That business performance is the aggregation of the performance of each individual in the organization impacted by situational factors outside of the control of the business. It is important to note that performance derives from employee motivation (and is exhibited in the form of discretionary effort) that is aroused by the organizational climate of the employees, which in turn is created by leadership style.

Put simply, the ability of business leaders to create a positive and energizing climate is a key determinant in the performance of the business – an excellent example is the highly successful search engine Google where even the corporate website (www.google. com/corporate/culture.html) reflects a laid back and relaxed 'organizational climate' that emanates from the company's leaders – original founders Larry Page and Sergey Brin, along with CEO Dr Eric Schmidt.

Considered in the context of electronically enabling a business, leadership style and its impact on organizational climate is a key determinant of the success or otherwise of 'E'enablement.

2. *Infrastructure* – Issues that must be addressed here span the technology spectrum from the requirement of a single internet file server connected to a commercial *internet service provider* (ISP) all the way to the information-intense online transaction processing of a large company. Infrastructure issues need to be considered on three levels:

   - *Strategic level.* Here the focus is on determining the impact future technologies will have on the market and the organization.
   - *Organizational level.* The challenge at this level is to align work practices, process flow and the structure of the organization to execute the strategic goals effectively.
   - *Physical level.* At this level, issues surrounding the hardware and software of the computing environment in conjunction with the telecommunications infrastructure must be addressed.

   Not all organizations have the ability to be nimble in responding to these challenges. Successful organizations work towards a fluid and flexible architecture that allows for change, whether that change be of organic growth through corporate acquisition, or of streamlining through divestiture, or of a complete strategic turnaround due to the pressures of new technology – for example, it is easier to create a brand new value chain that is based upon an internet-based pipeline philosophy than it is to change an established value chain which has inertia built into its practices and processes. The lessons for managers include:

   - Recognize that there is a need to create a flexible infrastructure that can act as the 'shock absorber' of change.
   - Recognize that infrastructure creation requires open levels of communication at and across all levels of the organization.
   - Ensure that a technology solution that is created is scalable, secure and robust.
   - Maintain awareness of all standards as they evolve and attempt to influence their development. Plan for standards integration as soon as is feasible, so that actual integration will not occur in a pressurized environment.
   - Recognize that managers and executives cannot divorce themselves from technological understanding: the techno-CEO is the leadership model of the future.

3. *Organizational learning* – The ability of established organizations to react, understand and deploy an electronically enabled business solution is very dependent upon the ability of an organization to effectively manage change and leverage its organizational learning. Two types of organizational learning are recognized, associated with the type of change the organization is undergoing:

   - *Adaptive learning* which takes place in reaction to changed environmental conditions
   - *Proactive or generative learning* which creates organizational changes that are made on a more wilful basis and are beyond the simple reaction to environmental changes.

   Adaptive learning tends to come from a lower degree of organizational change than proactive learning and is seen as a process of incremental changes. It is also seen as more automatic and less cognitively induced than proactive learning (Senge, 1990; Dodgson, 1991). Many companies pride themselves on being a *learning organization* which, according to Senge in his famous book 'The Fifth Discipline: The Art and Practice of the Learning Organization', (1990:3), is an organization where:

...people continually expand their capacity to create the results they truly desire, where new and expansive patterns of thinking are nurtured, where collective aspiration is set free, and where people are continually learning to see the whole togetherness...

Such organizations have, at their core, knowledge in the form of five disciplines – each discipline provides a vital dimension. Each is necessary to the others if organizations are to 'learn'.

- *Systems thinking.* This approach comprehends and addresses the whole, and allows an examination of the interrelationship between the parts. The systems viewpoint is generally oriented towards the long-term view. Systems maps which are diagrams that show the key elements of systems and how they connect are a useful tool for studying the organization.
- *Personal mastery.* Personal mastery is the discipline of continually clarifying and deepening one's personal vision, of focusing one's energies and attaining a special kind of proficiency that goes beyond competence. People with a high level of personal mastery live in a continual learning mode – 'the journey is the reward' (Senge, 1990:142).
- *Mental models.* These are deeply ingrained assumptions, generalizations, or even pictures and images that influence how we understand the world and how we take action. Reflection on past learning experiences is the key to these mental models – on one hand, entrenched mental models can inhibit change, on the other, visionary models promote change.
- *Building a shared vision.* When there is a genuine shared vision (as opposed to the common or garden 'vision statement'), people excel and learn, not because they are told to, but because they want to.
- *Team learning.* This is the process of aligning and developing the capacities of a team to create the results that its members really want. It builds on personal mastery and shared vision and the ability to act together.

Dunne (2004) discusses how post industrial agribusinesses must develop their core competencies and core processes to create a learning organization which then adds to the organization's intangible assets. Most importantly, he reiterates that the development of trust and collaborative commitment both within an organization and between organizations in a chain is imperative if the added value of creating a learning organization/chain is to be achieved. A good example of an agribusiness addressing change in the market and realizing added value through developing a learning organization culture, is the cannery company Golden Circle – see Byte Idea this chapter.

Ultimately then, organizational learning occurs through shared insights, knowledge, and mental models and builds on past knowledge and experience – that is, on memory. Knowledge is seen as being of economic use in regard to organizational outputs. The key challenge is how to capture and translate the accumulated knowledge, expertise and experience held across the organization and chronologically over its historical development (see Chapter 3).

Developing a Winning 'E' Strategy

Developing a winning business, 'E' or otherwise, requires, as discussed, an awareness and a knowledge of both internal and external issues to the company. However, given that these are taken care of, there are some basic practical 'to do' things that can boost the likelihood of success:

Byte Idea – Golden Circle
www.goldencircle.com.au

State-of-the-Art Learning Organization

Golden Circle is a fresh produce (mainly fruit and vegetable) processing and canning enterprise. It is an unlisted public company based in Queensland, Australia, which is owned by 700 Australian farmers who supply more than 180,000 tonnes of fruit and vegetables every year to the factory.

Golden Circle manufactures more than 800 products including shelf stable fruit and vegetables (in cans and glass jars), fruit juices, cordials, soft drinks, jams, conserves and baby food but is best known as a tropical fruit specialist which traditionally has handled much of the pineapple processing in Australia.

Golden Circle, like many other grower-based cooperatives, has seen some tough times in recent years and has had to rethink and restructure itself to meet the ongoing changes in the international business world. The company has invested significantly in new processes, benchmarking itself against 'world best practice'. In recent years the company has invested more than A$130 million in its operations including:

- State-of-the-art pineapple processing technology
- A $7 million can supply system
- World's largest steam peeler and new beetroot processing lines
- Coldroom and packaging technology
- A paperless computerized inventory management system
- A $20 million food hall which is a fully computer controlled manufacturing plant capable of producing value-added products using different packaging styles such as glass and plastic and enabling a move into more innovative packaging and development of new products such as baby food
- A modern tetra brik plant that produces more than 42 million litres (19 Olympic-sized swimming pools) of fruit juices and drinks every year
- Electronic fruit sorting and grading facilities
- A robotic blow-moulding facility capable of producing 50 million plastic cordial and other beverage bottles every year.

In addition, Golden Circle has actively adopted a learning organization approach to developing and maintaining core corporate knowledge – particularly in relation to supply chain management. This approach is supported by senior management and is delivering dividends to the company which in little over a year has managed to reduce costs and up production significantly.

Information sourced from the Golden Circle website

- Ensure the project is backed by a senior executive – such a person can act as a mentor and/or can provide support when the going gets tough. Getting senior executives on board an electronic enablement project is discussed in Chapter 8.
- Develop a strategy by focusing on technology, branding, marketing and service. This is most important – most of the dot.coms that have died over recent times had no strategy plan and thus had nothing to fall back on when the bottom fell out of the market.
- Develop a strategy before developing a web presence. A web presence by itself is not what is going to make the company money – a web presence backed by a good business strategy plan might.
- Develop an IT infrastructure capable of matching the strategic objectives (align the company's IT to support the achievement of the business objectives not the other way around).
- Identify and use knowledge in the organization.
- The strategy must add value for customers, and it must change as the requirements of those customers change (a company must be agile in its approach to business in the 21st century – see Chapter 4).

Unfortunately it must be added for the unwary – a winning 'E' strategy is not necessarily the answer to a CEO's prayers. If the company is not 'E' ready, either technologically or culturally for electronic enablement, then it will be difficult to create success.

EReadiness

EReadiness is a phrase coined to describe the state of play in terms of a country's readiness or capability to actually support and undertake electronically enabled business (GeoSINC, 2002). However, the concepts and metrics devised to evaluate a country's eReadiness apply just as well to a single business or industry chain.

While eReadiness has often been seen really only as the ability to access adequate network infrastructure and technology, the broad definition of eReadiness encompasses the full range of electronic information and communication technologies (ICTs) as described in Chapter 1. It should be noted that while network access was a significant impediment to eReadiness in the 1980s and 1990s, in the 21st century there are a broad variety of other factors involved which are now coming to the fore – for example, some or all of the following have impact on the eReadiness of a country or business:

- Technology and ICT infrastructure
- Technology competencies
- The political environment
- The macroeconomic environment
- Market opportunities
- Policy towards free enterprise and competition
- Policy towards foreign investment
- Foreign trade and exchange controls
- Taxes
- Financing
- The labour market and infrastructure
- Access to education and general education levels.

In effect, there has been an evolution of issues in terms of eReadiness over time from technology driven imperatives to knowledge driven imperatives (Fig. 7.9) which is in line with the 'New Economy' focus of knowledge driven commerce (see Chapter 3).

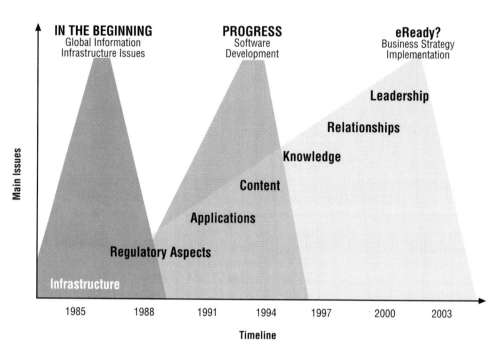

Fig. 7.9. Evolution in eReadiness imperatives over time – developing and maintaining relationships and having appropriate leadership are all significant in determining eReadiness

EReadiness Metrics

There have been a number of projects and related reports that have looked at defining the metrics associated with measuring eReadiness – the following six metrics are used by the Economist's Economic Intelligence Unit (EIU) (www.eiu.com). These six metrics are also quite appropriate to use for an evaluation of business eReadiness.

1. *Connectivity.* Electronically enabled business cannot function without adequate telecommunications and internet infrastructure. 'Connectivity' measures the access that individuals and businesses have to basic fixed and mobile telephone services, including voice and both narrowband and broadband data. Affordability and availability of service (both a function of the level of competition in the telecommunications market) also figure as determinants of connectivity. In rural areas this is a major issue.

2. *Business environment.* There are 70 indicators covering criteria such as the strength of the economy, political stability, the regulatory environment, taxation, and openness to trade used to evaluate the business environment. The resulting 'business environment rankings' measure the expected attractiveness of the general business environment over the next five years using a standard analytical framework. The rankings are designed to reflect the main criteria used by companies to formulate their global business strategies, and are based not only on historical conditions but also on expectations about conditions prevailing over the next five years.

3. *ECommerce consumer and business adoption.* Payment and logistics systems form the backbone of this set of criteria (e.g. the extent of credit card ownership, the existence of secure, reliable and efficient electronic payment mechanisms, the ability of vendors to ensure timely and reliable delivery of goods, and the extent of website development by local firms).

4. *Legal and regulatory environment.* The legal framework governing electronically enabled business is a vital factor than can enhance or inhibit the development of electronic trading. It includes the extent of legal support for virtual transactions and digital signatures, the ease of licensing and the ability of firms to operate with a minimal but effective degree of regulation.

5. *Supporting electronic services.* No business or industry can function efficiently without intermediaries and ancillary services to support it. For electronically enabled markets, these include portals, web-hosting firms, application service providers (ASPs), as well as website developers and eBusiness consultants. Access to these services is vital.

6. *Social and cultural infrastructure.* Education and literacy are necessary preconditions to a population's ability to navigate the web and drive future internet-based business development. Because entrepreneurship and risk taking play such an important role in building new electronically enabled business models, both nationally and internally, business attitude towards business innovation, risk and receptiveness to internet-based content should be measured. Basic 'E' competencies whether in house or available in the marketplace, are also necessary for businesses to be able to develop an electronic enablement capability (see Chapter 8).

According to the EIU the countries that are leaders in the eReadiness stakes are the USA, the European Union and Australia with Japan, Spain and Thailand following. Malaysia, Brazil and South Africa are getting there while some African nations, Russia and China are regarded as late starters. However, perhaps most importantly when attempting to deliver or evaluate the eReadiness of a country – or for that matter a business – it should be recognized that in reality the issues depicted in Figure 7.9 are overlapping and cannot be addressed in isolation (Lanvin, 2002).

EReadiness, the Internet and Strategy

The internet is an enabling technology that has, in recent years, become a basic infrastructure component of almost every business in just about every industry. The key question, however, is not whether to use it – there can be no doubt that to stay competitive in a global market which is based around information flow, using such technology is now mandatory – but *how* it can be employed to add value to the existing business (see Chapter 8).

Michael Porter (2001) makes the point that the creation of true economic value is the bottom line of business success and is nothing more than the gap between price and costs (i.e. profit). He believes that to simply use the internet to create revenue or to reduce expenses is not a value creation mechanism for a business in the long term – rather, the development of a strategy to provide such long-term value creation needs to look beyond the internet itself to its impact on an **industry structure** and **sustainable competitive advantage** of businesses within that industry.

1. *Industry structure.* Traditionally, the vertically integrated corporation has been the business strategy of success. In this model companies have performed most tasks associated with their product themselves rather than take the risk of partnering with others or outsourcing tasks. Whilst many organizations are still grounded in this strategy, the internet is slowly changing how businesses see themselves and how they see vital business activities being accomplished. For example, Boeing Company (www.boeing.com) no longer sees itself as an aircraft manufacturer but as a systems integrator of companies that make aircraft components – and IBM is a computer company that no longer makes its own computers but has a partner network that does instead.

    Such networks have been defined by Tapscott *et al.* (2000) as business webs or B-webs which are a system of suppliers, distributors, service providers, infrastructure providers and customers that use the internet for business communications and transactions. Participants of such B-webs have found significant reductions in search, coordination, contracting and other transaction costs between firms while increasing reach and links to partners never before available to them. A specialized form of these B-webs is known as an economic web and is detailed below.

2. *Sustainable competitive advantage of the business.* The internet impacts on operational effectiveness by increasing the speed of information flow which allows improvements throughout the value chain. However, this does not necessarily create a sustainable competitive advantage unless companies can achieve and sustain higher levels of operational effectiveness than their competitors. One way to do this is to use the internet to create unique products for the company or provide a unique customer service.

Successful 'E' Strategies

The 'New Economy' is fast, highly interconnected and knowledge- and relationship-intensive. Underlying the emergence of this new economy is a trend towards greater levels of complexity, a trend which seems to be based on the collapse of time and distance in business relationships which is dramatically changing the economic landscape. Just as biologists redefined the interconnections between species in a particular ecological framework as food webs rather than food chains, economists are building theories around knowledge-based *economic webs* or *value webs*, and *networks* rather than physical resource-based supply chains. The next two sections describe economic webs and the network orchestrator strategies, both of which have proven to be very successful.

Economic Webs – An Emerging Strategy for Business in the 'E' World

Economic webs have existed throughout the centuries in one form or another (e.g. craftsmen's guilds and in some manufacturing industries), but it has only recently been recognized that the interconnections are closed networks rather than linear chains. Since companies that have traditionally been separated by time and distance have operated under a value system based on physical resources (requiring linear production processes), these webs have been perceived as supply chains or value chains (see Chapter 4). An economic web approach, however, transforms the value chain into an adaptive value web of knowledge producers, solution providers and end users.

Essentially, economic webs are clusters of companies that collaborate on a particular technology or particular issue (Hagel, 1996). The technology the current web members promote eventually becomes the defacto standard through sheer weight of numbers (market preference). Economic webs are a natural response to highly risky and uncertain environments – a reason why they are prevalent in high tech areas. For example, in the desktop computing technology web, two major webs compete – the Microsoft and Intel web which promotes PC standards defined by the use of the Intel processor and Microsoft operating systems, and the Apple web which promotes the Macintosh as the standard desktop computing platform. Other economic webs may be organized around the behaviour and spending patterns of specified customer segments (*customer webs*) or organized around a specific type of transaction (*market webs*) (Tapscott *et al.*, 2000; Tapscott, 2001).

Companies in an economic web use a common architecture to deliver independent elements of an overall value proposition that grows stronger as more companies join. A good example of a technology-based economic web is the Microsoft and Intel personal computer web in which hardware and component makers, software developers, channel partners and training providers combine to deliver the overall value proposition of a Windows PC.

Webs operate without any formal relationships among participants. All companies in a web are wholly independent; they price, market and sell their products autonomously. Only the pursuit of economic self-interest drives them into weblike behaviour. Companies must, however, be aware and capable of managing the fact that there is a more dense information flow between participants in an economic web than in a more traditional linear supply chain and that a company's 'borders' thus become more porous (Hagel and Armstrong, 1997). Being able to capitalize on the information available yet prevent a loss of internal company knowledge capital to potential competitors is a significant management issue (Hagel and Singer, 2000).

Strategy and Economic Webs

As indicated earlier in the chapter, companies can have two different strategic postures in dealing with uncertainty, and thus with economic webs: adaptation and shaping. As an *adapter*, a company will be looking to stay ahead of the pack and must therefore be an early participant, be capable of competing for market share and be capable of diversifying its position. A *shaper* focuses on how to determine or influence outcomes of situations as they arise and can thus reap large returns by speeding adoption of core technologies and expanding the range and number of web participants. A web shaper's source of competitive advantage is thus within the economic web itself – not the company.

Whether a company chooses to be an adapter or a shaper has much more to do with its senior management's degree of ambition and willingness to take risks than with objective market conditions. Table 7.3 outlines the major strategies employed in economic webs by successful shapers and adapters using Microsoft and Compaq as examples.

Web strategies both narrow and broaden the focus of management. They narrow it because they encourage unbundling and the outsourcing of undifferentiated business activities. They broaden it because the context for defining strategy expands from maximizing value for the enterprise to maximizing value for the web. If the web does not maximize value, neither can the enterprises within it.

Table 7.3. Economic web strategies of a shaper and an adapter

	Web Formation (Get into the flow)	Web Mobilization (Build momentum)	Web Evolution (Encourage lock-in)
SHAPER E.g. Microsoft *Concentrates on occupying key leverage points on architectural grounds and using these to shape what others do. Marketing differentiates on the basis of the overall technological web rather than individual products*	• Ownership of key platform technology (Intel processor, MS OS) • Enter market quickly • Accelerate adoption	• Manage perceptions actively • Create economic incentives for others by unbundling the business • Actively sell the opportunities of being involved in the web (e.g. tied contracts with peripheral players)	• Frequently enhance platform technology • Promote standardization among other players
ADAPTER E.g. Compaq/HP *Forms alliances to boost responsiveness by improving access to technologies and markets. Marketing stresses product quality*	• Identify winning web early • Focus on near-term profit opportunities • Establish dense information links with other web participants	• Compete aggressively for web share • Link up with web shaper's strategy	• Exploit customer lock-in • Undermine supplier, shaper lock-in or diversify into new webs • Link and leverage

The Network Orchestrator 'E' Strategy

An alternative to the economic web, but no less important – indeed some would say more successful – strategy model in the 'E' world is the networked orchestrator model (Hachi and Lighton, 2001). In this model a company creates a network of contract suppliers and manufacturers connected to one another on its own proprietary extranet. Network participants exchange information about customers, products, inventories, schedules and costs pertaining to their commercial relationship with the orchestrator company, but are not owned or the core business controlled by that company.

An excellent example is eBay – one of the original dot.coms that has been phenomenally successful – see Byte Idea this chapter. EBay is the 'network orchestrator' for a group of companies that it considers necessary for it to run its online auction business. In the network, eBay supplies the communication platform via their auction software which enables buyers and sellers to come together as an online community to exchange information. EBay also provides the product management and the distribution links. Its specialist partners, including NEC, Yahoo, AT&T Wireless, America Online, Freightquote.com, UPS, Visa and PayPal handle direct payment, shipping, communication and other essential services (Fig. 7.10).

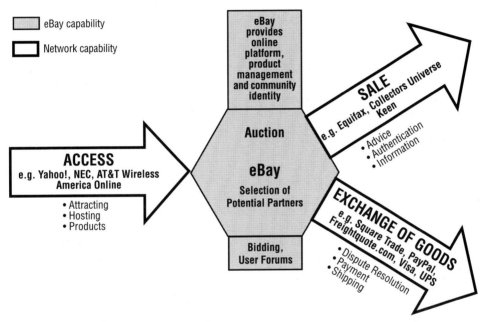

Fig. 7.10. EBay as an example of a successful network orchestrator

Network orchestrators control their circle of companies and choose them with care in order to match their own productive capacity and requirements – unlike the economic webs where the weight in the marketplace that comes with many partners is the important issue. There are five key characteristics of successful network orchestrators resulting in a highly transparent and integrated company which can respond quickly and appropriately to changes in supply and demand requirements:

- They maintain uniform standards for exchanging information.
- They rigorously monitor and evaluate performance standards.
- They share benefits generated by the network amongst all partners.
- They move key business processes online.
- They develop and test products across the network.

The result is a competitive, highly integrated company driven by network-wide access to critical market information and sources of knowledge. The internal transparency involved allows an organization to respond quickly and appropriately to shifts in supply and demand. In addition, it turns suppliers and subcontractors into network partners – and ultimately, close collaborators.

Byte Idea – eBay
www.ebay.com

Online Auctioning

eBay Inc. is the world's largest personal online trading community located at www.ebay.com. When eBay was created in 1995 as a vehicle for founder Pierre Omidyar to help his then-girlfriend collect Pez dispensers, a new market was also created: an efficient one-to-one trading in an auction format on the web.

eBay has developed a web-based community in which buyers and sellers are brought together to buy and sell items such as collectibles, cars, high-end or premium art items, jewellery, consumer electronics and practical and miscellaneous items. The eBay auction-style format permits sellers to list items for sale, buyers to bid on items of interest and all eBay users to browse through listed items. eBay's service is fully automated, topically arranged and easy to use. eBay features a variety of international sites, specialty sites, categories and services that aim to provide users with the necessary tools for efficient online trading in the auction-style and fixed price formats. Individuals use eBay to buy and sell items in more than 1,600 categories. Around the world, eBay runs over two million auctions each day – and more than 250,000 new items join the 'for sale' list every 24 hours.

eBay's success is mainly due to its business model which is a blend of garage sale, classified ad and auction. For example, if you have a fishing rod you want to get rid of, list it on eBay. The site accepts bids for three or maybe five days, then sells the item to the highest bidder. eBay gets a small fee for listing the rod, then a tiny commission from the sale. The beauty of eBay's model is that the company just facilitates the listings and the sales; it doesn't have to make or transport any goods or carry any inventory. As a result, eBay may collect as little as 6%, but most of that is profit. It is the sort of business that couldn't exist offline.

eBay is a publicly traded company under NASDAQ EBAY with headquarters in San Jose, California, USA.

Information sourced from Wikipedia (en.wikipedia.org/wiki/eBay), Forbes.com (www.forbes. com), NASDAQ (www.nasdaq.com) and eBay website

Internal Electronic Enablement as a Part of an 'E' Strategy

Internal electronic enablement is the oil that lubricates the engine of business – it provides the infrastructure and processes that enable efficiencies and time saving to be generated. In earlier chapters the hardware and some of the business systems involved have been discussed. In the following sections, three particularly important applications in relation to internally enabling the business are discussed – **executive information systems**, **'E' human resources** and **'E' occupational health and safety.** All are significant components of managing a business.

Executive Information Systems

Executive information systems (EIS) of all kinds are based on the concept of portals, and in particular the enterprise information portal (EIP) which was discussed in Chapter 3. In that chapter some emphasis was placed on the fact that EIPs come into their own when they are used to integrate various tasks and applications, including content management, data warehousing and business intelligence analyses as well as data external to these applications, into a single system that can 'share, manage and maintain information from one central user interface' (Luce, 2002).

In other words, the corporate portal is more than simply a gateway. It is a gateway to an application that may provide comprehensive support for the end user's job role. Executive information systems (EIS) are the next stage of development in the portal game providing management and executives with an integrated snapshot of the company's performance so that timely decisions can be taken and appropriate strategies elaborated (Varon, 2002). Ease of use is an important feature so that enquiries can be made without a detailed knowledge of the underlying data structures. Graphical user interfaces (GUI) make it possible to request reports and queries without resorting to programming.

Why have EISs developed? Edwards (2002) makes the point that the major trends in information technology in recent times have consisted of automating procedures, developing more appropriate IT-based processes and replacing analogue with digital information. Further, he notes that companies are growing weary of the *'effort, complexities and time required to implement ERP systems'* and to train staff in the quite complex use of them, and so are now looking for a *'value-centred IT return on investment'* (Edwards, 2002:34). EISs fulfil this latter requirement exactly, providing a cost-effective way to deliver enterprise information access in a graphically rich, application-independent interface (generally known as an executive or managerial dashboard) and to provide a user-centric single point of integration through the enterprise (Kirtland, 2003). Table 7.4 shows the key performance criteria you should expect from an EIS and the associated benefits.

Perhaps one of the more important forms of an executive information system is one dealing with the company finances (Reason, 2001). Chief financial officers (CFOs) need a system that gives them completely transparent access to the full array of supply chain, tax and accounting applications found in their companies to accurately keep track of the dollars coming in and going out of the business, as well as to give them capability of disseminating this information among employees as required.

A good financial EIS helps drive corporate strategy. It focuses managers' attention on areas that have the most impact on results and it unifies an organization behind its business goals by clearly linking goals to measures and tracking these measures on a real-time basis. It also provides the basis for accountability throughout an organization (Colkin, 2002).

Table 7.4. Key performance criteria and expected benefits associated with an EIS

Key Performance Criteria	Expected Benefits
Drill-down capability into various different data repositories in the organization	Fast and targeted action due to focused information delivery
Online and as close to real time data and information availability as ERP allows	One source of truth for all
A good analytic engine	Improved operational productivity, cost-effectiveness and decision making
User-friendly customization Mobile commerce management	Minimization of channel communication breakdown in strategy to execution
Project and innovation tracking and auditing	Online initiation of idea or project and standardized templates for idea submission and justification, review and approval of project/ideas
	Online tracking of project performance to timeline and budget

Electronically Enabled HR (E-HR)

HR or human resources is the term now used as a moniker for the old 'personnel management department' of most companies. E-HR refers to electronic human resource management and addresses the broad access to human resources data, tools and transactions available directly on the internet/intranet in most workplaces today (Lengnick-Hall and Moritz, 2003). It describes the 'internet effect' of the explosion in web technologies on HR and the dramatic impact this growth has had on the way employees now receive employment-related information through integrated self-service applications (Sullivan, 2003). It also includes the variety of new technologies available that help connect multiple systems, tools and databases, both inside and outside organizations.

What is the need for electronically enabled HR? Well, there are a number of reasons, primarily based around what HR is in the 21st century. Sullivan (1998) discusses the HR section being essentially a *solver of business problems* directly impacting the business, its products and its profitability. Its primary role will be to act as a *productivity consultant* that helps managers recruit and retain the best workers and to develop and motivate all employees so they are the most productive they can be (per dollar spent return on investment). HR will need to be able to monitor the environment and anticipate business opportunities and problems, and because HR is metric and reward driven it will need to be able to measure and reward the business impact of everything it does (Miller, 2003). Finally, a good HR group will shift ownership of people issues to managers and employees and will both coach and influence managers to make hard people decisions. Watson Wyatt (2005) in their renowned series of studies on human capital management show that superior human resource management is both linked to improved financial returns *and* a leading indicator of shareholder value.

What this means is that there is a growing requirement that HR management personnel become active participants in business strategy development and that HR itself becomes part of the company's strategic focus. This in turn means that HR managers must reduce the amount of time spent on administering the more mundane tasks of day-to-day HR such as:

- Payroll administration
- Salary packaging and benefits administration
- Expense reimbursements
- Leave management and approval
- Forms administration
- Occupational health and safety
- Recruitment
- Rostering
- Training.

Technological change is a key driver for HR transformation, in particular, web and internet technologies have already given workers direct access to each other, to HR, and to business information with such ease and intelligence that every worker can contribute more directly to business results. Further, the general comfort level using the internet has increased so steadily that it has become the preferred media in the workplace – making the introduction of E-HR a relatively 'painless' and cost efficient welcome event in almost any organization.

For the last decade, HR departments in large organizations have invested heavily in systems from interactive voice response systems to implementing enterprise-wide HR/payroll systems. Until recently, costs were high and progress was slow. Internet technology, however, has dramatically accelerated the transition to a more strategically orientated HR function with new resources becoming available to enable today's employees and managers in any sized company to complete HR related transactions on their own. In other words, electronically enabled HR has provided the foundation for a flexible workplace where employees have easy access to communication tools.

The use of internet and web technology has also freed management to re-assume its abdicated role of day-to-day people management without the endless wait for HR to eventually process requests for information, and/or provide vital reporting information needed to make swift people management decisions.

The Transition

For HR to complete the transition to E-HR, it must become the gateway (portal) to information that employees and managers need. New, evolving standards such as XML (Appendix 3) are increasing the ease with which HR can integrate various systems and databases into an electronically enabled HR environment. These technologies also help present information in a user-friendly manner that is available to employees and managers at all times and from all locations.

As discussed in earlier chapters, the internet reduces the tyranny of distance and time and provides the ability to transfer information across networks, both public and private. Internally it also provides the opportunity to transfer many common administrative tasks to employees by facilitating information flows, electronically enabling employee data changes and other transactions that have monopolized the HR department's time. Employees and managers are embracing this 'self-service' direction and the functionality of the web is appealing for many reasons beyond cost. Two of the most common are listed below (Sacht, 2003):

- A more mobile workforce needs access to HR information at various locations and times (e.g. electricity line construction crews working in a rurally isolated area)
- Changes in work style resulting from the proliferation of more functional wireless devices (phones, pagers, handsets, laptops and palmtops).

Implementation

Nearly all E-HR systems are based around the internet and an intranet and thus, to implement an E-HR system that is strategically linked to the company's business objectives, it must be tied in with the **internet** and **intranet strategies** of the company, as well as legacy system upgrades and integration since these older systems probably contain a lot of HR related information.

1. *Internet strategy.* Because the internet has, and will continue to have, such a major impact on companies and industries, it is essential that the internet planning process is not divorced from the corporate strategic management process, but is integrated into each stage of the company's processes. According to Solvik (2001), some basic points about setting up an internet strategy are:

    - Strategic use of the internet within a business will not happen unless the senior executives, particularly the CEO, get actively involved and drive its development.
    - The internet strategy must have measurable and reportable results. To do this it is important to find a benchmark company outside the business which can be used as a 'best in class' for each desired function. Once found, a comparison of the strategy being developed by the organization against that of the benchmark company can be made to determine if the business is on track to create a successful initiative.
    - It is also important to create an *internet capabilities scale* for the indusry sector of interest (Fig. 7.11). This can then be used to compare the business against other companies in the sector.
    - The company's proposed internet initiatives should be mapped to an *internet value map* (Fig. 7.12) so as to understand the effort required and the predicted returns.
      Most IT projects fail because of cultural and organizational gaps either between internal units or across external links. The internet era is forcing change in the way companies do business – in particular, in the way partnerships and collaborations (both within and external to the business) are being developed. The internet strategy development must be jointly developed by the IT and various business groups in the company (including HR), and integration issues must be planned for – both internally with legacy systems (see next section) and externally if business partners use different technology.
      Iterative reviews of the internet strategy need to be undertaken and change planned for if something does not work as well as it might, or if one area has shown more potential than was originally thought possible.

2. *Legacy system integration.* Legacy systems are part of a business's information system architecture that are old and technically obsolete but still exist because they perform essential data processing such as accounting and customer billing – and contain a lot of HR related information. They are difficult to change or to integrate with more modern systems because they are very often hard wired by rigid process flows making integration with CRM and internet-based business applications such as EIS almost impossible. According to Zoufaly (2002), maintaining and upgrading legacy systems is one of the most difficult

and challenging tasks that face businesses today. Despite constant technological change in the last ten years, up to 80% of IT systems in businesses are still running on legacy platforms!

Essentially the problem is cost. Driving the need for change is the cost of maintaining such systems and the constantly developing new business information systems being employed (for example, it is becoming almost impossible to hire developers who are familiar with applications written in languages that are no longer taught such as Cobol or Fortran), versus the business value of the legacy systems. Unfortunately, most legacy systems have had a huge investment made in them over time and CEOs are rarely inclined to approve major expenditure if the only result is lower maintenance costs rather than additional business functionality. There are other reasons why legacy systems are still existing and why they need to be upgraded – and Zoufaly's work details these – but today the most important driver of change in relation to legacy systems is when a company is wanting to implement a new business model based around the internet such as an internet-based procurement or E-HR.

The issues in this type of scenario focus around analysing and assessing the legacy system particularly the core business logic behind the legacy system, its source code, and the databases involved. Database migration can be the most complex and time consuming component despite most of the larger vendors (e.g. Microsoft, SUN, Oracle and IBM) providing tools to help automate the process. In essence, if all source database components are not mapped entirely accurately to the new target database then the migration process can be a total failure. However, despite the challenges, legacy migration is crucial if the business is to operate in the rapidly evolving electronically enabled world.

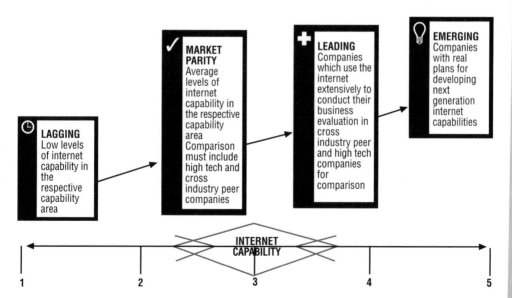

Fig. 7.11. An example of an internet capabilities scale – a score of 3-4 is optimum indicating the company is matching its peers in being ready and using the internet to maximize its business

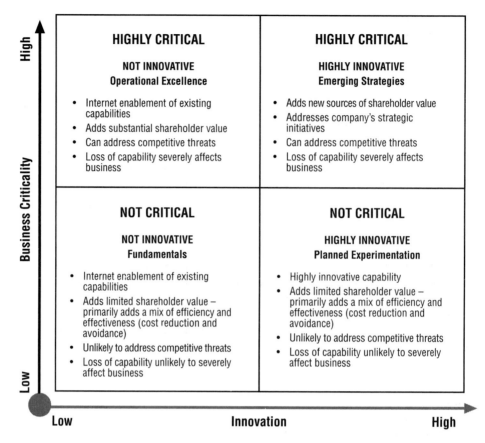

Fig. 7.12. An example of an internet value map – the vertical axis is Business Criticality (low to high) and the horizontal axis is Innovation level (of practice, strategy, management, etc.) (low to high)

3. *Intranet strategy.* An intranet is an internet-based technology within the organization which is an integrating mechanism for people, processes and information within it. The company's intranet is generally one of the HR section's 'babies'. An intranet is essentially an internal website that functions as a corporate network and uses the same protocols as the internet (e.g. http (hypertext protocol) and HTML (hypertext markup language) protocols). An intranet can also enhance the user's ability to find, manage, create and distribute information such as:

   - Manuals and procedures
   - Policies, production schedules and forms
   - Employee information
   - Schedules, calendars and timelines
   - Company databases
   - Product information and availability
   - Marketing data and sales goals

- Personal web pages
- Company telephone directories
- Company newsletters and news updates customized to user profile
- Hotlinks to other systems (e.g. company credit union site, expense system)
- Online ideas box.

For both internet and intranet strategy development the planning process should include investigating the development of new business strategies which incorporate electronically enabled business, and the associated need for the correct technology platform, integration with interdepartmental business systems, benchmarking against other sites, maintenance and ongoing management and development of site relationship management with customers, suppliers and allies and the legal/security implications of trading online (see Chapter 9).

E-HR Best Practice

Best practice organizations try to create corporate home pages, or portals, that serve as the single destination for employee and manager access to communications inside and outside the organization, including websites with job relevant information, data and tools, shared data and knowledge throughout the organization, work tools and applications from basic word processing and email to total retirement planning. However, to establish even the above, it is necessary to:

- Establish solid executive support for an E-HR strategy. There is an initial investment of money and resources required to be successful so the return on investment needs to be clear.
- Be aware of what the core HR system does currently and how far down the track to full electronic enablement it is. A technology audit will be of value here.
- Be aware of the company's eReadiness status – particularly in relation to its internet and remote computing strategy.
- Develop a formal strategy for your HR web that improves and integrates existing content and provides a framework for seamlessly providing access to users of the systems.
- Get the most effective specialized support possible (from the IT function and vendors) to deal with complicated issues such as legal, communications, data integrity and data security.
- Finally, as with other systems, Think Big (have a vision), Start Small (gain rapid returns) and Scale Fast (grow when needed).

Functionally, developing an E-HR system is a complex task, dealing as it must, with people. The following is a list of the types of processes that should be implemented:

1. Intranet-based workflow processes to automate paper-based processes – this is a priority if an organization is to enable online approval and authorizations
2. A standardized performance management process that can be monitored by HR for compliance (i.e. are current performance agreements in place by the required date)
   i. Provides easy review by one-up line management
   ii. Contains required fields that must be completed before closure
   iii. Has online sign-off by employee, manager and whoever else may be required
   iv. Is easily updated
   v. Allows last year's agreement to be copied into this year and edited to suit

3. An online leave management process which
   i. Enables requests and approval of leave online
   ii. Makes leave entitlements available online
4. Online training management
   i. Allows requests, approvals, bookings and confirmation of training online
5. Online recruitment
   i. Request/justification and approval for additional staff online
   ii. Online initiation of recruitment process
   iii. Online initiation and approval of termination and redundancy processes
6. Payroll automation
   i. Must capture hours worked for wages, employees via electronic timesheets
   ii. Automatic calculation of overtime and leave payments via defined rules
7. Expenses and petty cash
   i. Speedy online submission and approval of expense statements
   ii. Based on corporate credit card where the user is personally responsible for late fees and the appropriate use of the card
   iii. Facilitates automated audit for 'unusual activity'
   iv. Demands compliance with time requirements (otherwise user pays the late fee)
   v. Does away with petty cash as cash is withdrawn from ATMs using the card and accounted for as part of routine online expense statement
   vi. HR kiosks (terminals) on shop floor for blue collar worker access
8. Facilitation of collaborative project groups
   i. Links to specified shared databases and/or information repositories for groups with a common interest such as project teams or senior management
9. Rostering (workforce management)
   i. A roster is a schedule of personnel activities over a defined period and has population rules (who is the best person to fill a shift), costing or payment rules based on hours worked, payroll costs, etc. and self rostering or request rules
10. Integration with other databases is necessary – particularly in relation to costs, shift lengths, task specifications, etc.

In essence all of the above implementations are based around the premise that electronically enabling the HR side of a business will increase efficiencies (and thereby reduce costs) by avoiding unnecessary handling of the same information multiple times. Further, that it will enhance the quality of service provided (e.g. by allowing easy benchmarking, performance management and timely follow-up), as well as providing better communications within the business and a greater reach in terms of delivering business intelligence and information required for workers to more effectively do their jobs. An Australian company that has certainly taken advantage of this trend, developing HR software for many of Australia's leading organizations, is Aurion which is featured in a Byte Idea in this chapter.

Byte Idea – Aurion
www.aurion.com.au

E-HR

Aurion builds HR software with the goal of delivering innovative, value for money solutions that can be successfully implemented within customer organizations. The Australian company was formed in 1991 and the founding development team are still working together providing continuity for the products developed. The Queensland State Government is the major shareholder with Business Management Limited (BML) and the founders holding the remaining equity.

Aurion provides a comprehensive corporate system for personnel administration, leave, payroll, superannuation, salary budgeting and training administration. It incorporates a self service system for workers where staff can access their own personnel data over the internet using the Employee Self Service (ESS) module of Aurion, as well as intranet-based workflow and reporting mechanisms. Built around a modular format organizations can implement HR and payroll systems when they are developmentally ready to do so.

Aurion software streamlines human resource management by capturing data at the point of entry allowing the processes of the organization to then be applied to it. Automated processing provides timely reporting and analysis capability. Real time integration with other business applications including leading financial packages is another key feature of this software, enabling managers to drill down to view the source transactions. The following modules make up the Aurion software suite of programs:

- Self Service
- Workflow
- Query Tool
- Salary Packages and Benefits
- Payroll, Payroll Accounting
- Recruitment, Workforce Budgeting
- Reporting
- Leave
- Organization and People, Learning and Development, Career Manager
- Occupational Health and Safety.

Aurion has a number of business partners that provide aspects of their solutions – these include CITEC (www.citec.com.au) to deliver ASP and payment processing; SPHERION (www.spherion.com.au) and United KFPW HR Services (www.unitedkfpwhrs.com.au) to deliver outsourced HR/Payroll and Compuware (www.compuware.com.au) who are the developers of Uniface, the development tool used to build the Aurion software.

Information sourced from Aurion website

Occupational Health and Safety Management (OH&S)

Occupational health and safety (OH&S) is a major issue in any workplace and generally is the responsibility of the HR section. The reality is that no workplace is absolutely free of injury or illness and these are a drain on the organization and impact on the life of employees. It is thus most important to minimize the risk of accidents, emergencies, disease, injuries and chemical and manual handling hazards while at the same time maximizing safety management in organizations. While this may sound like common sense, particularly in terms of maximizing productivity, it has not always been at the forefront of the business community's agenda. Today, in Australia, as in most countries, OH&S is seen as an essential component of modern business with health and safety in the workplace being governed by a framework of Acts, Regulations and support material including codes of practice and standards (Fig. 7.13).

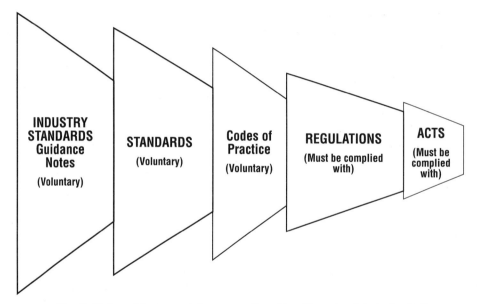

Fig. 7.13. Legal framework for occupational health and safety in Australia

According to the National Occupational Health and Safety Commission of Australia, it is only obligatory for an organization to comply with Health and Safety Acts (which set out legal rules that govern workplaces to ensure persons in workplaces do not suffer injury or illness) and Government Regulations supporting an Act (which outline how the general obligations of an Act will be applied in a workplace and are usually made in relation to a particular type of health and safety issue, such as asbestos, first aid, or a dangerous chemical) (NOHSC, 2003). Putting in place codes of practice and attaining and maintaining industry standards are not mandatory, but businesses and organizations wishing to be considered 'best practice' will adopt as many as is practicable.

OH&S Policies and Planning

The first step towards a safe workplace for a business is to develop an occupational health and safety (OH&S) policy and plan. There are a number of points that need to be borne in mind when developing an OH&S plan, and when electronically enabling the business. These include:

- *Ensuring management commitment to developing a company OH&S policy and plan.* CEOs also need to know that OH&S law aims to ensure that there are methods in place for identifying and transferring the knowledge and skills employees need to perform their job safely including an employee induction checklist.
- *Ensuring that visitors, contractors and others on site are covered.* A company's duty of care obligations under OH&S law extend to anyone entering the workplace including contractors, their employees and visitors. Information on safety procedures and processes needs to be readily available.
- *Having a safe workplace and buildings.* The main issues that need to be dealt with here are safe construction of the buildings, safe access and exit (particularly in response to an emergency evacuation), good housekeeping in terms of clearing up debris and mess from a day's work, and security.
- *Creating safe work procedures.* These need to be standardized and procedures outlined, developed and implemented.
- *General safety rules.* These need to be developed according to industry guidelines and should cover the company's code of conduct and safety rules.
- *Hazard reporting.* The main issues here are associated with spotting a hazard and reporting it correctly in a timely manner and the follow up procedures associated with the particular hazard.
- *Keeping good records and administration.* This is straightforward and involves procedures and processes associated with injury reporting, first aid and generally keeping appropriate records.

'E' Occupational Health and Safety

Where does electronic enablement fit in with OH&S? What are the benefits? As indicated in the last section, OH&S law aims to ensure that there are methods in place in organizations for identifying and transferring the knowledge and skills employees need to perform their job safely. This means providing information, education and training in the most appropriate manner to create a proactive workforce dedicated to minimizing the risks of accidents and maximizing safety in the workplace. Having this information available online is the best way to increase the richness and extend the reach of corporate OH&S information.

Since OH&S normally comes under the HR domain it is normally a component or module of most sophisticated E-HR systems running on corporate intranets. An electronic OH&S system needs to provide dedicated online easy access to:

- Full background information on key OH&S issues, policies and current plans and online access to procedures and templates
- Fully adaptable manual handling, hazardous waste, risk assessments and generic procedures
- Checklists compiled for specific requirements and best practice

- OH&S auditing at any level
- Simple multi-site operation (both domestically and internationally)
- Accident reporting, statistics and automated investigation
- Self auditing, designed to be used by all employees (e.g. online creation of OH&S improvement plans where progress can be monitored by line management and OH&S functionaries)
- Strategy planning for large and small companies alike
- Scheduling to ease management tasks
- A training matrix to help manage health and safety (and other) training
- Powerful reporting tools
- Access to corporation wide statistics and incident reports to facilitate shared learning and benchmarking of performance.

Having this type of system in place also provides better and more timely data to improve accountability and as a defence against prosecutions of the company by workers or visitors suffering injuries or illness due to workplace issues and reduced workcover claims (no win no fee). From a financial perspective good quality data and record keeping enable the company to control insurance costs, reduce sickness payments, reduce legal costs as outlined and reduce damage in the media should an incident be reported wrongly.

'E' Asset and Facilities Management

There are a number of definitions of asset management (AM) but according to Garvin (2001) a number of common themes in these definitions that can be extracted to produce a meaningful operational definition – which is:

> **Asset management is a methodology for programming infrastructure capital investments and adjusting infrastructure service provision to fulfill established performance and service objectives. In addition, the methodology takes a systemic and life cycle perspective of infrastructure, and its application is systematic. Finally, the methodology's decision process is scenario-driven and founded upon principles of engineering economy and management science.**

A shorter but no less encompassing description (Whitaker, 2001) is that asset and facilities management covers a range of responsibilities, which include:

1. Managing the provision of asset services to enable an organization to achieve its core business mission and associated goals
2. Maintaining the condition of owned assets to the required level to ensure provision of required services throughout their useful life
3. Acquiring, rehabilitating and disposing of assets to ensure cost effective provision of required asset services
4. Managing contracts and service level agreements where required asset services are provided through outsourced solutions and arrangements.

Essentially then, asset and facilities management involves making the best decisions on what is needed to be done with the assets throughout their life cycle (Fig. 7.14).

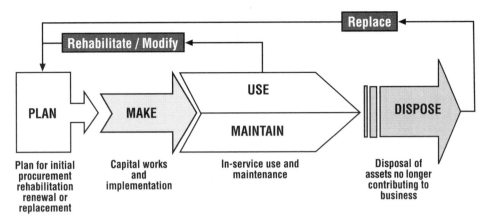

Fig. 7.14. The main phases of managing facilities and assets throughout their effective lives

Responsibility for managing an asset or facility usually begins when its construction is complete and the facility or asset is handed over for occupancy and use. Despite advances in the use of computer-based asset management systems, there is still considerable manual transfer of asset data and information from the construction phase to the commissioning into service phase, and associated inefficiencies and overheads.

There is no doubt that this juncture is a critical one, and it is imperative that full information disclosure is accomplished which may require a cultural and communications paradigm shift by both vendors and customers. Essentially, in an electronically enabled business environment where information is regarded as the gateway for a deal or development, there is a particular need for greater cross-functional collaboration in the decision making associated with creating, re-creating and maintaining facilities and assets. This is a departure from the old way in which such decisions were made and all stakeholders should be aware of the impact that such a change will have on the organizational culture and 'the way things are done around here' – issues which are discussed in more detail in Chapter 8. Operationally the main managerial issues associated with assets management in an electronically enabled business environment are about:

- Creating a balanced integration of technologies and physical elements of the new business environment
- Creating flexibility in the way end users can use the technology and the physical space it occupies
- Creating and maintaining relationships and communication between business infrastructure users and the asset managers.

Asset Management (AM) Systems

Traditionally, different aspects of AM (e.g. financial management, maintenance management and scheduling, strategic asset planning and management, drawing generation and management, and spatial management – an example of this would be electricity or telephone pole mapping and monitoring) have tended to be carried out in separate systems and applications. However, there is a growing need for more integrated functionality to support integrated facilities and asset management.

The large ERP systems (SAP, SAS, etc.) have attempted to provide total solutions covering all aspects of AM, but as discussed in the EIS section of this chapter, most companies have now given up on the idea of a total solution. This is primarily because it is recognized that such 'total solutions' are based around a specialist base system such as financial or maintenance management and, while having good functionality in the primary area, are generally not well balanced in other areas.

The emerging integrated asset management requirement needs a system that is fully integrated, truly modular in terms of implementation, and flexible and cost effective in implementation to meet the changing requirements of a more dynamic electronically enabled business environment. The key issue for these systems will be their flexibility in being able to support the changing nature of assets as impacted by electronic technology applications, and hence asset management approaches and the supporting information systems tools needed.

AM Systems Functionality

Most of the asset management systems available on the market originate from a core function, and have expanded their functionality with the addition of modules. They generally fall into two categories:

1. Systems developed from an asset maintenance management or works scheduling functional perspective. These systems generally have well developed asset maintenance management functionality and are able to support the management of more complex facilities and asset systems.
2. Systems developed from an accounting and financial management functional perspective. These systems generally are very strong in financial functionality but are usually weaker in maintenance management functionality. In general such systems do not handle as well the management of more complex facility and asset systems from the maintenance management perspective.

Typical modules that add functionality to a base system include:

*   Asset management – location, valuation, depreciation
*   Maintenance management
*   Work orders
*   Internet enabled job requests
*   Property management
*   Stock/Material management
*   Financial management
*   Contract management
*   Security access and management
*   Barcoding (see Chapter 6)
*   Reporting.

Strategic and Life Cycle Asset Management

The future business environment is without doubt one where facilities and assets and supporting equipment will be more complex and used in more complex ways, as well as changing more frequently. Hence, supporting facilities and asset management systems will need to be able to cope with increasing complexity, and frequent and short-term changes. There is also a growing need for systems to be able to interface with real time reconfiguration and automatic self-test and diagnosis information. Some of the common AM systems include:

- **ManageStar** which is a comprehensive and flexible software to manage assets, facilities, real estate, infrastructure, people and services. The product is based on an integrated web application (www.managestar.com).
- **ChubbCare** which is advertised as a system for both low and high tech situations (www. chubb.com.au/cams_asset.asp).
- **GeoDecisions** which is a spatially based asset management package (www.geodecisions. com).
- **PeopleSoft** Asset Management for Education and Government (www.peoplesoft.com.au/ en/us/products/applications/eg/fmeg/egasset.html).

The Challenges for Asset Managers in an 'E' World

There are many challenges to electronically enabling AM – Garvin (2001) provides a breakdown of these in terms of strategic, integration, measurement, analytic and institutional challenges (Table 7.5) while Whitaker (2001) discusses the major issues from an AM life cycle perspective (Table 7.6).

Essentially the challenges are all focused around the fact that constant innovation and development in the technology arena has created a reduction in the life cycle time of IT and associated assets, along with the follow-on effect this has on the assets cost structure, both during building and throughout the in-use life cycle. Extensive electronic enablement means that there is likely to be a significant impact on overall cost structures and budget profiles and allocations between capital investment and maintenance and operations costs, and managers must be aware of this, plan for it and be capable of dealing with it efficiently and cost effectively.

Table 7.5. The challenges for asset managers in an 'E' world (adapted from Garvin, 2001)

Challenge	Issue 1	Issue 2
Strategic	Procurement processes – current procurement practices are restrictive. Alternative arrangements for facility delivery, facility operation and maintenance, and system operation and maintenance are available and credible.	Defining system objectives – complex as it necessitates a paradigm shift in communication by requiring input from multiple sources that will certainly introduce competing objectives.
Integration	Integrating functional and condition-driven requirements – Current programming practices and decision support systems generally consider functional and condition driven requirements independently.	Integrating asset classes and sectors – generally, asset classes (such as computers, computer peripherals) and asset sectors (such as buildings) are separately managed and maintained. Existing databases and computer models are generally structured to support asset classes not asset sectors and cannot 'communicate' effectively with one another.
Measurement	Identifying appropriate indicators. There has been little progress made towards identifying indicators of deterioration, obsolescence, capacity and function.	Establishing comparative baselines. Benchmarks for evaluating alternatives during planning and monitoring performance during operation are not widely developed or deployed.
Analysis	Improving data collection and management. Data collection, particularly in support of condition assessment, is time intensive and periodic. In addition, protocols for and linkages to data users at different organizational levels need development.	Constructing and evaluating alternative life cycle scenarios. The emphasis upon life cycle principles is recent, so models, methods and tools to construct and analyse life cycle tradeoffs are still developing. Incorporating flexibility. Events over a life cycle are uncertain, so programming analyses must incorporate flexibility to preclude ill-advised technological or capital commitments.
Institutional	Affecting coordination between or reconfiguration of business 'silos' – asset management is holistic, so it depends upon comprehensive coordination and communication. Many enterprises are functionally segregated into divisions such as 'Engineering' and 'Construction' and this needs to change.	Defining roles of public and private sectors – as alternatives for delivery of infrastructure projects and services continue to emerge, the roles played by the public and private sectors will change and a rethink of public planning and procurement processes will need to happen.

Table 7.6. Asset management life cycle challenges (adapted from Whitaker, 2001). Life cycle stages are as represented in Figure 7.14

Life Cycle Stage/ Challenge	Plan / Make	Use	Maintain	Recycle/ Dispose
Capital Procurement	A key pressure on the procurement of replacement and new assets is the shortening life cycles of some of the technologies that are being introduced. This has implications for the building assets that provide the framework for carrying the infrastructure, where technology is regularly removed and installed.		Parts	Recycling possibilities
Finance	There is continuing pressure to be innovative in finance sourcing in budget constrained environments.	Yes	Yes	Reselling opportunities
Inventory Management	Electronic technologies as a business infrastructure mean a more complex and demanding environment for inventory management. The key factors include a greater number of items that need to be managed that are increasingly geographically spread across a greater number of locations. Greater sophistication and automation of inventory management will be required by asset managers to maintain effectiveness and efficiency levels.	Drives a need for more sophisticated functionality from asset management systems, with coordinate-based databases (e.g. GIS systems) and technology enabled automation such as barcoding.	Yes	

Life Cycle Stage/ Challenge	Plan / Make	Use	Maintain	Recycle/ Dispose
Fitness for Purpose / Functionality and Standard		Indicates the suitability of the facility or asset for its intended end use. 'E' techs will require managers to review, adapt and change their traditional view of 'fitness for purpose' of facilities and assets, and establish new performance indicators for both functionality and standard.	Traditional relatively low-tech buildings are likely to be inadequate for new 'E' enabled high-tech functions and will require greater maintenance than updated facilities.	
Maintenance Management		Yes	New facilities and designs decrease maintenance costs but the new environment drives the need for a higher skill set often resulting in the need to outsource.	

Strategic Alignment of 'E' Enablement with the Business Strategy

In order to support a balanced 'E' strategy, some quite traditional drivers are also required to bond or align the organizational strategy and the IT strategy together (Fig. 7.15). The most important among them are:

- Active support by the organization's content owners (i.e. groups and individuals that have a direct stake in the positional eStrategy mix – leaders in the corporation's technology, marketing, service and branding groups)
- The ability to climb the learning curve quickly. The companies that make the best use of electronically enabled business are identifiable by the speed at which they develop online projects and the wealth of future online options that they can then consider
- Belief that R&D for online activities is a strategic investment
- Adoption of a sourcing option that reflects the mission-critical nature of the internet. This includes partnering (i.e. working with a set of specialist providers) and traditional outsourcing to more effectively use internal resources while maximizing the efficiency of vendors.

The planning associated with electronic enablement development in an organization is a continuous, iterative and sustained process and is meant to deliver business benefits to an organization. It involves understanding what the business goals are and identifying how electronic enablement can support those goals by delivering benefits. A manager involved in this type of planning must understand:

- What the nature of the organization is, its goals and objectives, where it is going, what its culture is, and how it thinks
- What is the business about? What is its strategy and purpose?
- What the environment is, what influences the organization, such as legislation, market, technology, and media
- What are people's roles within the organization? What are their objectives and motives? How do they influence business processes?
- What information systems does the organization already have? Are they effective? Are they meeting business needs?
- What technological opportunities exist?
- How could new IT support the business?

This involves the *alignment* of electronic enablement planning with the business planning process and involves developing action plans to accomplish the requirements as shown in Figure 7.15.

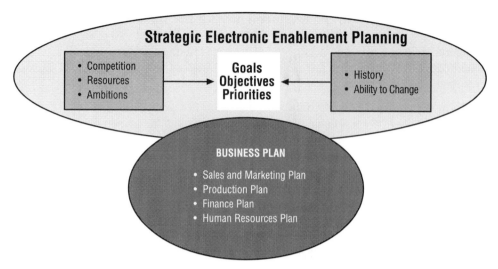

Fig. 7.15. The electronic enablement plan as part of the business plan

Summary

This chapter has addressed what some people consider the most fundamentally important issue in creating and running a business – developing the business strategy. The business strategy is the 'blueprint' of how a business will run. Without such a plan, the business can rapidly dive from the height of a good idea with huge potential value, to the depths of one with no value and potentially many debts. This has been shown to be particularly true in the dot.com arena where many companies started up on an idea only, and, with little thought of anything other than making money, got into difficulty when they lacked a strategy to counteract the increasing business pressures in the highly uncertain economic environment of the early 2000s. Thus this chapter has dwelt at some length on discussing 'E' strategies under uncertainty and the eReadiness of the business and surrounding environment as an area of importance for managers. Strategic alignment of the electronic enablement strategy with the business strategy was also discussed.

Meg Whitman, CEO of eBay as discussed earlier is the Smart Thinker for this chapter.

Smart Thinker – *Meg Whitman*

President and CEO e-Bay

Meg Whitman runs e-Bay, the international eCommerce giant which is the world's largest online marketplace for the sale of goods and services by a diverse community of individuals and businesses. Today, the eBay community includes 157 million registered users, and is the most popular shopping site on the internet when measured by total user minutes, according to Media Metrix.

With degrees from Princeton and Harvard, she had senior positions at a number of companies before joining eBay in 1998. Her first task was to make eBay more corporate, focusing on profitability, returns-based investment, and on growing revenue faster than costs. And that focus was maintained even when it wasn't fashionable, because it was felt that it was crucial to have predictable earnings and revenue growth. Most importantly she felt very strongly that sound business practices including sound risk management had to be embedded into eBay's culture from the beginning.

Since then Whitman has guided eBay to be the internet's number-one seller of consumer electronics, ahead of Amazon.com and Buy.com. It is also number two in books, movies, and music. And as a result, she says that eBay is far and away the largest e-marketplace because sellers want to be where the buyers are, and buyers want to be where the sellers are and if you are the largest player around, that is where all the action will be.

How does she do it? Her management style is about creating a collaborative network in partnership with the community of users. The key is connecting employees and customers in two-way communication. She calls it 'The Power of All of Us'.

Meg Whitman stands as the most successful internet executive of all and among a plethora of awards, Fortune Magazine ranked her the most powerful woman in business in 2004 and 2005.

Source: USNews.com (www.usnews.com), Time Magazine (www.time.com) and the eBay website

The Review and Critical Thinking Questions below have been provided for you to revise and reflect on the content of this chapter.

Review Questions

1. What is a business strategy and why is it important in the long term for a company?
2. What is economic uncertainty and what effect does it have on a business?
3. Is an electronically enabled business strategy any different to a bricks-and-mortar business strategy? Give reasons and examples.
4. Choose an agri-food chain of your choice and discuss with examples whether the businesses involved are eReady.
5. Compare and contrast economic webs and networks. Explain how each is facilitated and why they are of importance in today's business world.
6. What is E-HR? Discuss with reasons, two examples of HR processes other than OH&S that benefit from it.
7. Discuss the key points of developing both an internet and intranet strategy.

Critical Thinking Questions

1. Michael Porter emphasizes that a business strategy is about creating a unique place for a company. What does he mean and why is this important? Illustrate your answer with examples.
2. Discuss the difference between a 'shaper' and 'adapter' company and relate these back to an agribusiness of your choice, discussing the strategies and tactics that it needs to follow to ensure success, particularly in an uncertain business environment.
3. Discuss Meg Whitman's obviously successful business strategy for eBay and compare it to Golden Circle's.
4. What are the major internal processes of a company that would benefit (why and how) from electronic enablement?
5. Compare and contrast eBay and The Seam.

Chapter 8

The Practicalities of 'Making It Happen'

You quite often don't know what you don't know

This chapter is about the nitty-gritty of making an electronic enablement project work successfully to ensure a return on the investment made in it. The chapter looks at the major issues involved in determining the resources and management strategies required in a business (including the key competencies necessary) to develop and undertake such a project. The 'Think Big, Start Small, Scale Fast' mantra from Chapter 7 is revisited with some insights from industry as to what is required in such projects, and there is a significant section on *outsourcing* as a managerial tool in the electronic enablement approach.

As alluded to in earlier chapters, there are very real challenges to the business presented by existing *organizational culture* and difficulties of creating *organizational change* when introducing something new like electronic enablement of business processes. These issues are looked at in some depth – again with some industry insights and tips on how to put together a good business case for the task in order to get senior executives on side and supportive.

Chapter Objectives

After reading this chapter you should understand and be able to describe:

1. What electronically enabling a business or project entails
2. What a 'resource utilization analysis' is, and when and why it should be undertaken
3. The differences between a key competency and a capability and which are required for a successful electronic enablement of a business/project
4. The concept and practices of outsourcing
5. The main issues associated with managing the change necessary in an organizational culture to facilitate electronic enablement
6. How to deliver a business case for electronic enablement and the key features that need to be addressed.

Electronically Enabling a Company

To successfully electronically enable a company requires a thorough understanding of the current business environment including the market, customers, competitors and culture of the organization, the people involved, the current business and organizational models, the application models being used, and the management processes and technology involved (Hoque, 2000). It also involves anticipating the changes that must occur in every aspect of the organization when one of these variables changes (Fig. 8.1).

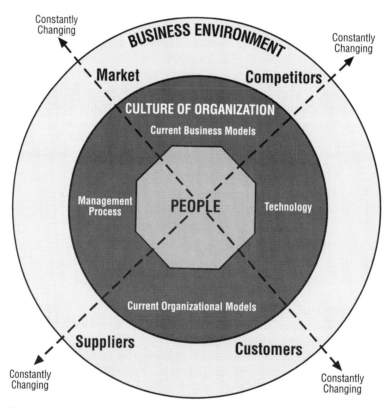

Fig. 8.1. To successfully electronically enable, constant monitoring of the internal and external worlds in which the business 'does business' is necessary

Since the heady days of the dot.com era when 'anything went', the current approach to addressing these issues is a more sober one focusing on integration and cooperation in order to minimize risks to the business. Table 8.1 compares the approaches taken in the dot.com era and the current electronically enabled business world.

Table 8.1. The change in approach to electronically enabling a company during the dot.com era and the current day (IPO: initial public offering – a company's first sale of stock to the public)

Issue	Dot.com	'E' Enabled
Culture *Intangible and essential in shaping the character and effectiveness of the enterprise*	• Chaotic • Dot.com mania • Rare business model • Reactive • Individualistic control • Hype	• Enterprise focus • Dot.com mania settled down • Business model focus • Proactive • Process creation and integration, cooperation • Branding
People *Intangible asset and store of knowledge capital*	• Technology and implementation focus • Specific 'hot' skills (e.g. webpage design, Java)	• Strategy and process design and development focus • Core competencies and knowledge-based • Business analysis / ROI
Business Model	• Startups • Content aggregation • Venture capital funded • Pure net play • IPO focused	• Mission-critical business • Process and people aggregation (process-orientated teams) • Business unit or spinoffs • Hybrid and inter-enterprise integration / consolidation
Organizational Model	• Functional focus • Isolated • Founder driven • Loose organizational models • In-sourcing	• Cross-organizational and extended enterprise focus • Collaborative with a blurring of corporation boundaries • CEO-driven • Tight organizational models • Outsourcing and partner-sourcing
Application Model	• B2C focused • Online marketplaces for hard goods	• B2B procurement and B2B community focused • Marketplace for soft goods • Mission critical inter-enterprise applications
Management Processes	• Situation driven • Project orientation • Disconnected	• Process, and model / ROI driven • Programme and enterprise solution orientated • Connected, integrated and extended
Technology	• Catalyst for business opportunity • Client driven • Unique	• Driver to automation • Server-network driven and distributed systems • Accepted as the norm

The 'Think Big, Start Small, Scale Fast' Mantra

The preparation and resources required for a smooth development or transfer to an electronically enabled situation must not be underestimated and involve planning, choosing the appropriate organizational and business models and implementing these in a timely and structured fashion. Some organizations have a culture of striving to achieve 'perfection', however that may be defined. Admirable as that aspiration might be, it has serious flaws in the context of electronically enabling a business. The cost of shooting for perfection, particularly in terms of design and implementation time, is significant. A company that both epitomizes the mantra and helps others to do so is MYOB – see Byte Idea this chapter.

In setting out on the 'E' journey you don't know what you don't know; therefore, whatever is developed will require modification once it is in use. So instead of squandering resources on trying for the 'right' solution, a better path is *shooting for the 80% solution* at 'go live' and have in place capacity to quickly change things as lessons are learnt and shortcomings identified.

In essence, as was emphasized in Chapter 7, the 'Think Big, Start Small, Scale Fast' mantra (Fig. 8.2) captures one of the most crucial concepts in electronically enabling a business – *have a vision, build a solid foundation and then grow when the time is right.*

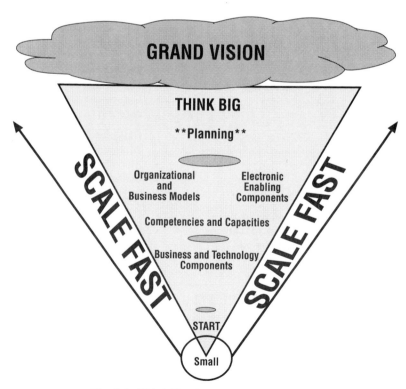

Fig. 8.2. 'Think Big, Start Small, Scale Fast'

In the context of electronically enabling a project or business, the 'Think Big, Start Small, Scale Fast' concept can be practically implemented by three actions:

1. *Don't be constrained by current technology.* The propensity for IT technologies to develop and expand in capability is well known. So it is important that in crafting the 'vision' for the future that the guiding principle should be to imagine what the technology should be able to do and not to be constrained by what can be done now. Chances are that by the time you get close to achieving your vision the technology will have caught up with you.

2. *Avoid the Big Bang.* It is very easy to 'think big' and then act big. History is littered with businesses that had grand and quite often appropriate aspirations about what their company could achieve then set about designing the 'perfect system' to deliver that vision in a very short timeframe. This approach inevitably involves large investments of capital up front with the promise of significant returns on the big play. However, what it overlooks is that at the start of a new venture like the journey to electronically enable the business, you quite often don't know what you don't know. It is often said that ignorance is bliss, but in this context it is a significant problem. As you, your staff and your value chain partners embark on the 'E' journey there are going to be unforeseen issues that will arise. These will range from detailed technical IT issues through to attitudes of individual stakeholders proving more intransigent than expected. So instead of making the big play and implementing the perfect solution up front, be humble, recognize that you and your experts don't know all there is to know and start out modestly, limit early downsides, be prepared to be flexible and if necessary completely change direction in order to move towards your vision.

3. *Learn by doing.* 'Learn by doing' (the 'know how' discussed in Chapter 3) sounds trite but again this is a very powerful concept. As you embark on a new venture like electronically enabling a business, theory is no substitute for real life experience. Pay a lot of attention to the feedback being given on the project and take every opportunity to refine the approach, craft the next step and if necessary reshape the vision as new insights help to understand what the future can look like.

Byte Idea – MYOB
www.myob.org.au

Mind Your Own Business

MYOB is a classic case of 'Think Big, Start Small, Scale Fast'. The company was set up in 1991 by CEO Craig Winkler, and was one of the first businesses to recognize the unique needs of small-to-medium-sized enterprises in the electronically enabled world. They set about developing and implementing powerful, accessible and affordable business management systems for small to medium enterprises (SMEs) and are most well known for accounting software. From the beginning, MYOB solutions have been aimed at minimizing an increasing administrative burden associated with the changing business environment and in giving business owners the necessary insights to run their businesses more successfully.

In 1996 MYOB acquired all intellectual property related to the MYOB name around the world, expanding operations to include North America and Europe. In the same year MYOB received the prestigious Telstra and Australian Government's Small Business of the Year Award.

In 1999 MYOB was successfully listed on the Australian Stock Exchange and by 2003 it had entered the top ten on the Australian Corporate Image Honours List (National Business Bulletin). In 2004 the company merged with accountants' system provider Solution 6 and CEO Craig Winkler became the National Winner of the 2004 Technology, Communications and E-Commerce Category of the Ernst and Young Entrepreneur of the Year Awards.

In 2005 MYOB is one of the most valuable listed information technology companies in Australia, delivering business solutions to more than 500,000 businesses and over 10,000 accounting firms worldwide.

Information sourced from MYOB website

Resource Utilization and Electronic Enablement

In Chapter 1 the notion of the Resource Based View (RBV) of a company was introduced. It has also been referred to at various stages in this book – mainly in relation to how you add value to the company and create a competitive advantage. However, there is a less lofty but no less important issue that is directly related to RBV theory when electronically enabling a business or project, and that is: knowing what resources are actually available for the task (and what needs to be brought in/sent out), how such resources may be used, where and when. This means undertaking a *resource utilization analysis (RUA)*.

To start at the beginning – what are we talking about in terms of 'resources'? According to the Miriam Webster Dictionary, a resource is a) a source of supply or support: an available means, or b) a natural source of wealth or revenue, or c) computable wealth, or finally d) a source of information or expertise. Thus we can define *'resource utilization'* in a business sense as:

> **the use of available means of supply or support in the organization to most efficiently address the requirements of carrying out the business of that organization to add value and create competitive advantage.**

For an electronic enablement project, a significant focus in an RUA is obviously going to be in the IT arena for, as companies develop and expand their global networks and enterprise infrastructures to incorporate the internet, intranets, and other forms of electronic enablement, IT resource utilization increases dramatically – as does the need for integrated and centralized resource management. IT departments are expected to maintain applications and services outside the traditional data centre with the same level of predictability that users have come to expect from services that were originally inside the data centre.

Other than in the very large organizations such as the utilities (water, telephone, electricity), and major multinationals, the management of network systems and storage resources is still relatively piecemeal in most organizations, with separate systems, which typically do not work well together, being used to manage each element. However, even in smaller organizations, this situation is beginning to change quite rapidly as the importance of having seamless information flows on efficiencies and costs is recognized – and as more and more businesses 'doing business' with others need to exchange information electronically. The typical questions therefore that need to be asked in relation to undertaking an RUA include:

- Who are the biggest consumers?
- Are costs allocated appropriately?
- What is the impact on resources from the organization's electronic enablement initiatives?
- What resources are really being used?
- How long will resources last?
- Is the project delivering an acceptable level of performance?
- How can all the resources be accounted for?

Given that there are risks associated with developing any new business model and in particular electronic enablement in all its forms, there are, in addition to the above questions, certain issues associated with resource allocation and utilization that should be audited prior to undertaking any new venture. These include the size of the project, the type of technology required for the project, the vendors that are most likely to be supplying the technology and services associated with the project, the resource availability throughout the proposed project life cycle, the stability (or otherwise) of the organization's management, market conditions and the culture of the organization and its customers.

- *Project size.* The bigger the electronic enablement project the more susceptible it is to risk. Bigger projects are more difficult to scope. Small projects are more tangible, and it's easier to predict how long they'll take to complete.
- *Technology.* Every time you put a new technology in place you run the risk that it won't work as planned – for example, will that new CRM package really do everything that it is expected to? This also happens when modifying an enterprise's legacy technology – indeed, software integration projects are notorious for this.
- *Vendors.* Vendors regularly modify the services they offer, by either shifting strategy or through acquisitions, mergers, or simply going out of business. All of these can have serious effects on availability and support for enterprises.
- *Resources.* Necessary resources aren't always available throughout a project's life cycle, which can result in expensive delays. This is particularly true of staffing resources.

- *Management.* Enterprise leadership isn't static and this can have a dramatic effect on costs. Management can change business direction or there can be changes in management itself. Leadership and change management are discussed later in the chapter.
- *Market conditions.* Changing market conditions are probably the most infamous risk. For example, since the events of 11 September 2001, the travel industry has plummeted while demand for security products and services has grown dramatically. In the face of such changes, IT projects are forced to dramatically scale up or down. Also, some markets are particularly influenced by changes in government regulations – new ones are introduced, or existing regulations are modified. The effects of these changes need to be considered.
- *Culture.* When new technology is introduced it can meet a chilly reception if the users don't want to change the way they are used to working. More than one eBusiness project has gone nowhere because of resistance to change. Even if enterprise users are receptive to change, there's always a risk that a new technology will be ignored or used improperly or incompletely. Training costs must be factored in but the risk that some staff won't effectively grasp the training they receive should also be considered.

Efficient resource utilization allows an organization to maximize resources, streamline core business processes, and maximize returns on skill investments by optimizing resource capacities. A good example of how this has been accomplished in a company via electronically enabling significant components of the company's business is in Elders, the large multifaceted Australian agribusiness – see Byte Idea this chapter. It also allows the organization to:

- Reduce time consuming manual tasks and identify training needs
- Forecast future costs and revenues
- Coordinate all project and resource information, reducing poor information flow between areas
- Identify resourcing problems, minimize redundancies of effort and prevent imbalances in work distribution
- Reduce project overheads and maximize chargeable revenue
- Increase global visibility of all programme and project data.

Byte Idea – Elders Limited
www.elders.com.au

Managing resources smartly in the electronically enabled rural sector

Elders Limited is one of the oldest, and one of the largest, agribusinesses in Australia. It has been existence since 1839 and has not only grown with the times but has also seen changing fortunes before becoming, in the 21st century, a company with national and international offices in locations from within all the States of Australia to Germany, Japan, China, Korea and South East Asia.

As a multifaceted enterprise Elders provides a wide range of agricultural services to the rural community of Australia and more recently internationally, including livestock marketing, merchandise, financing, insurance, real estate, wool brokerage and sales. With over 400 locations, 85,000 product offerings, 1,600 suppliers, 100,000 clients and a sales staff of 1,000, Elders has traditionally generated an enormous amount of paperwork and is a very transaction-based company.

So, it only seemed logical that as electronic enablement became the norm and businesses started looking for ways to extend their competitive advantage in the early 2000s, Elders took the plunge and decided to automate their supply chain. The project was outsourced in the main to IBM Services and Software who worked closely with Elders. Of major importance in the supply chain management system development was the fact that a large variety of platforms were used by suppliers and buyers and these needed to be taken into account. To address this need, XML was/is used wherever possible, to ensure that new suppliers and business requirements can be supported into the future. Other elements of the solution included Microsoft BizTalk, which served as the eCommerce gateway for supporting secure document transaction between Elders and their suppliers, and as a mapping tool to transform suppliers' business documents to Elders' business documents and vice versa.

Chris Ferguson, IT General Manager for Elders viewpoint is 'We have a large back office, with over 1.2 million merchandise invoices going through the system in one year...we are achieving a very rapid return on our investment with IBM...and it [the system] should pay for itself completely through increases in productivity in less than a year, with additional benefits for our customers and suppliers in improved lead times and increased accuracy.'

As it happens – during the past five years, Elders', turnover in sales of merchandise has grown from $A450 million per annum to $A1000 million. Rural business has grown from $A400 million to $A1,300 million and livestock business has grown from $A1,000 million to $A2,000 million. And to cap it off – with the addition of the electronically enabled SCM system – to the start of 2005 Elders had experienced a 70% productivity improvement in accounts payable, a 70% reduction in manual paperwork and a significantly decreased rate of error...you couldn't hope for a better situation really!

Information sourced from iStart (www.istart.com.nz) and the Elders website.

Planning an Electronic Enablement Project 1: Organizational Models

A necessary component of planning for electronic enablement is to address the current organizational model and to project what it is likely to become in an electronic environment. Organizational models have undergone a number of fundamental revisions over time. The following six trends are likely to impact heavily on the design of organizational models in the future:

1. *Vertical and/or horizontal integration.* Vertical integration is when a company expands its business into areas that are at different points of the same production path – for example, a car company expanding into tyre manufacturing. A company such as this is often referred to as being 'vertically integrated'. Horizontal integration is when a company expands its business into different products that are similar to current lines – for example, a cafe selling coffee and cakes, expanding into selling full meals. Currently there is a lot of activity in the corporate world in relation to defining organizational structures because of the impact that electronic enablement has on 'opening up' the business, and the impact that this has on managing the business processes that extend across key business groups such as the supply chain, customer relationship management, global alliances and IT (Mougayar, 2002a and b).

2. *Extension of enterprise relationships.* In electronically enabled business, the extended enterprise (Hoque, 2000) is a network of relationships encompassing customers, suppliers, partners and their own internal activities. Such a network covers all key business functions and thus there is a need to integrate the inward-facing business applications such as enterprise resource planning (ERP), supply chain management (SCM) and legacy systems, with applications that provide outward-facing connectivity to suppliers, partners and customers.

3. *Creation of new roles for executive management.* New roles for executives are developing around specialist areas directly related to doing business electronically – for example, Chief Knowledge Officer, E-Procurement Manager.

4. *Partner sourcing vs outsourcing vs insourcing.* Partner sourcing refers to the practice of using a partner's expertise or resources, or of buying from a partner the expertise or components for product and/or service production. *Outsourcing* which is discussed in a later section of this chapter in some detail is the practice of purchasing a significant percentage of intermediate components for the business's products from outside suppliers that are generally not partners in business. *Insourcing* is where the business contracts out a complete area of operations to an external company (e.g. the IT area), who runs it on your behalf, providing the labour, premises, vehicles, communications, tooling, management and supervision while maintaining a seamless integration with your business, thus enabling staffing and administration costs to be reduced.

5. *Employee incentives and accountability.* An incentive program is a planned activity designed to motivate people to achieve organizational objectives through improved performance. Incentives provide concrete rewards for quality performance that is consistent with short- and long-term organizational goals – for example, extra financial rewards, share packages in the company, travel and holiday rewards over and above the statutory salary and entitlements. Accountability involves assessing how well the employee and/ or the organization is integrating and responding to the needs and requirements of its stakeholders.

6. *Employee empowerment and participative management.* Empowerment is about fostering improvement and self-determination. The most important concept of empowerment is for senior managers to delegate responsibility to the lowest levels in the organization. The idea is that the decision-making processes within the organization should be to a high degree decentralized and individuals or work-designed teams should be responsible for a complete part of the work process. Participative management has become a key word in empowerment and it has been shown that there is a positive link between participation and satisfaction, motivation and performance (Hollander and Offerman, 1990; Keighley, 1993).

Planning an Electronic Enablement Project 2: Electronic Enabling Components

The driving factors behind the new economy are *flexibility* in terms of sensing and managing change in the business environment, *agility* in terms of having people with core skills that can be used to address a number of different problems that may develop with constantly changing business requirements, and finally a reduction in *time-to-market* which is the speed at which a company can respond to market changes by rolling out new, iterative versions of technology and application platforms (Scott, 2000). In such a business environment, **processes, people** and **technology** within and around the business must be assessed as part of the enablement planning.

- *Processes* – The velocity and immediacy of information in the 'E' world demands that business processes which are at the heart of a business must be robust, effective and well documented in order for a business to take full advantage of the opportunities that electronic enabling brings. In the 'E' world, human intervention and problem resolution cause unacceptable delays and may in fact generate more problems.
- *People* – Often the most challenging aspect of electronically enabling a business is aligning the people and managing them to effectively adopt the changes in business operation that are inevitable. For instance, a business not used to maintaining data quality is likely to experience resistance from staff as new standards are imposed. Part of the challenge here lies in the ease with which the human aspects of electronic enabling may be overlooked. The quantifiable nature of the business case or the technical appeal of the technology and process implementation can cause the less sophisticated manager to overlook the impact of the changes on the people who at some point are critical to the success of the business. It is important to maintain a focus on the fact that electronically enabling the business is often a significant change for those involved, particularly those with a key stake in the existing 'system'.
- *Technology* – 'Fit for purpose' as discussed in Chapter 7 in relation to asset management issues captures the essence of the technology selection task in electronically enabling projects. It is very easy to become enamoured with the promise of what a particular software platform holds for the future, but it is important to maintain perspective on where the business is in its electronic enabling journey and what is the right balance of functionality now versus capability for the future. Legacy IT systems are probably the biggest technological challenge in electronically enabling a business. It is important to appreciate from the outset that electronic enabling involves some non-negotiable conditions for success. Systems need to be configurable enough to interface with a variety of technology alternatives. Server and communications band width must have

the capacity to handle data at required speeds. The software that is used for any specific task must have the capacity to deliver the appropriate level of automation to achieve the specific objectives of the strategy.

Key Competencies and Capabilities in an 'E' World

The other aspect of an electronically enabled business is the 'bench strength' of the organization to implement the often significant information technology requirements. The availability of IT project managers, business analysts and business operatives – all of whom can bring together an understanding of how the new business processes are supposed to work along with an intimate knowledge of how the business operates at a very detailed level – is critical to the success of a project. In fact, with rapidly evolving applications software in the electronically enabled business space, a challenge that arises from time to time is the availability of specialists to implement the specific software to be implemented. There is little point in a business selecting a 'bleeding edge', 'latest and greatest' software package if there is no one available to implement it. In this situation delays and poor quality programming will erode the business value promised by the technology.

Part of a good RUA is to assess the competencies and capabilities of the organization, its people and those of its business partners as a key part of electronic enablement. *Competencies* are general descriptions of the abilities necessary to perform successfully in specified areas. They are, by definition, an abstraction. They are meant to represent the potential to perform and can never therefore be absolute. As such, competency profiles synthesize skills, knowledge, attributes and values and express performance requirements only in behavioural terms. *Capabilities* on the other hand are a quality or state of being able to do something.

Prahalad and Hamel first used the term *core competencies* in their 1990 article entitled 'The Core Competence of a Corporation' and they described them as the collective learning and coordination skills that lead to a firm's value creation and revenue accruing capabilities. They also believe that the intersection of market opportunities with the core competencies of a business forms the basis for launching new businesses.

When a business is looking to electronically enable itself or a process, what core competencies are going to be needed? In a project undertaken in 1991, the US Departments of Labor and Education investigated the kinds of competencies and skills that workers would have to have to succeed in the year 2000 with five main categories of competency being identified: **resource management, interpersonal skills, information handling, a systems approach** and **technology skills**. A summary of the findings is provided below.

1. *Resource management.* Includes identifying, organizing, planning, and allocating resources – for example:

    - *Time.* This involves being able to select goal-relevant activities, rank them, allocate time and prepare and follow schedules
    - *Money.* This involves being able to use or prepare budgets, make forecasts, keep records and make adjustments to meet objectives
    - *Material and facilities.* This involves being able to acquire, store, allocate and use materials or space efficiently
    - *Human resources.* This involves being able to assess skills and distribute work accordingly, evaluate performance and provide feedback.

2. *Interpersonal skills.* Is about being able to work with others – for example:

- As a team member which involves being able to participate and contribute to the group effort
- As a teacher which involves being able to teach others new skills
- With services clients/customers which involves being able to work to satisfy customers' expectations
- As a leader which revolves around being able to communicate ideas to justify a position, being able to persuade and convince others, and being able to responsibly challenge existing procedures and policies
- As a negotiator which involves being able to work towards agreements involving exchange of resources, and resolve divergent interests
- Being multiculturally secure which means working well with men and women from diverse backgrounds.

3. *Information handling.* The ability to understand the value and use of information. Skills include:

- Being able to acquire and evaluate information
- Being able to organize and manage information
- Being able to interpret and communicate information
- Computer literacy in processing information.

4. *Systems thinking.* The ability to understand complex systems and interrelationships including:

- Understanding how social, organizational and technological systems work and to be able to operate effectively with them
- Being able to monitor and correct performance and distinguish trends, predict impacts on system operations, diagnose deviations in systems performance and correct malfunctions
- Being able to improve or design systems as well as being able to suggest modifications to existing systems and develop new or alternative systems to improve performance.

5. *Technology skills.* The ability to work with a wide variety of technologies and includes:

- Being able to choose appropriate procedures, tools or equipment including computers and related technologies for specified tasks
- Understanding the intent and proper procedures for setup and operation of equipment and applying technologies to particular tasks
- Being able to troubleshoot and/or maintain equipment (i.e. being able to prevent, identify, or solve problems with equipment, including computers and other technologies).

These individual competency requirements are as valid currently as they were in 2000. However, in addition, within the business as a whole, the understanding of the role of technology as a business infrastructure and the efficiencies created by transparent information flows within and between businesses in the industry is becoming a necessary competency – as is the ability to recognize the opportunities that these two features in combination open up in terms of value adding for a business or industry.

Planning an Electronic Enablement Project 3: Electronic Business Components

High-level business services are known as business components – they facilitate critical business rules and functions and consist of business objects (each business object represents a particular entity, process or business event), data structures, published application programming interfaces or APIs, and configuration and administration tools. APIs are particularly important since they are a series of functions that programs can use to make the operating system do their dirty work. For example, a program can open windows, files and message boxes, as well as perform more complicated tasks, by passing a single instruction. Example electronic business components include:

- *Security access control* – establishes trust between parties
- *User profiling* – automatically logs and collects profile information
- *Search engines* – allow users to easily locate information via parametric or content searching
- *Content management and cataloguing* – must be able to handle multiple authors with multiple roles to manage complex catalogues
- *Payment* – B2C and B2B have different requirements but both need the involvement of financial institutions, online payment service providers and third party brokers
- *Workflow management* – acts as a vehicle to pass a process from one stakeholder to another according to a predefined set of business rules – consists of the workflow definition module, the business rules definition module and the workflow engine
- *Event notification* – must be capable of monitoring status of processes and notifying users of important occurrences
- *Collaboration* – sharing of information and provision of negotiation and collaboration functionality between combinations of buyers, sellers, intermediaries and other parties
- *Reporting and analysis* – delivers critical information in user-defined reports
- *Data message integration* – must allow definition of rules and criteria to construct and pass message types and data that are understandable by disparate application infrastructures.

Planning an Electronic Enablement Project 4: Electronic Technology Components

Technology components focus on accessing, facilitating and performing system functions (Boey *et al.*, 1999). They are cross-application, cross-industry and application-specific. They consist of:

- *Client components.* These are programs or applications which manage the user interface. They send messages to the server components to perform user-specific tasks (e.g. JavaBeans or MS Active X).
- *Server components.* These are programs or applications which implement the server side of the business logic such as creating a purchase order (e.g. CORBA, Enterprise JavaBeans).
- *Application servers.* These are deployment platforms that provide application management, distribution, scalability and security over the inter-, intra- and extranet.
- *Application frameworks.* These are a collection of business functionality components used to provide a framework for an overall system.
- *Enterprise application integration (EAI).* This is necessary given the new set of integrated systems that have been created by combining existing legacy applications, custom applications and packaged applications with new web functionality (Leclerc, 2000).

- *Technology standards* – must be focused on cross-system issues. They include:

    - *UML.* Which is the 'unified modelling language' for visualizing, specifying, modelling and documenting the architecture of a system
    - *Public key infrastructure (PKI).* Which uses a standardized set of transactions that use asymmetric public key cryptography for limiting access to digital resources
    - *Digital certificates.* Which verify the sender's identity, place a tamper-resistant seal on a message and provide proof that a transaction has occurred
    - *Digital signatures.* Which are electronic signatures that can be used to authenticate the identity of the sender of a message or of the signer of a document
    - *Workflow technologies.* For example, XML (extensible markup language)
    - *Web audit.* There are a wide variety of legal issues presented by websites and it is important to pay careful attention to those issues before they become problem (Hoffman, 2002).

Outsourcing

If key competencies for electronic enablement are not available in house, and it is determined that it would be too expensive, difficult or impractical to develop them in house, an organizational decision may be to *outsource* the development. The term *outsourcing* defines the use of external agents or outside business relationships to perform one or more necessary organizational activities in lieu of internal capabilities. In addition, according to Corbett (2001), a leading outsourcing specialist consulting house in the USA:

> **Outsourcing is nothing less than a basic redefinition of the corporation around core competencies and long-term outside relationships. These core competencies and outside relationships are chosen to bring the greatest value to the ultimate customer and the greatest productivity to the corporation itself. Outsourcing is more than purchasing, and it is more than consulting. It is a long-term results-oriented relationship for a whole business activity over which the provider has a large amount of control and managerial discretion**

Thus, what defines outsourcing is more the circumstances of the relationship than the nature of the work performed and there are three main models:

- *Outsourcing versus suppliers.* Outsourcing relationships are characterized by replacing or substituting the services of an external provider for internal capabilities. Importantly, outsourcing applies to an activity an organization did do, or would have done, itself. Outsourcing is not about 'supplying' commodities or totally unrelated products or services.
- *Outsourcing versus consulting.* Consultants advise on how to do something: outsourcing providers actually do it. Sometimes a consultant will deliver a business service or product, and that's when they are acting like a provider. At other times an outsourcing provider will advise, but generally the distinction is easy to see. The reality is that many firms are both consultants and providers, but play different roles with different clients – and at different times.

- *Outsourcing versus jobbing and out-tasking.* Outsourcing relationships are high value add, durable and ongoing – they are not a one-time-only deal. Hiring a provider to set up your technology, or a manufacturer to handle production when demand exceeds capacity, or using IPEX to deliver overnight packages is not outsourcing. Outsourcing relationships are high level, contractual relationships for a fixed period of time, usually measured in years, but they are assumed to be continuous. Provider and user often work to define the service delivered. There is frequent interaction between user and provider and a lot of communication. The outsourcing service is customized to the needs of the user.

Those who provide outsourcing are referred to as outsourcing partners, suppliers and providers. Those who are purchasing the outsourcing services are called buyers or users.

The History of Outsourcing as a Managerial Tool

Since 1990 there has been a huge growth in the use of outsourcing. There have been three distinct stages in its evolution (Corbett and Associates, 2000):

1. *Tactical outsourcing* – In this first stage, the reasons for outsourcing were usually tied to specific problems the firm was having and outsourcing was seen as a direct way to address either the lack of financial resources to make capital investments, inadequate internal managerial competence, an absence of talent, or a desire to reduce headcount. Outsourcing often accompanied large-scale corporate restructuring. Many tactical relationships were forged to create immediate cost savings, eliminate the need for future investments, to realize a cash infusion from the sale of assets and to relieve the burden of staffing.

   The focus of tactical outsourcing is the contract, constructing the right contract, and holding the vendor to the contract. The expertise for constructing these arrangements emerged from purchasing. Frequently the contract was simply a fee for services. Much of the value stemmed from the discipline of spending dollars externally. When managers created successful tactical relationships the value of using outside providers was clear: better service for less investment of capital and management time.

2. *Strategic outsourcing* – Strategic outsourcing is the redefinition of the corporation around its core competencies and the creation of strategic, long-term, results-oriented relationships with service providers. It is about fundamentally redefining the business and separating the core from non-core activities and making decisions about how to get the work done.

   Strategic outsourcing relationships are about building long-term value. The working relationships with providers evolve from adversarial vendor-supplier relationships to long-term partnerships between equals where the emphasis is on mutual benefit. In this more strategic model, corporations work with a smaller number of best-in-class integrated service providers and the managerial mindset about the nature of the relationships has matured from one between buyer and supplier to one between business partners.

3. *Transformational outsourcing* – This is the term used to describe the third stage of the outsourcing evolution which is the use of outsourcing to redefine the business by recognizing that the real power of outsourcing is in the innovations that outside specialists bring to their customers' businesses. Outsourcing can be used to radically change the definition of the business – open new markets, deliver new customers and create new products. Outsourcing is leverage. Here outsourcing is a vehicle for changing the firm's relationships with customers, employees and business partners by working with best-in-world partners. Outsourcing enables growth and is a way to grow by alliance.

The Range of Outsourcing Options

The range of outsourcing options gets bigger by the day – however, in general terms the following are the main categories of outsourcing (CIO Magazine, 2002):

1. *Human resources.* Management uses outsourcing as a way to meet short-term demand – for example, the use of contract programmers and personnel who are managed by company employees or hiring of a staffing company to monitor and manage all the non-exempt hiring for the company.
2. *Project management.* Management outsources for a specific project or portion of a project work using vendors to develop a new system, support an existing application, handle disaster recovery, provide training, or manage a network. For example, a university hires an information technology company to manage all of its desktop PCs and staff the user help desk. In this case, the vendor is responsible for managing and completing the work.
3. *IT and business process outsourcing.* This has grown tremendously in recent years and includes: user support, disaster recovery, data network management, data centre operations, support services, application hosting, voice network management, software development, software maintenance, IT strategy and planning, internet services, business process modelling.
4. *Total outsourcing.* The vendor is in total charge of a significant piece of work – for example, a pharmaceutical firm hires a contract manufacturing organization to manufacture a product, or a company totally outsources the entire hardware and software support functions to an outside vendor. In this instance, outsourcing represents a significant transfer of assets, leases and staff to a vendor that now assumes profit and loss responsibility. It also allows a third party to operate, manage and control information systems functions (which may create problems – see the section on criticisms of outsourcing below).
5. *Offshore outsourcing.* This allows organizations to successfully overcome the worldwide shortage of qualified resources and provide their users with critical time-to-market solutions. It requires setting up an offshore development centre and due diligence procedures for selecting and monitoring the right vendor.

Drivers of Outsourcing

The drivers for outsourcing depend very much on the evolutionary stage of the managerial mindset (Corbett and Associates, 2000) and include tactical, strategic and transformational issues.

1. *Tactical issues.* The top five tactical reasons for outsourcing include:

   i. *To reduce and control operating costs.* The single most important tactical reason for outsourcing is to reduce or control operating costs. Access to the outside provider's lower cost structure is one of the most compelling short-term benefits of outsourcing. One study, conducted in 1993 found companies reporting an average 9% reduction in costs through outsourcing.
   ii. *To free up investment dollars.* Outsourcing reduces the need to invest capital funds in non-core business functions. This makes capital funds more available for core areas. Outsourcing can also improve certain financial measurements of the firm by eliminating the need to show return on equity from capital investments in non-core areas.

iii. *To create a cash infusion.* Outsourcing may involve the transfer of assets from the customer to the provider. Equipment, facilities, vehicles and licences used in the current operations all have a value and are, in effect, sold to the provider as part of the transaction resulting in a cash payment. This cash can then be used in other parts of the operation. Similarly, some contracts involve an up-front payment by the provider to the customer of anticipated savings during the first few years of the contract. Similarly, this cash is then immediately available for investment elsewhere.

iv. *When resources are not available internally.* Companies outsource because they do not have access to the required resources – human, capital or intellectual. For example, if an organization is expanding its operations, especially into a new geographic area, outsourcing is a viable and important alternative to building the needed capability from the ground up.

v. *When a function is difficult to manage or out of control.* Outsourcing is certainly one option for addressing these types of problems. Outsourcing does not, however, mean abdication of management responsibility, nor does it work well as a knee-jerk reaction by companies in trouble. However, introducing the best-in-class management and business processes that a provider offers can be a quick way to bring control to a situation.

2. *Strategic issues.* The top five strategic reasons for outsourcing include:

   i. *To improve business focus.* Outsourcing lets the company focus on broader business issues while having operational details assumed by an outside expert.

   ii. *To access world-class capabilities.* Outsourcing providers bring extensive worldwide, world-class resources to meeting the needs of their customers. Partnering with an organization with world-class capabilities can offer access to new technology, tools and techniques that the organization may not currently possess; better career opportunities for personnel who transition to the outsourcing provider; more structured methodologies, procedures, and documentation; and competitive advantage through expanded skills.

   iii. *If accelerated reengineering efforts are taking place.* Outsourcing is often a byproduct of business process reengineering. It allows an organization to immediately realize the anticipated benefits of reengineering by having an outside organization take over the process.

   iv. *To share risks.* There are tremendous risks associated with the investments an organization makes. When companies outsource they become more flexible, more dynamic and better able to adapt to changing opportunities.

   v. *To free resources for other purposes.* Every organization has limits on the resources available to it. Outsourcing permits an organization to redirect its resources from non-core activities towards building knowledge sets with long-term pay back and impact on innovation.

3. *Transformational issues.* The top five transformational reasons to oursource are:

   i. *To bring new solutions to customers, faster.* The pace of business change is accelerating. Today, customer expectations and preferences can shift at the speed of a mouse click. The specialized capabilities of an outside expert are increasingly important in building that next generation solution.

ii. *To respond to shortening product life cycles.* Product life cycles that recently measured in years, today measure in months. The 1999 Strategic Outsourcing Study (Corbett and Associates, 2000), found that companies facing this type of growing external turbulence are more and more turning to specialized outside suppliers to effect the necessary changes.

iii. *To redefine relationships with suppliers and business partners.* Outsourcing can transform the entire supply and customer chains in a host of different ways. New, integrated providers can shorten the chain and provide additional services that transform markets or transform the customer experience.

iv. *To leapfrog competitors.* Amazon.com fundamentally changed retailing thanks to a different philosophy, new technology and a cadre of outsourcing providers. Everything changed when Amazon.com partnered with the providers who could manage call centres, process orders, fulfill orders and warehouse books in ways previously untested.

v. *To enter new markets with reduced risks.* Firms are moving into new markets in response to new competition or opportunities, sometimes with little experience. This is most evident in the areas of eCommerce and eBusiness. Providers like IBM Global Services are helping bricks-and-mortar firms create entirely new customer bases online.

Examples of companies using outsourcing are numerous but include Du Pont (see Byte Idea this chapter), and Mission Foods – a major American food processor which has outsourced its deliveries to Penske Logistics and is operating at 99.9% on-time with deliveries because of it. Penske is also working to reduce the handling and optimize the routes to move the product faster (thus making the product fresher at market as well as being damage-free). As a result, Mission Foods is seeing the development of greater competitive advantage.

The Disciplines of Outsourcing

The disciplines necessary for successful outsourcing are similar to any other major business project undertaking with perhaps an additional issue associated with making sure the knowledge associated with the project stays in house once the project is completed. The framework (Firmbuilder, 2001) for a successful outsourcing project includes:

- *Understanding outsourcing's value proposition.* Outsourcing can help to redefine the corporation and it can build long-term, sustainable competitive advantage thus building shareholder value.
- *Identifying core competencies.* Outsourcing is the tool that allows organizations to focus on their core competencies and allows businesses to adapt to marketplace opportunities. As an organization outsources those activities that are non-core, it frees management time and resources to focus on core activities.
- *Selecting the best candidates for outsourcing.* Assess the marketplace and consider bundling several activities into one outsourcing contract.
- *Forming successful project teams.* Project teams are usually necessary for any large and complicated undertaking. Outsourcing often covers many activities requiring a number of skills.

Byte Idea – Du Pont
www.dupont.com

Outsourcing

Du Pont is in its third century as a company – two hundred years ago, it was primarily an explosives company. One hundred years ago, it turned to global chemicals, materials and energy. Today, Du Pont with revenues exceeding US$27 billion delivers innovative science-based solutions in areas as diverse as food and nutrition, health care, construction, electronics and transportation.

Du Pont has over 1,100 IT professionals and a substantial IT budget, but this is not enough to supply all the necessary services to Du Pont's 94,000 workers in 70 countries. The company uses a variety of managed services, including web hosting from Proxicom Inc, an internationally recognized internet consulting firm, which delivers custom-tailored web applications, industry-specific solutions and system integration services to Global 1000 organizations.

Du Pont's approach is to consider outsourcing when it contributes to value and growth for the company, which allows Du Pont to get IT to focus in areas where it can create competitive advantage. Those competitive areas include such things as process control and research and development computing where outsourcing is not normally considered – but even in the latter category, the company is now prepared to consider managed services for infrastructure, such as network monitoring and system backups.

Once the decision to outsource is made, a close eye must be kept on the new 'partners' no matter how much they are trusted – service level agreements are an absolute must. At Du Pont, everything is audited – from source code to security.

Information sourced from Business Weekly (www.businessweekly.co.uk) and the Du Pont website

- *Defining requirements.* It is an absolute necessity to define the requirements, goals and expectations of an outsourcing project.
- *Performing the financial analysis.* To make sure that the organization's financial resources are being wisely used and are adding value.
- *Selecting providers.* A multistage process that involves a number of activities that need to be addressed before deciding which provider will create the most value for an organization.
- *Negotiating price and contract.* Once an organization has selected its outsourcing provider, the deal's pricing structure needs to be determined. There are a number of key considerations: how will pricing work to align the interests of both parties and create what might be called alignment between the principal and the agent? How much variability in the pricing are both parties willing to accept? Where does management put forth the effort into measuring outcomes or performance?
- *Negotiating a win-win relationship.* Every agreement is unique and the goal is to achieve the best price and performance reasonable given the responsibility and risks involved.
- *Managing the people dimension.* The process organizations use to manage the people dynamics will ultimately make or break the value of an outsourcing decision.
- *Managing the ongoing relationship.* At one time, the management of the relationship simply meant structuring the contract. Today it is much more along the lines of defining

and working towards mutually agreed to goals and creating a strong working relationship between the firms as keys to successful outsourcing relationships. The bottom line – both parties must maintain a win-win approach.

- *Making outsourcing a strategic management tool.* Industry convergence is a major force of change and it is changing the competitive landscape as well as shaping the way business will be conducted in the near future. The sooner organizations understand the culture of change and embrace it, the more successful they will become.

Criticisms of Outsourcing

Despite the extreme growth in outsourcing there are some major issues that still need to be dealt with (Quinn *et al.*, 2000; Firmbuilder, 2001):

- Outsourcing can result in a loss of control, particularly in relation to IT and information systems (IS) outsourcing.
- Outsourcing can foster the threat of opportunism from the supplier unless the relationship is well developed and nourished.
- Outsourcing can result in a loss of IT/IS expertise and corporate memory.
- Outsourcing can create a decline in morale and the performance of remaining employees.
- Outsourcing can create intellectual property problems in the new interconnected eLandscape, particularly in relation to technology transfers, the division of ownership interests in R&D, and in how to protect global branding and trademark portfolios.
- Anticipated cost savings may in fact be able to be more easily achieved internally.
- Anticipated cost savings do not eventuate.
- Unless good relationship management strategies are in place, the resultant disconnect between expectations and results is too great to maintain the relationship.

Managing Change during Electronic Enablement Projects

Electronic enablement is an evolution of the workplace that has impacted everyone by its speed and flexibility (Connor, 1998). It has changed how business has traditionally been conducted and threatens the behavioural norms of most organizations. This is especially true during the stage of transition when executives begin to outsource non-core competencies to electronic technology providers. There are seven major steps in minimizing the impact of new electronic enablement strategies. These include:

1. *Laying the foundation.* If a realignment of the workforce will be necessary to execute the electronic enablement strategy, communicating the changed vision is the best way to lay the foundation. Employees are better able to understand why certain activities (such as those associated with non-core competencies) are no longer part of the future strategy, and the business case for change appears less random and more objective.
2. *Starting early.* By sharing the changed vision early in the transformation process, employees have time to react and possibly transfer or build skills to positively impact the new strategy. A strong change vision lays the foundation for a smoother transformation.
3. *Clarifying roles.* Clearly communicating new goals, the actions required, and anticipated benefits, will clarify to the workforce how and what they will be contributing to the future success of the organization.

4. *Assessing change readiness.* A critical area that must be assessed in the early stages of transformation is the organization's readiness for the change. Based on that assessment, the company can select and develop a change strategy that addresses the specific needs of the organization. This change strategy must support the transformation, align with the company's vision, and be implemented concurrently with the change in strategic business direction.

5. *Integrating electronic enablement requirements into the organizational design.* Organizational design includes the work that goes into defining the reporting structure, roles, performance measures, workgroups and integrating mechanisms for a company moving into the eBusiness environment. It is critical that the employees in the new organization understand their roles, and that the systems by which they are measured and rewarded are aligned with the new processes.

6. *Managing the corporate culture.* Culture speaks to the behaviours that are required to succeed in the electronically enabled environment. It is the 'how' things get done to support the strategy. In any organization, developing, building, and constantly maintaining a culture that supports the evolving business strategy is important. Culture and managing change is discussed in the next section. The speed at which the marketplace moves currently makes 'change' a critical issue to be managed in electronically enabling a business successfully.

7. *Communicating to build commitment.* Communication can be the most critical aspect of whether a change strategy will succeed or fail. The degree to which information flows down through the organization can influence how well the change vision is understood and embraced. It can impact the values, behaviours and mindset of the workforce. Lateral communication and cooperation is important to building an innovative organization as it supports risk-taking and provides tools to take action.

A number of these points have been discussed in earlier chapters – however, the big issues of organizational culture and change management should not be forgotten when looking to electronically enable a business. The following sections give a brief overview of these two areas particularly as they relate to technological change dilemmas. For detailed study of these areas, you may access some of the classic readings within the general management discipline of organizational behaviour.

Organizational Culture

The concept of organizational culture is based on the proposition that organizations operate within a cultural/social context, and are also themselves culture-bearing entities (Frost *et al.*, 1985). There is no one definition of organizational culture but essentially there are three main elements (Schein, 1992):

• *Basic assumptions and beliefs* which are 'taken-for-granted' interpretations of the nature of reality which have become the basis on which people think and act. Basic assumptions are likely to exist in relation to the need for hierarchy, the basis of competitiveness, the identity of the market in which the organization operates, the importance of consultation.

• *Expressed beliefs and values* about how and why things are done.

• *Artefacts* which are the language, myths, ceremonies, rituals and norms involved in everyday life of the organization.

These combine in various ratios to create three perspectives on organizational culture:

1. *The integration perspective* which treats culture as a unified phenomenon – that is, it is shared by all organizational members and the various cultural phenomena are consistent with each other. The assumption underlying this perspective is that culture integrates and binds.
2. *The differentiation perspective* where rather than focusing on the organizations as a whole as the key unit, pre-eminence is given to the notion of sub-cultures, both those formed as part of the organizational structure such as departments, and those imported from outside, such as class, occupation and gender.
3. *The ambiguity / fragmentation perspective* where clear patterns are not typical and where consensus, divergent views and confusion exist. Unlike the first two perspectives, there is no shared meaning across the organization or within sub-cultures.

The need to understand culture and manage for it is based around the premise that it has an effect on organizational effectiveness. Under this premise, culture is both an asset and a liability. As an *asset*, shared beliefs ease and facilitate communication, and shared values generate higher levels of cooperation and commitments than is otherwise possible, which itself creates efficiencies. As a *liability*, efficiency does not necessarily imply effectiveness. The extent to which something is being done in an appropriate way is a question of effectiveness. If culture guides behaviour in inappropriate ways, ineffectiveness and inefficiencies can result.

Organizational Change

> *'It is not the strongest of the species that survives, nor the most intelligent, it is the one that is most adaptable to change'*

> Charles Darwin (1859) The Origin of Species

Organizational change is the implementation of new procedures or technologies intended to realign an organization with the changing demands of its business environment, or to capitalize on business opportunities. Change usually involves the introduction of new procedures, people, or ways of working which have a direct impact on the various stakeholders within an organization. Change is inherent in contemporary organizations and its management is not only critical to organizational success but to survival. There are three major types of organizational change recognized:

1. Incremental or adaptive change
2. Radical or generative change (i.e. paradigm shifts)
3. Continuous change.

In a business context, a paradigm refers to a way of doing business and has the connotation of conventional wisdom – 'this is how things have always been done and must continue to be done'. Paradigm shifts require disruptive breaks and are not achievable by smooth transitions such as the adaptive changes.

In terms of managing change in an organization, both evolutionary and revolutionary change are important for overall organizational progress. It is a mistake to regard revolutionary change as more important due to its higher visibility. The cumulative effects of incremental change are often far greater than the effect of radical change. The challenging task for management is to generate the correct balance of radical and incremental change.

The *Punctuated Equilibrium Model* of change (Tushman and Romanelli, 1985) emphasizes the discontinuous nature of change. Long periods of small incremental change are interrupted by brief periods of discontinuous, radical change. These occasional dramatic revolutions or punctuations overcome organizational inertia, which alters the organization frame.

Brown and Eisenhardt (1997), however, argue that organizational survival in today's environment of very rapid technological and associated economic change depends on the firm's ability to engage in rapid and continuous change, in contrast to the rare, episodic phenomenon described by the Punctuated Equilibrium Model.

Drivers of and Barriers to Change

Drivers to change can come from anywhere – in the era of electronic enablement the original drivers were the role of the internet as an 'information superhighway' and of doing business in a different way (Chapter 3). These are still core drivers, but, in addition, change in everyone else's businesses happening at the same time creates other impactors on individual businesses and thus new drivers such as:

1. Changes in the mission or 'reason to be'
2. Changes in the identity or outside image of the company (e.g. bricks and mortar vs clicks and mortar)
3. Changes in relationships to key stakeholders (particularly in relation to supply chains)
4. Changes in the way of working (e.g. the paperless office, mCommerce)
5. Change in the culture (e.g. traditional vs electronically enabled).

Barriers to change in an organization are often referred to as *organizational inertia* and can exist at four levels: corporate, divisional, functional and individual.

*   *Corporate obstacles.* Existing structure and strategy may be an obstacle to change and strategic change is often a slow process due to the huge amount of inertia that needs to be overcome. The type of structure within an organization may impede change – a matrix or decentralized structure may be easier to change than a highly functional, centralized or integrated structure, due to the greater flexibility and cooperative, cross-functional relationships involved. Corporate culture can also be an obstacle to change, and organizational cultures based on flexibility and openness to change would be easier to change.
*   *Divisional obstacles.* If divisions are highly interrelated and share resources, change will be more difficult as operational shifts in one division may significantly affect another division. Diversification strategies in related divisions are harder to change than in unrelated divisions since change often favours some divisions over others. Managers too have different attitudes to change, especially if their division is negatively affected by the change in terms of resources or status.
*   *Functional obstacles.* At the functional level, there may be different strategic orientations and goals and these functions may react differently to the implementation of change. This in turn may reduce a company's ability to respond to changes in the external environment.
*   *Individual obstacles.* Uncertainty, fear of the unknown and insecurity are symptoms of an individual's resistance to change. Potential layoffs as a result of restructuring and reengineering cause fear and insecurity for managers and staff which may result in reduced commitment to organizational goals and lowered morale.

Change Management

> *'There is nothing more difficult to take in hand, more perilous to conduct, or more uncertain in its success, than to take the lead in the introduction of a new order of thing'*
>
> Machiavelli, (1513).

The key to successful change management lies in understanding the potential effects of a change initiative on these stakeholders. Will employees be scared, resistant, pessimistic or enthusiastic about the proposed changes? How can each possible reaction be anticipated and managed?

Peter Drucker (1999) one of the 20th century's great management gurus comments that at the end of the day, change boils down to people. People make things happen, and company people will be the ones carrying out the changes. However, this is only likely to occur if a change initiative is properly communicated by management. The organizations that succeed at change are the ones that are able to consider the people who will be affected by the change and who will be crucial to its long-term chances of success.

In reality though, without the leaders of the organization recognizing the need for change and adopting an appropriate leadership role, it is almost impossible to have a smooth transition to an electronically enabled business model. It is therefore of the utmost importance that the CEO and other senior management lead the way forward and support change management strategies in the organization. FedEx Corporation (see Byte Idea this chapter) provides a good example of a company whose leadership recognized that great heights could be achieved on the back of electronic enablement of both itself and the product it delivers – and went for it.

Getting Senior Management Onside and Online

It is critical not to overlook the importance of the role of senior management in electronically enabling a business. Not only do they control the resources and money that will be required to implement the specific projects in the plan, but, depending on the nature of the business, senior management support will be crucial in determining the way in which the whole strategy is approached. 'Think Big, Start Small, Scale Fast' rolls easily off the tongue, but it can be quite challenging to an organization in its execution. The people aspect associated with electronic enablement also emphasizes the need for senior management support. Like any other significant change in a business it stands little chance of success if it is not led or at least staunchly supported from the top.

Getting senior management onside and online is thus a crucial part of ensuring successful electronic enablement. Even today, there are some senior managers who rarely touch their email, relying on secretaries to check their email for them, and some will never use the internet – ever. Why is this so and is it a sustainable situation, either for the managers concerned or for the organizations in which they work?

In short, no, it is not sustainable for the individuals themselves – both from the point of view of their accessing information and because, as email communication becomes more and more the 'norm', this attitude starts to impinge on staff workloads. From a company perspective, senior executives and managers who do not remain up to date and innovative in the way they communicate are probably unable to adapt quickly enough to the rate of change in the business environment around them, and are thus potential liabilities to the organization.

This resistance to technology may exist for a combination of reasons which are likely to include a fear of computers, a lack of knowledge, an innate fear of being 'shown up' by younger colleagues, laziness and organizational culture. As discussed above, this latter pervades everything an organization and the individuals in it operate under and is a major challenge when it comes to changing the way people 'do' things around the organization. There are many texts and many people working in this area – thus the next three sections which deal with organizational culture, organizational change and change management will only briefly outline the major issues associated with each subject to provide the context in which electronic enablement development has to operate.

Byte Idea – FedEx Corporation
www.fedex.com

Express Delivery via Electronic Means

In the last 30 years FedEx Corporation has created what we know as the modern air/ground express parcel delivery industry and was the first to provide online shipping and tracking. Today, FedEx Corporation leads the broadest array of transportation, eCommerce and supply chain solutions in the world and employs around 213,000 employees, has an annual revenue of US$21 billion and services 210 countries.

FedEx's business model has a two pronged approach: independently, each FedEx Corp. subsidiary focuses on what it does best – offering the best transportation and logistics services in its respective market. Collectively, FedEx Corp. provides the strategic direction for its independent operating companies to compete worldwide under the powerful FedEx brand.

The FedEx group of companies includes:

- FedEx Express – The world's largest express transportation company
- FedEx Ground – North America's second largest small-package ground carrier
- FedEx Freight – The leading provider of regional less-than-truckload freight services
- FedEx Custom Critical – The world's largest expedited, time-critical shipment carrier
- FedEx Trade Networks – Provides customs brokerage, consulting, information technology and trade facilitation solutions
- FedEx Services – Provides integrated sales, marketing and information technology support for the FedEx companies
- FedEx Supply Chain Services – Provides outsourcing services to establish competitive supply chains.

The company did a total site redesign in 2001 to take itself global, adding international forms, currency converters and 11 languages. New tools for online merchants include FedEx Net Return to handle product returns, and FedEx Insight, a new tracking tool for business shippers which proactively notifies customers of any shipping problems. Other new technology includes an internally developed routing innovation called Purple Lights – named for the violet hue its laser scanners cast – which helps make sure workers put letters and small packages in the right bins and bags. If a parcel is incorrectly sorted, the system lets workers know by literally telling them 'wrong bag', which has meant that FedEx has been able to recover over $75 million in profit that was otherwise being lost.

Information sourced from CIO Magazine, Financial Times and the Fedex website

A Solid Business Case Is the Key

So, to return to the question we posed at the start of this chapter – how is the business going to make money out of electronic enablement? The formulation of a cohesive *business case* to justify the strategy and each of its component projects is fundamental to answering this question. As the name implies, a business case is a collection of arguments not unlike a legal case in court, which justifies why the business should invest in the initiative.

The business case discipline is critical to the genuine success of the 'E' strategy. The dot. com days and their aftermath demonstrated how easy it is to become enamoured with a 'new way' of doing business and overlook the fundamentals of business success in the process. The business case provides the decision makers with information that they require to make the hard decisions about whether to support the new initiative – and it also provides those who are advocating the case *for* the new initiative with a tool to ensure that all the various justification bases are covered and that there is genuine value to be had from pursuing the project.

It is also important to understand that it is likely that electronically enabling the business is just one of many initiatives within a business competing for capital and resources. The business case must demonstrate to senior management (and/or the Board of Directors) the compelling reasons why the business will be better off through the implementation of the strategy.

Justifying the overall strategy is a fairly high level analysis that is focused on selling the *Think Big* picture. It provides the overall context within which each individual initiative can be judged. This business case should be revisited regularly as you refine your understanding of what is required to achieve your vision.

Individual projects must also be justified both on their own merits and in the context of the overall strategy. The early *Start Small* projects will generally be somewhat 'speculative' in nature but the business case should still quantify the likely benefits of these projects while at the same time recognizing that the analysis is somewhat 'rubbery'. As the project journey progresses, however, later projects can and should be supported by analysis that has a clear return to the business in terms of increased profitability, improved asset utilization, reduced risk or a combination of all three (Fig. 8.3). A good business case should:

- Demonstrate the alignment of electronic enablement with the overall strategy of the business
- Establish how the needs and issues of *all* key stakeholders are to be addressed
- Explain clearly how value will be created, and how increased revenue, reduced transaction costs, decreased inventory levels, etc. will be achieved
- Quantify the value to the business through a discounted cash flow analysis of costs and benefits over three to five years
- Highlight the business's capacity to execute specific projects and to exploit the opportunities that the initiative will generate (i.e. *Scale Fast* which is related to the business's 'E' readiness – Chapter 7).

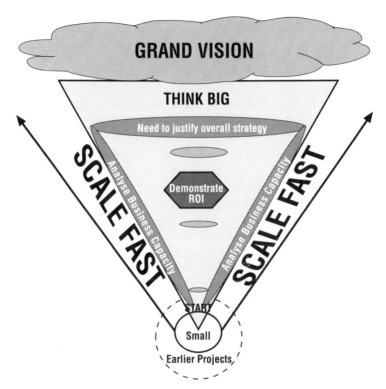

Fig. 8.3. Developing a business case using the 'Think Big, Start Small, Scale Fast' theme

Project Execution – The Decision Has Been Made, Get On With It!

A compelling business case supported by robust analysis will count for nothing if the business is unable to properly execute. While seemingly self evident, quality project execution is often overlooked. Key features include: ownership of the project by employees, availability of project champions, good project management, availability of appropriate resources, definition of the IT group's role, appropriate change management strategies and appropriate evaluation mechanisms.

1. *Business ownership.* Underpinning a successful project is ownership by those who will be charged with taking it forward after it has been implemented. A defining feature of a project is that it has a beginning and a defined end. After that end, if the people within the business do not genuinely feel ownership of a project, it is likely to fail. There will always be a need to 'fix' problems after a project has 'gone live' and it is the goodwill of those working with the fruits of the project that will determine whether problems are proactively addressed or whether 'I told you so' and 'it was never going to work anyway' become self fulfilling prophecies.

2. *Project champions.* You could be forgiven for surmising that as long as you follow the prescriptions of this book, 'E Nirvana' can be achieved with relative ease. The reality is likely to be somewhat different. Each organization will have its own unique issues,

sensitivities and roadblocks to be overcome. This comes into sharp focus as you set about executing a project. To manage these issues, it is critical to have a senior manager with significant organizational credibility and authority to act as a project champion. This person on the one hand can kick down roadblocks as they arise and, on the other hand, ensure that the project execution is aligned with project success within the specific context of the business and delivery of the promised business value.

3. *Project management.* Disciplined planning and measurement of progress against the project plan is critical to the success of any project. The elements of success include:

   - A steering committee to overview progress and alignment
   - A project manager who can deliver the project
   - A detailed plan of activities that addresses task dependencies and resourcing
   - A detailed budget that can be tracked in parallel with the task plan
   - Adequate contingency allowances and planning
   - Appropriate resources – This last point about resources has been discussed earlier but it is worth reiterating here – *just as critical as good project management is the need to resource a project properly.* These resources (mostly people) will be a mixture of 'in house' and external consultants. People from the business usually bring the all important context and specific knowledge of the way the business operates while the external consultants will bring specific technical skills to implement the project.

4. *The IT group's role.* Issues around the IT group's role are highly organization specific and relate to how their role is defined more broadly and how much expertise they have. The IT group must be involved in the project from its inception. They are the organization's technical experts on the systems that are already in place and are likely to have important perspectives on new technologies and their implementation. However, it is *critical* that this capability does not translate into the IT group owning the project: their input is a necessity but it is the people running the business *at all levels* who must own the project. The IT group should be an active partner and key provider of resources and advice, but the project must be driven from a business perspective not from a narrow IT view. It is an easy error for management to assume that because a project is predominantly implementing IT that the IT group should be in charge, and it is a project-killing mistake to make.

5. *Change management.* Change management must be an integral part of the project team's remit. A beautifully executed project will fail to deliver business benefit if the relevant people have not been 'brought along'. Change management is a critical part of the business ownership agenda. It is a joint responsibility of the project team and line management and it is often overlooked. It involves a complex mix of leadership and training and, like all issues associated with people, is highly specific to the business.

6. *Evaluation.* The final discipline of project execution is the evaluation stage. Once the dust has settled and the project has been operating as an integral part of the business it is important to review its performance, determine whether the project produced its specific deliverables and measure how much of the promised business benefit has been delivered. This evaluation must be driven by senior management and should be understood as a non-negotiable part of the process from the outset.

Summary

This chapter has addressed some of the practical issues associated with making electronic enablement happen successfully – either as a whole of business, or as individual projects within a business. Having read this chapter you *must* be able to answer the question *how is the business going to make money out of electronic enablement* to proceed with a project of any size. Along the way we have discussed some of the organizational models, electronic enabling components (including processes and people) and the major issues associated with implementing change within an organization – change that inevitably comes with electronic enablement.

A constant theme throughout this book has been that electronic enablement is just the use of a specific set of technologies and processes to improve business performance and as such the 'normal' rules of business apply. That is to say, electronic enablement must be able to contribute to return on investment of the business and that contribution must be credibly projected prior to management sign-off. To this end, the key components of a good clear business case have been discussed with the agenda that without such a document, no electronic enablement project should proceed further.

The Smart Thinker for this chapter is Michael Dell – the founder of Dell Computers who certainly knows about 'Making it Happen'!

The Review and Critical Thinking Questions below have been provided for you to revise and reflect on the content of this chapter.

Review Questions

1. What are the major differences in approach between the old dot.coms and electronically enabling a business in the 21st century?
2. What are key electronically enabling components?
3. Discuss the 'Think Big, Start Small, Scale Fast' mantra – why is it so important in an electronic enabling project?
4. Why is understanding the concepts of resource utilization important – give examples relating to electronic enablement development.
5. What is outsourcing and why is it a useful management tool?
6. Why does organizational culture present a challenge to management in a world of almost constant change?
7. List the key elements of a good business case and very briefly describe each element.

Critical Thinking Questions

1. Key competencies are abstract concepts yet are highly important in running a successful business. Discuss the conundrum with examples to reflect your argument.
2. Discuss and analyse the drivers and the criticisms of outsourcing with examples.
3. What was Michael Dell's vision for his company when he first started Dell Inc.?
4. What is so different at Dell that has allowed the company to be as successful as it has been?

Smart Thinker – *Michael Dell*

Founder, Executive Chairman and former CEO of Dell Inc.

Michael Dell founded Dell in 1984 with $1,000 with the idea of bypassing the middleman who adds little value to the products, and to sell custom-built computer systems directly to end-users.

By using this innovative direct-marketing approach and by pioneering the industry's first service and support programs, Dell has established itself as the No. 1 vendor of computer systems worldwide and is a premier provider of products and services required for customers to build their information-technology and Internet infrastructures. By 2002 the company's sales had grown from $6 million to $31.8 billion and it had opened sales offices worldwide.

With the addition of Dell to the Fortune 500 list in 1992, Michael Dell became the youngest CEO of a company ever to earn a ranking on the Fortune 500®; because of the phenomenal success of the company, he has been honoured many times for his visionary leadership and in 1997, 1998 and 1999, he was included in Business Week's list of 'The Top 25 Managers of the Year'. By 2001 such accolades as 'Chief Executive of the Year' from Chief Executive Magazine, 'Entrepreneur of the Year' from Inc. magazine, 'Man of the Year' from PC Magazine, 'High Impact CEO' from executive search firm Heidrick and Struggles, 'Top CEO in American Business' from Worth magazine and 'CEO of the Year' by Financial World and Industry Week magazines were commonplace.

In 1999, Michael Dell wrote the bestselling book 'Direct From Dell: Strategies That Revolutionized an Industry', which is his story of the rise of Dell Computer Corporation and the strategies he has refined that apply to all businesses – definitely a 'Must Read' book.

In 2004 he resigned as CEO of Dell but stayed on as Chairman of the Board. In 2005 he was listed as the 4th richest man in the United States and the 18th richest in the world.

Source: Time Magazine (www.time.com), USNews.com (www.usnews.com), Forbes 500 (www.forbes.com) and the Dell website

Chapter 9

EGovernance and Legal Issues in the 'E' World

Good behaviour and the law

Corporate entities – that is, corporations, trusts, foundations and partnerships – can be, and quite often are, misused for money laundering, bribery and corruption, shielding assets from creditors, tax evasion, self-dealing, market fraud and other illegal activities. This chapter considers how the rules and regulations associated with running a business should be applied and gives a broad overview of *corporate governance* – what it is, why it is necessary and how it is accomplished. It then goes on to focus in detail on issues associated with *corporate governance in cyberspace* and in electronically enabled business as well as the management of e-technology.

Chapter Objectives

After reading this chapter you should understand and be able to discuss:

- What corporate governance is and what are the two main global governance models
- The OECD Principles of Corporate Governance
- Why good governance is necessary and how it is accomplished
- The governance issues associated with the internet (eGovernance)
- The major governance issues associated with doing electronic business and how these differ from traditional corporate governance requirements
- The basic principles of intellectual property protection, particularly as they relate to electronically enabled business.

What Is Corporate Governance?

The Organization for Economic and Cooperative Development (OECD) Principles of Corporate Governance state that Corporate Governance relates to:

> **the internal means by which corporations are operated and controlled. While governments play a central role in shaping the legal, institutional and regulatory climate within which individual corporate governance systems are developed, the main responsibility lies with the private sector.**

Corporate governance practices should focus a company's Board of Directors (the Board's) attention on optimizing the company's operating performance over time. In particular, the company should strive to excel in specific sector, peer group comparisons.

OECD Principles of Corporate Governance

The OECD Principles of Corporate Governance (2004) which can be found at the OECD website (www.oecd.org) are the best known globally and are the most frequently cited principles of corporate governance. They cover the **rights of shareholders,** the **equitable treatment of shareholders,** the **role of stakeholders** in corporate governance, **disclosure and transparency** of information and the **responsibilities of the Board of Directors.** The OECD principles are 'non-binding' and do not aim at detailed prescriptions for national legislation. Rather, their purpose is to serve as a starting point for policy makers and market participants as they examine and develop their legal and regulatory frameworks for corporate governance to reflect their own economic, social, legal and cultural circumstances.

1. *The rights of shareholders.* The corporate governance framework should protect shareholders' rights which means that:

   - Major strategic modifications to the core business/s of a corporation should not be made without prior shareholder approval of the proposed modification.
   - Equally, major corporate changes which in substance or effect actually dilute the equity or erode the economic interests or share ownership rights of existing shareholders should not be made without prior shareholder approval of the proposed change.
   - Shareholders should be given sufficient information about any such proposal sufficiently early to allow them to make an informed judgment and exercise their voting rights.
   - The right and opportunity to vote at shareholder meetings is of paramount importance – but hinges in part on the adequacy of the voting system which may include the secure use of telecommunication and other electronic channels. As a matter of transparency, meeting procedures should ensure that votes are properly counted and recorded, and that an announcement of the outcome be made in a timely manner.
   - Any move away from a 'one-share, one-vote' standard which gives certain shareholders disproportionate power to their equity ownership is undesirable. Any such divergence should be both disclosed and justified.

2. *The equitable treatment of shareholders.* The corporate governance framework including the Board of Directors should ensure the equitable treatment of all shareholders, including minority and foreign shareholders. In other words:

- All shareholders should have the opportunity to obtain effective redress for violation of their rights.
- The concept of *one-share, one-vote* is based around the notion that national capital markets can grow best over the long term if they move towards the 'one-share, one-vote' principle.
- Capital markets that retain inequities are likely to be disadvantaged compared with markets that embrace fair voting procedures.

3. *The role of stakeholders in corporate governance.* The corporate governance framework should recognize the rights of stakeholders as established by law and encourage active cooperation between corporations and stakeholders in creating wealth, jobs and the sustainability of financially sound enterprises. Thus:

- The Board should be accountable to shareholders and responsible for managing successful and productive relationships with the corporation's stakeholders.
- Active cooperation between corporations and stakeholders is essential in creating wealth, employment and financially-sound enterprises over time.
- Performance-enhancing mechanisms promote employee participation and align shareholder and stakeholder interests. These may include broad-based employee share ownership plans or other profit-sharing programs (however, note the current move away from this in Australian enterprises, e.g. the Commonwealth Bank, toward senior executives being given options on company stock as part of their remuneration package as a means of minimizing the risk of inappropriate behaviour in relation to company management).

4. *Disclosure and transparency.* The corporate governance framework should ensure that timely and accurate disclosure is made on all material matters regarding the corporation, including the financial situation, performance, ownership and governance of the company. This means that:

- Corporations should disclose accurate, adequate and timely information – in particular meeting market guidelines where they exist – so as to allow investors to make informed decisions about the acquisition, ownership obligations and rights, and sale of shares.
- In addition to financial and operating results, company objectives, risk factors, stakeholder issues and governance structures, data on major shareholders and others that control or may control the company (including information on special voting rights, shareholder agreements, the ownership of large blocks of shares and significant cross-shareholding relationships) should be included.
- Corporations should disclose information on core competencies, professional backgrounds, other Board memberships, factors affecting independence, and overall qualifications of Board members and nominees upon appointment to the Board in each annual report, so as to enable shareholders to make an assessment of what value the Board members add to the company. Information on the appointment procedure should also be disclosed annually.

- Remuneration of corporate directors and key executives should be aligned with the interests of shareholders.
- Corporations should disclose in each annual report the Board's policies on remuneration, and, preferably, the remuneration break up of individual Directors and top executives, so that it can be judged whether corporate pay policies and practices meet the standard.
- Annual audits of corporations by independent, outside auditors, together with measures that enhance confidence in the quality and independence of the audit, should be undertaken.

5. *The responsibilities of the Board.* The corporate governance framework should ensure the strategic guidance of the company, the effective monitoring of management by the Board, and the Board's accountability to the company and the shareholders. *In other words:*

- The Board should ensure that the organization should adhere to all applicable laws of the jurisdictions in which they operate.
- The Board should be able to exercise objective judgment on corporate affairs independent from, in particular, management.
- The Board of Directors, or supervisory Board, as an entity, and each of its members, as individuals, is a fiduciary body for all shareholders, and should be accountable to the shareholder body as a whole. Each elected member should stand for election on a regular basis.
- Each Board should include sufficient independent non-executive members with appropriate competencies. Responsibilities should include monitoring and contributing effectively to the strategy and performance of management, staffing key committees of the Board, and influencing the conduct of the Board as a whole.
- Certain key responsibilities of the Board such as audit, nomination and executive remuneration, require the attention of independent, non-executive members of the Board. The Board should be composed predominantly of non-executives and, ideally, there should be no fewer than three independent non-executives.

The International Corporate Governance Network (ICGN)

The ICGN was founded in 1995 at the instigation of major institutional investors to represent investors, companies, financial intermediaries, academics and other parties interested in the development of global corporate governance practices to international bodies such as the OECD, the European Commission and the World Bank, in order to inform the decision making and influence policies and proposals where they relate to corporate governance (ICGN, 1999). It supports the OECD principles fully and promotes the adoption of codes of best corporate governance. The ICGN states that where codes of best corporate governance practice do not yet exist, investors and others should endeavour to develop them.

In addition, the ICGN advocates that if corporate governance issues between shareholders, the Board and management arise, they should be pursued by dialogue and, where appropriate, with government and regulatory representatives as well as other concerned bodies, so as to resolve disputes, if possible, through negotiation, mediation or arbitration. Where those means fail, investors should have the right to convene extraordinary meetings.

Governance Models

A governance model is a particular approach to governance and can be defined by a set of attributes which include a set of **structures, functions** and **practices** that define who does what, and how they do it (Coombes and Watson, 2000, 2001). These attributes typically relate to the role and relationships of the Board of Directors and the senior staff member of an organization (CEO or Executive Director).

- *Structure* refers to the parameters for selection and operation of the Board and CEO. These are typically established by legislation, regulations, bylaws and policies. Structure consists of the legislative framework under which the organization is created (its legal mandate), the letters patent, the bylaws and governance-related policies created by the Board to define how it will carry out its responsibilities, and the general rules under which it will operate.
- *Functions* refer to the 'what' of governance (i.e. what are the roles of different players involved in governance)
- *Practices* refer to the 'how' of governance and describe the activities or tasks which need to be undertaken.

These areas may overlap with each other to some degree but together they define the main parameters of how governance works in a given organization. Table 9.1 outlines in more detail the various issues that structure, function and practice cover in terms of an organization.

Understanding who does what in an organization and why is fundamental to running a successful business, particularly in terms of the authorizations and delegation of authority at the top of the organization among the corporate governance team. These include the Board of Directors, senior management and shareholders. Table 9.2 outlines the roles of the corporate governance team associated with a given organization.

Table 9.1. Governance structure, functions and practices

Structure	Function	Practice
Involves a statement of the organization's mission or purpose and should include: • Job descriptions for the Board, individual Board members and the senior manager (e.g. Executive Director or CEO) • Identification of Board committees and their terms of reference • Mechanisms for accountability to principal stakeholders in the organization including voting members, community groups or other entities with an important interest in the organization's activities, and the formal 'owners' of the organization • Governance policies which may include, for example: – Policies governing the interrelationship between the Board and staff, and the degree of Board involvement in management or operational matters – A statement of organizational values – A code of conduct – A conflict of interest policy – Policies related to communication (i.e. who speaks for the organization) – Formal rules for management of meetings – Procedures for handling complaints from volunteers, partners, staff, etc., and the role of the Board or Board Chair in this regard	The responsibilities of the Board may be described under the following headings: • Establishing the mission and planning for the future • Financial stewardship • Human resources stewardship • Performance monitoring and accountability to key stakeholders • Risk management • Community representation, education and advocacy • Managing transitional phases and critical events	This is the 'how' of governance (i.e. how functions are exercised and the organizational context in which they occur). They include: • Board development • Creation of Board work plans • Recruitment and nominations process • Contracts or agreements to serve as a director (including a confidentiality agreement) • Orientation process • Succession planning • Management of Board work and meetings, agenda management • Managing conflicts and conflicts of interest • Managing volunteer Board members (e.g. discussion dominators, attendance problems, failure to complete tasks, lack of preparation, undermining Board solidarity, breach of policies or rules of conduct) • Creation of methods for Board and director self assessment – Board work plans • Decision making (how decisions are made) – Encouraging input – Styles (consensus, near-consensus, majority) – Rules • Developing Board and organizational culture – History and traditions – Trust – Teamwork • How the Board actually conducts itself in relationship to the organization's statement of values (if it has one), code of conduct, or ethical behaviour and rules of confidentiality

Table 9.2. Who does what in the corporate governance team

Shareholder	Board	Individual Director	Chair of the Board	Board Committee and Committee Chair	CEO	Senior Management	Corporate Secretary
• Provider of purpose and capital • Governance role is limited to appointing directors, auditors and ensuring an appropriate governance system is put in place	• A link between owners and specialists • Represents shareholders to managers and managers to shareholders • Actively involved in long-term strategic planning and risk management • Must assure appropriate governance leadership and stewardship systems are in place • Reports to shareholders	• General Board roles • Inquirer • Mentor and influencer	• General Board and individual director roles • Achiever of diversity • Consensus builder • Promoter of solidarity	• General Board and individual director roles • Policy developer • Advisor • Planner	• General Board roles • Director • Initiator of new developments • Implementer • Mentor	• Day-to-day operations • Strategic leadership • Innovators • Recommend changes • Identify business risks • Manage intellectual property, capital and technical resources	• Communicator • Information manager • Team builder • Recorder • Mentor • Advisor to CEO and Board

Governance models vary according to how a Board is structured, how responsibilities are distributed between Board, management and staff, and in the processes used for Board development, management and decision making. However, in a global sense, there are two major governance models (Franks and Mayer, 1994) based on the way in which ownership and control are organized. These are the Anglo-Saxon Control Model and the Stakeholder/ Relationship Market Model (Figs 9.1a and b).

1. *The Anglo-Saxon Control Model* is predominantly found in the UK and USA, where the rights of shareholders reign supreme. This is because investors are a very large, dispersed class of people with no prior connection to the companies listed on the public exchanges. Such investors tend to insist on Boards that are similarly independent and demand a high level of financial and business disclosure.
2. *Stakeholder/Relationship Model* is found predominantly in Continental Europe, Asia and in many emerging markets, where family-owned businesses predominate. The senior managers of these companies are very often owners and they expect to have a Board presence. In addition, there is no anonymous investor class for them to worry about, and there is a feeling that more disclosure leads to more government interference.

While 'good governance' tends to be associated with the Anglo-Saxon model, primarily because of the emphasis on financial and business disclosure, recognized legal standards, shareholder value and the role of independent directors, this model by no means guarantees that investors are safe from unscrupulous practices. However, if an emerging economy wishes to attract investment, the market model is the one that most investors look to be in place. Reform at the institutional level and driven by governments, local stock exchanges and regulatory bodies is also very important in securing investor interest. Figures 9.1a and 9.1b show the major issues associated with the two main corporate governance models as outlined by Coombes and Watson (2001).

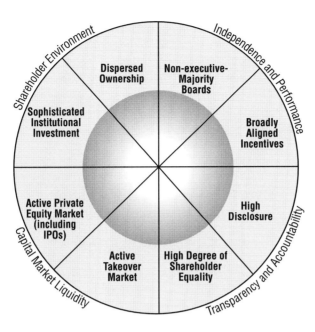

Institutional Context   Corporate Context

Fig. 9.1a. The Anglo-Saxon Control or Control Model of Corporate Governance found in the USA and UK (sourced from Coombes and Watson 2001, 'Corporate Governance in the Developing World', *McKinsey Quarterly,* No. 4 special ed., 89–93)

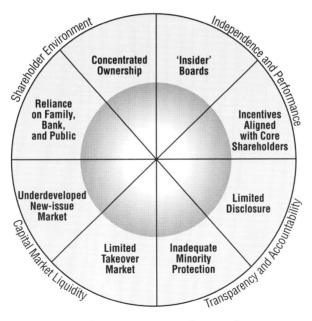

Institutional Context   Corporate Context

Fig. 9.1b. The Stakeholder/Relationship or Market Model of Corporate Governance found in Asia, Latin America and much of Continental Europe (sourced from Coombes and Watson 2001, 'Corporate Governance in the Developing World', *McKinsey Quarterly,* No. 4 special ed., 89-93)

Whatever the governance model used in an organization, it is only workable while the people involved in the corporate governance team take on board the responsibilities involved. Thus in corporate failure situations that have come to light in recent times such as the Enron, WorldCom, Parmalat, HIH and One Tel collapses (see Byte Ideas this chapter) the investigations of, and ramifications for, the Boards and senior management of these companies have been wide reaching and in some cases have led to jail terms.

Byte Idea – Enron Corporation
www.enron.com, www.ees.enron.com

Why Not Collapse Big Time?

Enron is the story of the largest bankruptcy in US history. It has cost thousands of employees their jobs and many more their superannuation and retirement funds. In addition Arthur Andersen, one of the 'Big 5' accountancy consultancy companies, was put on trial and found guilty of obstructing justice for shredding Enron documents while on notice of a federal investigation – thus causing the almost certain demise of that firm as well.

How did this happen? How does a multi-billion US$ company go bankrupt overnight?

Enron Corporation began in 1985 when Houston Natural Gas merged with Nebraska-based InterNorth to create the first American nationwide natural gas pipeline system. During the 1990s, Enron then created a trading business, beginning with natural gas and moving into electricity while in addition it created a number of subsidiary companies that delivered physical commodities as well as financial and risk management services to customers around the world. Innovation in business strategy development was a hallmark, and in creating Enron Online Trading, Enron ensured a major global trading presence that in the year 2000 netted the company 548,000 transactions with a US$336 billion notional value. By 2001, Enron was one of the largest traders of electricity in the world with 'real' revenues in excess of US$100 billion, a share price of US$81.39 on 25 January 2001, and a seventh place on the Fortune 500 list.

What went wrong?

Well, it appears that the executives and directors of Enron created a series of off-the-books partnerships that they used to hide millions in debt. Often employing complex financial structures, these partnerships, limited-liability companies, and other affiliates were used to inflate their profits and lower their debt and were so complex that they were not transparent to anyone.

In other words, a culture of rule breaking was allowed to flourish by management which was facilitated by their auditors (Arthur Andersen) through a series of accounting tricks so that investors, the general public and employees were misled about the company's financial condition. Once the off-the-books partnerships were exposed, the bottom dropped out of the game, with Enron's stock plummeting from a high of US$81 to 26c a share just before Enron filed for bankruptcy on 2 December 2001 – however, not before the Enron executives involved reaped millions through these partnerships by selling off stock before the stocks started rapidly plummeting.

Information sourced from a timeline of events which can be found at: www.cnn.com/SPECIALS/2002/ enron/ under Resources/TIME, and from the Enron website

Good Corporate Governance

According to the OECD (2004), there are seven primary characteristics of good corporate governance: **discipline** (good behaviour); **transparency** (it is easy to see what is going on); **independence** (there is little risk of conflicts of interest); **accountability** (investors can query and assess management actions); **responsibility** (management is held responsible for its actions); **fairness** (shareholders are all treated equally); and **social responsibility** (high ethical standards exist). These characteristics are explained further below.

1. *Discipline* – Corporate discipline is the commitment by a company's senior management to adhere to correct behaviour. This encompasses a company's, and in particular its Board's and senior management's awareness and commitment to the underlying principles of good governance.
2. *Transparency* – Transparency is the ease with which an outsider is able to make meaningful analysis of a company's actions, its economic fundamentals and its ownership. It is a measure of how good management is at making necessary information available in a candid, accurate and timely manner – particularly audit data but also through general reports and press releases. It reflects whether investors get a true picture of what is actually happening inside the company.
3. *Independence* – Independence is the extent to which mechanisms have been put in place to minimize or avoid potential conflicts of interest that may exist because the management of a company may be dominated by its largest shareholders. This ranges from the composition of the boardroom, to the appointments in various executive management committees and of external parties, such as the auditors. The decisions made, and internal processes established, should be objective and not allow for undue influences.
4. *Accountability* – Individuals or groups in a company who make decisions and take actions on specific issues need to be accountable for their decisions and actions. Mechanisms must exist and be effective to allow for accountability, thus facilitating both transparency and responsibility. This provides investors with the means to query and assess the actions of the Board and executive committee. The Board's role should be clearly defined and published so that responsibilities can be assigned appropriately. Worldwide standards that should be employed include GAAP (Generally Accepted Accounting Principles (US)) and the International Financial Reporting Standards (IFRS). In addition, most countries have their own Accounting Standards Board and accounting standards – in Australia that is the Australian Accounting Standards Board (AASB) which administers the Australian Accounting Standards (AAS).
5. *Responsibility* – Responsibility refers to Board and management behaviour that follows internal mechanisms to allow for corrective action and punishment of mismanagement. Responsible management would, when necessary, put in place what it takes to set the company on the right path.
6. *Fairness* – The systems that exist within the company must be balanced in taking into account all those who have an interest in the company's future. The rights of various groups have to be acknowledged and respected. Minority shareholder interests must receive equal consideration to that of the dominant shareholder.
7. *Social responsibility* – A well-managed company will be aware of, and respond to, social issues, placing a high priority on ethical standards. While this might have indirect economic benefits to the organization via improved productivity, a good corporate citizen is increasingly seen as one that is non-discriminatory and non-exploitative, as well as being responsible with regard to environmental issues.

Byte Idea – WorldCom
www.mci.com

Corporate Greed and VERY poor Governance

WorldCom was the USA's second biggest telecommunications company and by conducting business in 65 countries it had made itself one of the world's biggest "backbone" networks for internet traffic and electronic commerce. With 85,000 employees, revenues of US$35.2 billion in 2001 and share prices that had hit US$64 in 1999 – this was one big BIG company...

That is, until 25 June 2002 when the company announced that an internal audit had found accounting for expenses not in accordance with generally accepted accounting principles – to the tune of US$3.055 billion of expenses in 2001 and US$797 million in the first quarter of 2002. These amounts were wrongly listed on company books as capital investment, instead of expenses, which artificially inflated earnings before interest, tax, depreciation and amortization. Without the improper accounting practices, WorldCom would have posted net losses in 2001 and the first quarter of 2002.

The fraud is far bigger in money terms than Enron's misdeeds, and is a damning indictment of the auditor Arthur Andersen which was responsible for approving the accounts of WorldCom as well as Enron.

As one might have expected, a series of company-breaking events rapidly followed – the Chief Financial Officer of WorldCom (Scott Sullivan) was sacked, the share prices crashed to less than the dollar, and on July 21, WorldCom and substantially all of its active US subsidiaries filed voluntary petitions for reorganization under Chapter 11 of the US Bankruptcy Code.

Following this, a number of WorldCom executives including the former CEO Bernard Ebbers and Scott Sullivan were arrested and charged with securities fraud and filing false statements with the USA Securities and Exchange Commission (SEC). Federal prosecutors alleged they colluded to hide billions in expenses, while lying to investors and regulators. In July 2005 Ebbers received a 25 year sentence which he is appealing and Scott Sullivan finally received a 5 year jail sentence commencing in August 2005.

The USA is reeling under this, the second major corporate fraud of the 21st Century and US Attorney General John Ashcroft has stated that 'Corporate executives who cheat investors, steal savings and squander pensions will meet the judgement they fear and the punishment they deserve'.

As a postscript – WorldCom emerged from bankruptcy in 2004 with a new name MCI Inc although it has yet to pay off many of its creditors.

Information sourced from http://news.bbc.co.uk/1/hi/business/2166426.stm, www.FoxNews.com and Wikipedia

Why Is Corporate Governance Important?

Corporate governance is one of the critical issues in business today. For companies, good governance means securing access to broader-based, cheaper capital. For investors, a commitment to good governance means enhanced shareholder value. For both, following good governance equals good business. In recent surveys on corporate governance (Coombes and Watson, 2000; Newell and Wilson, 2002; Hoschka, 2002; Gouvea *et al.*, 2002) the following key findings were reported:

- Strong corporate governance is seen as crucial to assess and reduce investment risks.
- Stocks with good governance provide higher returns.
- Stocks with good governance can command as much as a 27% premium in some markets.
- Good corporate governance can bring down the cost of capital.
- In a well regulated market, businesses may only meet the bare standards of corporate governance.
- In bad market environments, there are usually some companies with good corporate governance.

What these surveys also showed was that after the Asian crisis in 1997 (Barton *et al.*, 2004), many Asian countries are making a concentrated effort to improve corporate governance. The different components of 'good governance' vary between countries (Fig. 9.2), and Newell and Wilson (2002) suggest that companies in different countries should address different components that might be generally below par in those countries.

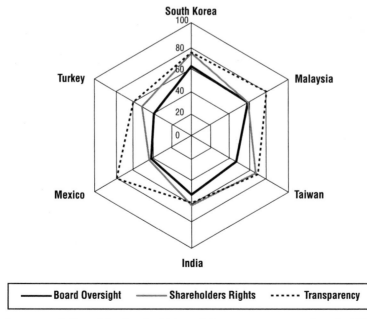

Fig. 9.2. A representation of the level which certain governance practices in 188 companies in South Korea, Malaysia, Taiwan, India, Mexico and Turkey attained when rated against key components of corporate governance (Board Oversight, Shareholder Rights and Transparency) in a recent survey by Newell and Wilson (2002)

Byte Idea – Parmalat
www.parmalat.com.au

How Can Things Go So Sour?

Once again a big multinational was caught out – or the senior executives were – for not ensuring that good corporate governance was maintained. Parmalat – the iconic Italian based dairy foods giant – managed to lose US$8.5 to US$12 billion in assets, in one of the largest financial frauds in history.

The trigger to the downfall happened when Parmalat defaulted on a US$185 million bond payment in mid-November 2003. In the ensuing investigation, it was found that managers simply invented assets to offset as much as US$16.2 billion in liabilities as well as falsifying accounts over a 15-year period, a state of affairs that forced the US$9.2 billion company into bankruptcy on 27 December 2003.

These events show how a lack of transparency in internal governance can be disastrous. Parmalat, like many other large European companies, is family controlled through a chain of holding companies, making corporate governance and supervision by regulators more difficult. In Parmalat's case it is founder and former Chief Executive Calisto Tanzi, whose family controls 51% of Parmalat, and his son Stefano and brother Giovanni as well as some 17 other individuals, including former Board members, who were arrested on suspicion of fraud, embezzlement, false accounting and misleading investors.

The trial started in September 2005 with all defendants claiming to be 'not guilty'. An interesting spin off is that prosecutors have also asked for the indictment of five global banks and securities companies for securities law violations. It is alleged that the banks provided false information on Parmalat's finances to investors and encouraged them to buy Parmalat stocks and bonds in the run-up to the crash.

The lesson is simple: keep it clean and stick to the Principles of Good Governance.

An article on company law that is worth reading on the ramifications of the Enron, Worldcom and Parmalat scandals can be found at www.infoconomy.com/pages/politics-management/group99613.adp

Information sourced from Forbes (www.forbes.com) and Business Week (www.businessweek.com)

EGovernance

The penetration of innovative communication technologies such as the internet into all facets of daily life, and in particular the business world, has created more freedom, flexibility and opportunities than there has ever been. The problem is 'opportunity' is not all good – the internet has gained a reputation as a largely unregulated realm, which in business terms provides a high-risk environment. There are three broad governance issues raised by the internet:

1. The management of the internet itself
2. How to legally govern activity on the internet
3. The corporate governance of the 'new enterprise' (B2C, B2B, etc.).

Managing the Internet

The internet is a technologically complex global communication network. The task is to balance competing interests around the world in the evolution of new technological standards. A main player in this area of technological standards is the World Wide Web Consortium (W3C), which develops interoperable technologies such as specifications, guidelines, software and tools for the effective delivery of web-based business.

The current environment associated with managing the internet is one of tension between four sectors: the **collaborative community**, the **deregulated competitive market,** the **established communications industry** and the **public regulator** of the day (Huston, 2002).

1. *The collaborative community.* In the beginning (early 1980s), the major characteristic associated with the governance of the network was that the internet was available through a collection of service providers, consumers, developers and researchers – all operating within a largely homogenous environment. The policy agenda was concerned with technological refinement, derived collaboratively. This approach enabled the internet to develop as a single environment of uniform connectivity rather than as fragmented islands of connectivity. In essence, this means that there was no formal regulation of the internet by an external body. Associated with this situation was the position of self-moderation of the internet by the internet community itself.

   Today, the collaborative community still has a strong influence over policy creation associated with the internet, although its level of influence over the technology and the resulting internet service environment is significantly reduced. The collaborative community's major contribution remains in the maintenance of the ethos of collaboration and active involvement by the community at large.

2. *The deregulated competitive market and communications environment.* The deregulated competitive market is now the dominating factor in the internet both in terms of usage (e.g. traffic volumes) and influence (e.g. investment levels). The internet service provider (ISP) industry has the view that the internet is an open and unregulated marketplace in which goods and services are freely traded based on perceptions of need and demand. This sector sees the internet as an opportunity for exploitation by market forces in the same way as other media do, such as radio and TV.

   The internet operates as a service and within many countries it operates under the terms and conditions of a value-added reseller rather than being an activity of an authorized common communications carrier such as a government owned entity. In most countries (including Australia – e.g. Telstra), the telecommunications industry is one that is in the process of complete deregulation.

   The motivation for deregulation is to encourage private investment in national infrastructure and to provide consumers with a range of services that are geared to efficient and effective service operation. Within such a framework, the internet then operates as any other value-added service in which service definition, pricing, policy of operation, and all other aspects of the service are defined through the operation of market forces within a competitive market. Additionally, policies are derived through a mix of market pressures and client expectation, and service and infrastructure roles are derived through attempts to generate both price and more critically service differentiation within the marketplace, rather than common collaborative agreement.

3. *The established communications industry.* Originally, the established communications industry had envisioned a data environment more on terms similar to telephony, in which the characteristics of the data service were an intrinsic attribute of the network rather than the host. This model gave many more elements of control to the network provider domain. Thus the established communications industry resisted the internet approach to supporting networking services for some time, offering an open-system-interconnection-based network service architecture as an alternative approach to the industry. However, the current approach from this sector is to quickly come to terms with the market recognizing that the internet is a significant market force. From a policy perspective, the initial motivation within this sector is to ensure that this activity does not erode current market position; the secondary motivation is to develop this market in a way that enhances the value of their total enterprise.

4. *The public regulator.* Public regulation pressure to provide international and national policies in relation to managing the internet to ensure that the stated social objectives are being achieved within the competitive marketplace is growing. There are three main areas of pressure in terms of a public regulatory resource: **fair trading, long-term social obligations and objectives** and **the avoidance of negative national outcomes**.

   i. *Fair trading* is a balance between a market trend towards the aggregation of supply channels, and the need to ensure that genuine and open competition exists that allows market forces to exert pressure on providers so that they operate in an efficient fashion. Examples of regulations include ensuring that current market dominance by any one player is not abused, that consumers are charged a fair price for the service, that new entrants are not artificially excluded from participating in the market, and that the provider environment is fair and open to all players.
   ii. *Long-term social obligations and objectives* recognize that the internet is a nationally important resource and one that would appear to have significance to the public interest in the future. Thus public resource management measures need to be imposed to ensure the efficient and effective use of the internet as a resource in the common national interest, and would essentially revolve around the current and likely future areas of uptake of the service.
   iii. *Avoiding negative outcomes from a national perspective* includes creating measures that will minimize the risks associated with companies using the internet as a major business tool to:
       • create an unregulated monopoly position over a strategically critical or widely used resource
       • disenfranchise areas of the population through the deliberate actions of service providers that operate outside any national policy framework
       • develop a situation where a nationally strategic resource is controlled by non-national interests.

The role of a national public resource manager has the objective of ensuring the maximal positive outcome of a public commodity or service for the benefit of the nation and the benefit of the national population. However, public policies must be effective within the context of the internet itself, and must reflect an adequate understanding of the current nature and environment of the internet. They should also make clear a coherent objective in terms of a future vision of the internet and its role in supporting broader social objectives. Because the internet environment is constantly evolving and expanding, the policy challenges are thus associated with estimating likely outcomes of further rapid growth in that environment and guessing the areas of opportunity that are presented within such a process.

ICANN

Internationally, there are a number of bodies that have been formed to look at specific internet governance policy issues – for example, ICANN (the Internet Corporation for Assigned Names and Numbers), which addresses directly the applicability of local and international law (Wilkinson, 2003). The ICANN Governmental Advisory Committee (GAC) has endorsed the scope of the public policy issues that should be within its remit including the general principle that the internet naming and addressing system is a public resource that must be managed in the interests of the global internet community.

ICANN is constantly confronted with significant differences between jurisdictions in certain areas of law – for example, respect for local and international law has in practice been interpreted as including competition laws, intellectual property laws, and data protection and privacy laws. However, ICANN currently functions as the sole regulatory coordinating organization for the internet infrastructure as well as being the creator of rules for the market for domain names and for the allocation of internet protocol addresses. This latter role is increasingly implemented via contracts between ICANN and the various registries and registrars around the world. In such situations, the contractual framework often affects a range of economic and legal issues, including prices, terms of conditions of registration, and dispute resolution and it is thus of great importance that ICANN operates in the public interest and is transparent to scrutiny.

In addition, ICANN determines the basis on which registration data is made available to the public, including (eventually) personal data and, through working with the IETF (Internet Engineering Task Force) and the IAB (Internet Architecture Board), the organizational framework in which the internet technical standardization process takes place.

How to Legally Govern Activity on the Internet

Legally governing activity on the internet remains the responsibility of the government in every country that is connected to the internet, and includes:

- The regulation of business transactions and securities trading
- The regulation of consumer protection (including minors)
- Fairness in advertising
- The protection of intellectual property (IP)
- Taxation on various goods and services
- Prohibition (in some countries) of internet gambling
- Prohibition in trafficking of alcohol and other banned substances across borders
- Food and safety regulations
- The protection of free speech
- Controls on indecent materials.

Each country has a different approach to these issues, which can cause fundamental problems in attempting to create good governance polices for the internet. These issues are dealt with later in this chapter.

The Corporate Governance of the New Enterprise

A new enterprise in this case means the B2C, B2B and eMarketplace online exchanges where information and information trading are the major commodities. As discussed in earlier chapters, information is power. The reality is that whatever business model a company uses, information plays a critical role in setting up governance and control mechanisms because access to quality and timely information and knowledge forms the basis of decision making and controlled action which can affect the security and profitability of the company.

Thus in addition to standard governance principles as detailed at the beginning of this chapter, measures to protect and control data privacy (both corporate and private information), taxation issues, content regulation, copyright, spam, technology deployment and more, are necessary, and are detailed in the next section.

Major Governance Issues for Conducting 'E' Enabled Business

Without a doubt, the global nature of the internet greatly complicates effective governance at the organizational electronically enabled business level. Transactions pass across borders without detection, thus any legal regime put in place to govern internet based activity must address issues of compliance and enforcement. Major conflicts can arise, however, in the area of **jurisdiction** which is about whose rules will apply to a transaction when the buyer and seller are located in different countries, and **sovereignty** which deals with how much domestic control over access to the global internet must be relinquished in order to promote global eCommerce and electronically enabled business.

PriceWaterhouseCoopers (1999) discuss these two issues in a document entitled 'Electronically Enabled Business Regulation: The Borderless Economy in a World of Borders' using the following points to create the basis of defining the major operational issues for conducting electronically enabled business:

- *Electronically enabled business is, by definition, global.* Business methods that are effective and acceptable in one jurisdiction may not be permitted or will not work well in other markets. Variations in practice across national lines, despite the best efforts of international rule-making bodies, should be expected.
- *Electronically enabled business shrinks the window for optimal regulatory action.* New business arrangements, with industry-churning impacts, can have impact in months not years. This rapid change means that regulatory issues must be addressed early on to avoid overly 'reactive' responses that could be counterproductive.
- *Electronically enabled business effectiveness depends on a regulatory environment that is both supportive and predictable.* While onerous rules can be stifling to business interests, regulatory indecision can be similarly disruptive. In order for electronically enabled business to work best, business must accept equal responsibility with governments to point the way.

Using the above guidelines, the major operational governance issues that electronically enabled businesses have to deal with can be defined and include:

- Trade and tariffs
- Data security
- Encryption
- Infrastructure and access
- Intellectual property rights
- Liability: choice of law and jurisdiction
- Content
- Antitrust legislation
- Self-regulation.

Operational Governance Issues for Electronically Enabled Business

Regulatory issues are inherent for electronically enabled business and both a knowledge of the issues, and an active involvement in the policies associated with them, will ensure better run and more profitable businesses. The following provides a brief overview of the major operational governance issues that are involved in managing an electronically enabled business:

- *Trade and tariffs.* The internet is changing who buys and sells in the international marketplace and what is bought and sold. It also potentially changes where the trade took place and thus where tax and tariff liabilities are due. Customs duties or tariffs are the most basic barrier to international trade. Currently, customs duties are not applied to electronic transmissions. However, the *product* that is ordered or exchanged via electronic means *is still* subject to customs duty and any other tax (e.g. GST in Australia) that is normally applied in the country in which the product is imported into. Organizations need to know what law prevails where.
- *Data security.* This is one of the most difficult international electronically enabled business governance challenges and includes issues of electronic authentication, digital signatures, data encryption and security of personal data (see Chapter 8). Companies are vulnerable at many levels from simple theft to credit-card and banking breaches of security, to the stealing of IP and damage to their brand and reputation. Many countries have amended national laws to allow the provision of electronic signatures as a legal alternative to written ones on a document.
- *Encryption.* The challenge with encryption technology is to ensure that legitimate users can make use of it to provide security for their customers, but that illegal activity using unbreakable encryption algorithms is not fostered.
- *Infrastructure and access.* This is primarily a broader telecommunications issue and varies from country to country. In Australia one of the major barriers to electronically enabled business uptake is access appropriate infrastructure – particularly in rural areas.
- *Intellectual property rights.* Copyright, patent and trademark concepts and laws are well established in the non-online environment. However, the internet provides many opportunities for piracy of intellectual property and protecting IP on the internet poses challenges both for individual businesses and governments since digitized information can be copied and distributed internationally with ease. IP is covered in more detail later in this chapter.

- *Liability: choice of law and jurisdiction.* As discussed above, this is a major area of concern. 'Choice of law' refers to the determination of what country's law should apply in the event of a dispute resulting from a cross-border electronic transaction. Jurisdiction refers to which country's legal system will hear the dispute. Companies can subject themselves to international liability without realizing it. Currently there is no universal agreement on rules governing jurisdiction or choice of law for transactions conducted on the internet.

- *Content.* Content management is addressed in Chapter 3. However, while content can enhance a potential customer's visit to a website, it can also be dangerous because imported material can infringe a third party's IP rights, defame someone or be illegal in some countries. A particular problem area is in the recording industry where the internet and associated technologies have made it very easy to distribute quality copies of music illegally. The main issues in regulating content concern who should regulate the information available online and the question of where does free speech end and appropriate regulation begin.

- *Antitrust legislation.* Antitrust legislation protects trade and commerce from unlawful restraints and monopolies or unfair business practices. Competitive practices of companies have long been subject to scrutiny by government regulatory bodies. In the non-digital era, a large market share was evidence for antitrust action: this has not been the case in relation to businesses using eCommerce to extend their reach, although there is in place a World Trade Organization (WTO) principle to prevent suppliers of internet access, network and certification services engaging in non-competitive practices.

- *Self-regulation.* Most western governments encourage the private sector to take the lead in establishing the rules governing electronically enabled business. For the individual business, clear policies that underpin an organization's electronically enabled business processes are a good measure to prevent the unwanted attention of a regulator.

ePharmacy (see Byte idea this chapter) is a good example of a company that has had to address all of these issues when deciding to go into the online environment to sell drugs and other pharmaceutical products.

Byte Idea – ePharmacy

www.epharmacy.com.au   www.health.gov.au/tga/

Regulated eCommerce

Buying and selling of therapeutic goods via the internet can be a dicey business. There are a number of questions (other than the standard ones of security related to payment online) that immediately spring to mind – from a buyer's perspective, and from the seller's perspective.

From a buyer's perspective:

- What are you actually buying, from whom and from where?
- What are the quality controls associated with the production, storage and distribution of the product?
- Is the drug/product legal – both in the country (apparently) of origin, and where it is going?
- What are the regulations in relation to export and importation of the product and who governs this?
- If a drug is called 'XX' in one country and 'XV' in another country is it the same drug and which regulator/regulations cover it?
- Are there any quarantine or tax/customs duties related issues?
- Will my personal information be kept private by the vendor?
- Can I buy prescription drugs online?
- If a drug is a prescription drug in one country and not in another, can I buy the drug online in the country where it is freely available and use it legally in the country where it is normally only available via prescription?

And from the online retailer's perspective:

- Does doing business via electronic means create any additional issues – if so what are they?
- What are the regulations associated with buying and selling therapeutic drugs in the country of origin and the country where I do business?
- Given I now have a global presence via the internet, what does this mean in terms of my IP protection, IT security and information privacy issues?
- What are my business risks, legal liability issues and insurance requirements?

ePharmacy is one of Australia's first true 'online' pharmacies and operates under the same Australian laws and standards that all other Australian retail pharmacies practise. Savings on products are due to aggressive pricing and larger volumes of trade. No product that has not been approved for sale in Australia by the relevant authorities is sold online. With strict privacy and security arrangements in place ePharmacy endeavours to maintain good corporate eGovernance and to always practise within the regulatory framework of the Australian Therapeutic Goods Act of 1989.

Information sourced from ePharmacy and the ePharmacy website

The International Regulators

Electronically enabled business is a global phenomenon and the internet as a communication technology that is available 24 x 7 ensures that. While governments in general are looking to the private sector to lead the establishment of governance criteria and standards, there are a number of organizations around the world that already act as regulators in terms of international trade governance issues. These include the World Trade Organization (WTO), the Organization for Economic Cooperation and Development (OECD), United Nations Commission on International Trade Law (UNCITRAL), the International Chamber of Commerce (ICC), Asia Pacific Economic Cooperation (APEC) and the World Intellectual Property Organization (WIPO).

The World Trade Organization (WTO) (www.wto.org)

The WTO, with 148 member countries, is the biggest and most powerful international organization regulating trade. It is a rules-based, member-driven organization (i.e. all decisions are made by the member governments), and the rules are the outcome of negotiations among members, and are backed by binding dispute resolution mechanisms. A major player in terms of agricultural product and food trade, electronically enabled business issues that the WTO is interested in include (www.wto.org/english/thewto_e/thewto_e.htm#intro):

- The classification of internet products and transactions
- The application of customs duties on electronic transactions
- The valuation of products that are transmitted electronically
- The scope of the WTO Agreement on Telecommunications in relation to the internet.

Organization for Economic Cooperation and Development (OECD) (www.oecd.org)

The OECD is a group of 30 member countries (including the USA, UK, Australia, Japan, Korea and many European countries) that share a commitment to democratic government and the market economy (www.oecd.org/about/). With active relationships with 70 other countries, it has a global reach. The OECD plays a prominent role in fostering good governance in the public service and in corporate activity. It helps governments to ensure the responsiveness of key economic areas and produces internationally agreed instruments, decisions and recommendations to promote good governance in areas where multilateral agreement is necessary for individual countries to make progress in a globalized economy. The OECD's electronically enabled business governance interests include:

- Taxation
- Consumer fraud and protection
- Privacy of personal and corporate information
- Security.

The United Nations Commission on International Trade Law (UNCITRAL)
(www.uncitral.org)

UNCITRAL is the core legal body of the United Nations system in the field of international trade law. It is made up of 60 member states elected by the General Assembly and has a mandate to modernize the rules on international business conventions, model laws and rules which are acceptable worldwide, and to create legal and legislative guides and recommendations of practical value.

In 1995 UNCITRAL created a Model Law of eCommerce which covers general legal principles governing electronic commerce and provides standards by which the legal value of electronic messages can be assessed. The 1995 Model Law also contains rules for electronic commerce in specific areas, such as carriage of goods. In 2001 the Model Law on Electronic Signatures was created which in particular deals with electronic contracting and the use of digital signatures.

International Chamber of Commerce (ICC) (www.iccwbo.org)

The ICC is *the* world business organization. It was founded in 1919 and today it represents thousands of member companies and associations from over 130 countries. The ICC promotes an open international trade and investment system and the market economy, and it also develops rules that govern the conduct of business across borders. These rules are voluntary, but they are observed everywhere transactions take place and have become part of the fabric of international trade. Business leaders and experts drawn from the ICC membership establish the business stance on broad issues of trade and investment policy as well as on vital technical and sectoral subjects. These include financial services, information technologies, telecommunications, marketing ethics, the environment, transportation, competition law and intellectual property, among others.

The ICC through one of its member groupings – the Alliance for Global Business (AGB) – has recently published the third edition of the Global Action Plan for Electronic Business (July 2002). The objective of the action plan is to provide an inventory of fundamental business views on the issues that government must deal with in terms of electronically enabled business governance and to give a clear overview of business action in those areas where market-driven, industry-led solutions are most likely to be found. Table 9.3 outlines the strategic objectives of the Global Action Plan and lists the issues that need addressing to achieve these objectives.

Table 9.3. The ICC Global Action Plan for Electronic Business, Third Edition, July 2002

Strategic Objectives	Issues to Be Addressed
Maximizing the benefits – economic and social impacts	• Economic and social impacts • Small and medium-sized enterprises • Skills development • Ensuring global participation • Infrastructure deployment • Government as a model user
Electronic business and the information infrastructure – trade aspects, standards, internet names, numbers	• Competition and trade-related aspects of electronic commerce • Telecommunications competition • IT equipment • Trade-related aspects of intellectual property • Convergence, standards • Internet names and numbers, domain name system
Building trust for users and consumers	• Protection of personal information • Privacy and trans-border flows of data, internet privacy • Consumer trust and content • Online alternative dispute resolution (ADR) • Content • Marketing and advertising ethics • Unsolicited commercial communications • Issues relating to cybersecurity • Cryptography for confidentiality • Information sharing • Cybercrime • Business monitoring of its own communications and networks • Data storage requirements • Searches and subpoenas of computer records • Access to public domain information • Legal government interception of telecommunications and electronic communications • Electronic authentication, legal validity of electronic signatures, interoperability of certificates and electronic signatures • Accreditation • Availability of certification practice statements
Establishing ground rules for the digital marketplace	• Contractual and other legal issues • Removing legal/regulatory obstacles • Create a new uniform legal framework • Jurisdiction and applicable law • Incorporation by reference • Transparency and availability of proprietary and best practice • Legal terms, model contracts, etc., dispute settlement • Fraud and other commercial crime, liability • Taxation and tariffs, customs duties • Taxation • Trade facilitation and customs modernization • Private/public sector interface • Capacity building • Intellectual property, copyright and neighbouring rights, trademarks • Databases

Asia Pacific Economic Cooperation (APEC) (www.apec.org)

The Asia Pacific Economic Cooperation (APEC) was established in 1989 in response to the growing interdependence among Asia Pacific economies. APEC started as an informal discussion group, but has since become the primary regional vehicle for promoting open trade and practical economic cooperation.

APEC is currently participating in encouraging eCommerce in the region by establishing the Electronic Commerce Work Program with the aim of building trust and confidence, enhancing government use, intensifying community outreach, promoting technical cooperation, directing work towards eliminating impediments to its uptake, and developing seamless legal, technical, operating and trading environments to facilitate its growth and development.

Developments include eCommerce Best Practice Guidelines for SMEs in the APEC Region (2001), the creation of B2C eCommerce/eBusiness Market Liasions with other international organizations to facilitate cyber security development, RFID Advocacy and Consumer Trust (2004), and most recently Public Key Infrastructure (PKI) Guidelines: Guidelines for Schemes to Issue Certificates Capable of Being Used in Cross-Jurisdiction eCommerce (2005).

The World Intellectual Property Organization (WIPO) (www.wipo.int)

WIPO is one of the 16 specialized agencies of the United Nations system of organizations. It administers 23 international treaties dealing with different aspects of intellectual property protection. The organization counts 179 nations as member states.

In 2000 WIPO published a 'Primer on Electronic Commerce and Intellectual Property' (WIPO, 2000) that addressed the main issues of electronically enabled business as they impact upon intellectual property, as well as the role of intellectual property in facilitating electronically enabled business. While this Primer is still well regarded globally, WIPO recently produced a follow-up document that deals with the results of a survey of intellectual property on the internet as it stands in 2005 (WIPO, 2005). Many of these issues will be dealt with in the next section on cyber law.

Cyber Law

It is not the intention of this section to give great detail on legal matters associated with using the internet – however, it is important that managers are aware of the main areas of potential concern under law when a company becomes involved in electronically enabled business. An excellent reference (as discussed above) is the WIPO Primer on Electronic Commerce Law and the more recent WIPO Survey on Issues of Intellectual Property on the Internet (both of which can be downloaded from the WIPO site).

According to WIPO, there are three main areas under which legal challenges can take place in relation to electronically enabled business – and these are:

1. The Paperless Environment – Electronic Contracts
2. The Internet – Jurisdiction and Applicable Law
3. Digital Technology – Issues of Enforcement and Privacy.

The Paperless Environment – Electronic Contracts

Electronically enabled business generates interactivity and transactions between parties that may have had no previous contact. These dealings can occur in real time over the internet between businesses, or between businesses and consumers. These transactions need rules to govern the relationship between the parties. The main instrument for these rules is the agreement itself – that is, the contract – which provides a flexible yet legally enforceable mechanism.

UNCITRAL's 1995/2001 Model Law of ECommerce covers this area of electronic contracting mainly in relation to electronic data interchange (EDI), which until recently was considered among the most common forms of online contract. In Chapter III, Article 11 the Model Law states that:

> *'In the context of contract formation, unless otherwise agreed by the parties, an offer and the acceptance of an offer may be expressed by means of data messages. Where a data message is used in the formation of a contract, that contract shall not be denied validity or enforceability on the sole ground that a data message was used for that purpose.'*

The accompanying Guide to the Model Law's Enactment (Chapter III, Article 11 (76)) clarifies this further by stating that:

> *'The Model Law is not intended to interfere with the law on formation of contracts, but rather to promote international trade by providing increased legal certainty as to the conclusion of contracts by electronic means and that contracts in electronic commerce should continue to meet with the traditional, technology-neutral principles that are necessary for validity.'*

An exchange of consideration (i.e. something of value) is also necessary to transform the agreement from merely a set of promises into a binding and enforceable contract. Electronic business raises questions as to how an offer, acceptance and exchange of consideration in the online environment can be achieved. It places a premium on the clarity and transparency of the contractual terms and conditions, particularly as electronic contracts may involve parties from different parts of the world who may have had little or no interaction with each other apart from their communications online.

With respect to contractual and evidentiary formalities, the UNCITRAL Model Law on Signatures (Chapter II, Article 7) says that:

> *'...where the law requires information to be in writing, that requirement is met by a data message if the information contained therein is accessible so as to be usable for subsequent reference. With respect to a legal requirement that information be in 'original form', this requirement is met if there exists a reliable assurance as to the integrity of the information from the time when it was first generated in its final form, as a data message or otherwise.'*

In addition, the Model Law (Chapter II, Article 7) is clear regarding a signature requirement, stating that digital signatures are sufficient if the method used in an electronic communication to identify a person and indicate that person's approval of the information contained in the message:

'...is as reliable as was appropriate for the purpose for which the data message was generated or communicated, in light of all the circumstances, including any relevant agreement.'

Many contracts in electronically enabled business implicate the intellectual property rights of one of the parties to the contract. Licences, distribution and franchising agreements, and joint venture arrangements are some of the most common forms of contract that protect intellectual property rights. For example, a licence is a contract that authorizes the licensee to do something that, in the absence of the licence, would normally constitute an infringement of the licensor's intellectual property right. When consumers on the internet access a musical composition, they are allowed to do so subject to a licence agreement. The business distributing the music, in turn, may hold licences from the copyright owner and the producer of the sound recording (all of whom are likely to be in different countries). Thus when setting up such contracts, questions of jurisdiction, applicable law and enforcement should be carefully considered – at the time of contracting – to bring added certainty, and where possible to limit potential exposure, for businesses and consumers.

The Internet – Jurisdiction and Applicable Law

Users can access the internet from almost any place on Earth and using worldwide telecommunications infrastructure, digitized information can travel through many countries and jurisdictions, each with its own legal system, in order to reach its destination. The internet can thus be thought of as *multi-jurisdictional*, and in consequence the jurisdictional issues associated with law are complex, particularly in relation to intellectual property, contracts, fraudulent behaviour of all kinds, consumer protection, taxation, and the regulation of online content relating to obscenity and criminal law. In the context of private international law, the following cross-country issues are important to be aware of:

- Jurisdiction to adjudicate a dispute at a particular location
- The law applicable to the dispute (choice of law)
- The recognition and enforcement of judgments in courts in foreign jurisdictions.

Such issues are complicated in electronically enabled business situations because one or more of the parties involved (or processes used) in the commercial activities – including internet users, service and content providers, buyers, sellers, businesses (and their assets), technology systems and computer servers – may be located in different countries. In such situations, it may not only be unclear as to *where* the relevant activities are taking place, but the activities themselves can have both intended and unintended consequences around the world, resulting in uncertainty when it comes to questions of localizing the dispute, determining the applicable law, and the practicalities of pursuing enforcement or adequate dispute-settlement alternatives.

Owners of intellectual property seeking to manage their rights through licensing agreements, or to enforce them against infringement, are confronted with complex issues. In the case of a licence to cover rights on the internet, you must consider which laws in which countries may have a bearing on the agreement, including laws addressing electronic contracts, consumer protection, intellectual property, disclaimers and privacy aspects.

The difficulty in terms of determining the jurisdiction in which to adjudicate stems from a reliance, in private international law, on physical 'points of attachment', such as 'the State *in which* the act or omission occurred' or '*in which* the injury arose', for determining jurisdictional competence. This approach does not address the multi-jurisdictional character of the internet and the activity conducted on it.

An example illustrating this dilemma is the case of an internet-based copyright infringement. When a user in one country or users in multiple countries download an allegedly infringing copy of a copyrighted work from a foreign website, has an offence (the unauthorized copying or distribution of the work) occurred in the user's location (i.e. by copying the work into the memory of the user's computer or other digital storage device), thus triggering a jurisdiction issue – or does the offence occur in the foreign State where the computer server hosting the website is located, with only an impact in the location of the user?

Similar questions may arise in terms of an alleged infringement of a trademark on the internet. If a company in one country operates a website using a sign that has not been registered there by any third party, and offers commercial services in connection with the use of the sign in an allegedly infringing manner in other countries where a corresponding trademark is registered, where has the infringement occurred, and where does jurisdiction lie?

Without an appropriate regulatory framework, someone using a trademark on the internet may potentially be sued in court in any country of the world, and courts will have to determine whether a sufficient connection exists to justify the exercise of jurisdiction.

Digital Technology – Issues of Enforcement

The internet has created new challenges in relation to issues of enforcement because all content reduces to digital data on the internet. Text, music and images are converted to strings of binary code which have potential for indefinite storage in computer memory and network devices. Vast amounts of information and intellectual property can therefore be transmitted in digital form to anyone with access to the network, thus hugely increasing the ease with which intellectual property can be infringed, either inadvertently or through piracy and counterfeiting. There are two main issues:

- Detecting infringements
- Effective enforcement.

Detecting Infringements

Detecting infringements associated in particular with intellectual property rights is not a simple matter. IT users can duplicate, manipulate and change content – perfectly, instantly and continuously – in ways that may be largely undetectable, thus increasing the opportunities for confusion, fraud and infringement of intellectual property rights. For example, one publication of protected content on the internet can lead to its proliferation around the world through rapid copying by third parties, making enforcement difficult in an international arena.

In addition, digital data is transient and infringing material may be on the internet for only a very short period of time, as files and/or content can be deleted within a matter of hours after their posting on the web, email, newsgroup, bulletin board or chat room. Sites that have been ordered to close down in one jurisdiction may easily reappear in another, or may be mirrored across multiple jurisdictions, thereby frustrating the effects of local enforcement proceedings.

Once an infringement has been detected, the party causing the problem needs to be identified in order to set in motion the enforcement process. However, the internet, by its very nature, makes anonymity possible, and tools are available, such as strong encryption technology, that can make it virtually impossible to detect who is at the source of a particular communication, particularly if a deliberate effort is made by the user to remain unknown.

Many companies that offer services for posting and distributing content on the internet (e.g. domain name registrars, bulletin board operators and commercial web page hosts) require their customers to identify themselves and specify contact details, but there are new restrictions on the collection, storage and public availability of this type of data as the issue of information privacy becomes more central. While privacy is emerging as a new area of public policy, appropriate solutions will also need to accommodate the recognized need for IP protection and enforcement.

Effective Enforcement

Effective enforcement pre-supposes an underlying legal framework that is conducive to the enforcement of the rights concerned in or on the medium where enforcement is sought. Unfortunately, IP law and its accompanying enforcement mechanisms are fundamentally physically based and territorial in nature. The scope of the rights established in each country is determined by that country and the effect of these rights, as well as their protection, are in principle confined to the territory of the country. As such, the internet is less of a threat to the rights of intellectual property holders than it is to the means by which intellectual property has been traditionally managed in the physical world. Therefore, the real issue is how to ensure that enforcement can be made less dependent on the concept of territory.

There are a number of ways this can be achieved including the use of technological measures of protection such as **encryption** and **watermarking** which are intended to permit owners and rights-holders to control access and manipulation of their works, and to track them on the internet.

- *Encryption* allows content to be transmitted through the internet in a scrambled, illegible format, which can only be decoded by means of a decryption key, the receipt of which may be conditioned upon payment to gain access to the work.
- *Watermarking* consists of embedding in a work and its legitimate copies data that permits the identification of rights-holders. The same technique can also be used to prevent a work from being modified (e.g. removing the watermark), because any tampering can be made to result in a visible or audible rearrangement of the data.

Other technologies allow works to be licensed online, eliminating many of the transaction costs involved in traditional forms of licensing.

The Impact of 'E' Enabled Business on the IP System

The intellectual property (IP) system covers issues of copyright and related rights, patents and trademarks. According to IP Australia (2005), ownership of IP rights is the legal recognition and reward received for creative effort. IP is a business asset and protecting it should be an integral part of the business process. In Australia, copyright issues, patents and trademarks have been Commonwealth Government functions since Federation and are administered by IP Australia, the Attorney General's Department and Plant Breeders Rights Australia. As discussed in previous sections of this chapter, the IP system can be severely compromised by the internet and business using it as the main medium for information dissemination and communication.

Copyright and Related Rights

Under the most important international copyright convention, The Berne Convention (Paris Act 1971), copyright protection covers all 'literary and artistic works'. This term covers diverse forms of creativity, such as writings, both fiction and non-fiction, including scientific and technical texts and computer programs; databases that are original due to the selection or arrangement of their contents; musical works; audiovisual works; works of fine art, including drawings and paintings; and photographs. Copyright does *not* cover the ideas themselves, only the original expression of the ideas.

Related rights protect the contributions of others who add value in the presentation of literary and artistic works to the public: performing artists, such as actors, dancers, singers and musicians; the producers of phonograms, including CDs; and broadcasting organizations. In Australia, copyright is free and is automatic under the *Copyright Protection Act 1968.*

Digital technology enables the transmission and use of all of these protected materials in digital form over interactive networks. Materials protected by copyright and related rights span the range of information and entertainment products, and constitute much of the valuable subject matter of electronically enabled business. Given how easy it is to copy material over the internet, if legal rules are not set and applied appropriately, digital technology has the potential to undermine the basic tenets of copyright and related rights. The result could be the disruption of traditional markets for the sale of copies of programs, music, art, books and movies.

The definition of rights and what exceptions and limitations are permitted is the key issue for policy makers and the legal fraternity, as intellectual property is no more or less than the sum of the rights granted by law. The development of digital technologies, which allow the movement of copyrighted works over networks, has raised questions about how these rights apply in the new environment. According to WIPO, these include:

- When multiple copies are made as works traverse the networks, is the reproduction right implicated by each copy?
- Is there a communication to the public when a work is not broadcast, but simply made available to individual members of the public if and when they wish to see or hear it?
- Does a public performance take place when a work is viewed at different times by different individuals on the monitors of their personal computers or other digital devices?

Similar questions are raised about exceptions and limitations to rights. Some exceptions, if applied literally in the digital environment, could eliminate large sectors of existing markets. Others may implement valid public policy goals, but be written too restrictively to apply to network transmissions.

Enforcement of Rights

In order for legal protection to remain meaningful, rights-holders must be able to detect and stop the dissemination of unauthorized digital copies, accomplished at levels of speed, accuracy, volume and distance that in the past were unimaginable. And for electronically enabled business to develop to its full potential, workable systems of online licensing must evolve in which consumers can have confidence.

In 1996, two treaties were concluded at WIPO: the WIPO Copyright Treaty (WCT) and the WIPO Performances and Phonograms Treaty (WPPT) (commonly referred to as the 'internet treaties'). These treaties address the issues of the definition and scope of rights in the digital environment, and some of the challenges of online enforcement and licensing. They recognize the emerging role of technological protection measures, and online management and licensing systems. In Australia, the *Copyright Amendment (Digital Agenda) Act 2000* was passed to address these issues.

Patents

The patent system provides a framework for innovation and technological development by granting an exclusive right to the owner of a patent to exploit an invention (whether products or processes, in all fields of technology, provided that they are new, involve an inventive step and are capable of industrial application) for a limited period, while at the same time balancing this right with a corresponding duty to disclose the information concerning the patented invention to the public. This information, which is classified and stored in the patent documentation, is available to anyone and increasingly is accessible through online, internet-based systems. This mandatory disclosure of the information associated with the invention adds to the available pool of technological knowledge, facilitates technology transfer, and enhances the opportunities for creativity and innovation by others.

In the past, the market exclusivity established through effective patent protection (an application has to be made to the Patent Office both in the country of origin and wherever else you wish it to be protected) has provided a reward for investment and has justified the expenditures on research and development to achieve further technological progress. However, digital technologies and electronically enabled business pose challenges to the conventional legal scheme for the patent system.

Patents have recently been granted to certain inventions concerning financial services, electronic sales and advertising methods, business methods consisting of processes to be performed on the internet, and telephone exchange and billing methods. Although some patent offices have established examination guidelines for computer related inventions, including software related inventions, very little international agreement has been achieved in this area. In the field of information technology, the value of intellectual assets often resides in the 'content' of the information. In the past, software has often been sold as an integral part of the computer system, while today software products are often marketed in the form of computer readable media – for example, diskettes and CD-ROMs or directly over the internet.

Software-related inventions are thus stored in such media, and commercialized separately from the computer hardware. It is necessary, therefore, to claim such software-related inventions as 'a computer readable medium storing the software that performs the claimed functions' which is known as a **Beauregard-type claim**, or 'a computer readable medium storing a data structure, which data structure is interrelated to the medium structurally and functionally' which is known as a **Lowry-type claim**, or a claim to 'a computer data signal that is embodied in a carrier wave' known as a **propagated signal claim** (WIPO 2005).

Enforcement of Rights

Patent protection is provided on a country-by-country basis, and the patent law of each country has application only within its borders in accordance with the traditional principles of territoriality. The internet thus poses significant problems for the enforcement of rights associated with protecting the patent.

For example, where patented software is sold and delivered over the internet internationally, any infringement action would require a consideration of the jurisdictional and choice of law issues. Moreover, the first practical issue may be that of detection, since the unauthorized importation of such software by means of the internet, unlike the importation of tangible goods, cannot be detected and stopped by customs authorities.

Trademarks

The identity of business goods and services is a major business asset and should be protected. A trademark enables consumers to identify the source of a product and to link the product with its manufacturer in widely distributed markets and is thus an important tool in commerce. The exclusive right to the use of the mark, which may be of indefinite duration, enables the owner to build goodwill and reputation within the expression of its identity, and to prevent others from misleading consumers into wrongly associating products with an organization from which they do not originate.

Trademarks are essential in electronically enabled business because organizations need to build recognition and goodwill, and inspire confidence in themselves and in their brands. This is even more important in the virtual world of online trading than in the offline world because there is little or no opportunity to inspect goods or services before purchasing them. Therefore, in the online domain, consumers will reward trusted and branded sources who are offering competitive products over and above unknown sources irrespective of price. In such circumstances, a company's mark or brand becomes a vital means of identifying and distinguishing itself.

Thus there is a growing international consensus that trademark protection should extend to the internet, and that it should be neither less, nor more, extensive than that which exists in the physical world. As soon as a trademark is used on the internet, it is immediately visible to a potentially global public and might be considered to have a global effect. This particular feature of the internet makes it extremely difficult for businesses to foresee in which countries their business activities might become legally relevant. Even within the boundaries of a single legal system, it is often difficult to fit the 'use' of a trademark on the internet into traditional legal concepts.

Given the rapid and continuing development of electronically enabled business, it is almost impossible to give an exhaustive list of ways in which trademarks can be used on the internet and to project what new forms of use might raise questions in the future. For the present, some of these practices such as *hyperlinking* or *metatagging* are currently indispensable for an efficient use of the web. Nevertheless, they pose potential threats to trademark owners since they facilitate the creation of associations, thus increasing the danger of confusion, dilution or other forms of unfair exploitation of trademarks.

Use of Trademarks as Metatags

A metatag is a keyword embedded in a website's HTML code as a means for internet search engines to categorize the contents of the website. Metatags are not visible on the website itself (although they can be made visible together with the source code of the page); however, a search engine seeking out all websites containing a particular keyword will find and list that particular site. The more often a keyword appears in the hidden code, the higher a search engine will rank the site in its search results. In various jurisdictions, trademark owners have challenged the unauthorized use of their trademark as a metatag.

The problem is that in such cases the trademark is not primarily used to distinguish particular goods or services, instead it is being used in a way that is invisible to the human eye to make a search engine list a particular website in response to a search. The user then has to click on one of the listed search results if he or she wants to view the content of that website itself.

Some courts have regarded this as a trademark infringement because of 'initial interest confusion' which could be created by consumers looking for the products of the trademark owner who might wrongly be directed to the website of someone else. If this is the website of a competitor, consumers might simply stop there and use the competing product, even though they are no longer confused when viewing that website.

Sale of Trademarks as Keywords

The websites of internet search engines are very attractive to advertisers because they are visited so often. Some of these search engines 'sell' keywords to advertisers who want to target their products to a particular group of internet users. This results in the fact that whenever the keyword is entered into the search engine an advertisement appears along with any search results. Thus retailers, for example, can purchase keywords so that their advertisements are displayed whenever it appears that products bearing a particular trademark (sold as the keyword) are being sought. This has been challenged by trademark owners who are concerned that such advertisements might divert customers from their own website, or from the websites of their preferred or authorized web retailers.

Acceptable Unauthorized Use

Legal systems may provide exceptions for the 'fair use' of a sign that is protected as a trademark. Such exceptions often apply when a sign is used fairly and in good faith in a purely descriptive or informative manner. Such use should not extend beyond that which is necessary to identify the person, entity or the goods or services, and nothing should be done in connection with the sign which might suggest endorsement or sponsorship by the trademark holder. Such exceptions may be equally applicable when a sign is used on the internet.

Use of a Sign on the Internet

The use of a sign on the internet can only infringe a trademark if such use can be deemed to have taken place in the country where the trademark enjoys protection. The question arises under what conditions might the appearance of a mark on the internet constitute *use* in a particular forum and give rise to infringement.

Unfair Competition

Business means competition and where there is competition, acts of unfair competition are liable to occur. Electronically enabled business is no exception. Protection against unfair competition supplements the protection of intellectual property rights. Without this type of protection, companies are likely to face the risks of damage to their reputations, loss of customers and liability from engaging in electronic commerce, with the threatened consequence that innovation and freedom of competition is stifled.

Protection against unfair competition provides a legal framework for all forms of marketing, and it supplements the protection of IP through statutory rights. So far, national or regional laws apply together with international provisions contained in the Paris Convention (which is the Agreement for the Protection of Industrial Property 1883, last revision, Stockholm, 14 July 1967) and the TRIPS Agreement (which is an Agreement on Trade-Related Aspects of Intellectual Property Rights, and is an Annex 1C of the Marrakesh Agreement Establishing the World Trade Organization, signed in Marrakesh, Morocco on 15 April 1994).

However, organizations have found that they cannot simply continue their normal marketing efforts in cyberspace. They have to adapt to and use the particular technical features of the internet, such as its interactivity and support of multimedia applications. Problems may therefore arise with regard to the following issues:

- *Interactive marketing practices.* Because electronic commerce relies on interactive contacts with prospective customers, attracting their attention is a core issue. Online marketing often uses strong incentives such as lotteries, free gifts or rebates, and tends towards more aggressive practices, such as comparative advertising or unsolicited emails (often referred to as 'spamming'). Under a number of legal systems, such inducements may be considered contrary to honest trade practices.
- *Transparency and privacy concerns.* As previously discussed, safeguarding transparency and maintaining privacy is of paramount importance when doing business over the internet. Unfair competition law may have to include rules requiring a clear distinction between informative text and advertising, and protecting consumers against the unauthorized collection of personal data for commercial purposes. Spamming is once again a related problem.
- *National versus international standards of 'unfair' marketing practices.* Marketing practices in electronic commerce are often directed at a public in more than one country. What might be perfectly clear in one country can be misunderstood in another.
- *Trade secrets.* The protection of trade secrets is in many countries covered by unfair competition law. The protection of trade secrets in a network environment relies heavily on technological measures for information security, especially because after a trade secret has been stolen and posted on the internet, courts sometimes experience difficulty finding the 'secrecy' element of a trade secret (i.e. it can no longer be regarded as a secret once in the public domain of the internet).

Domain Names

Domain names are a simple form of internet address, designed to serve the function of enabling users to locate sites on the internet in an easy manner. They may be registered in spaces known as *generic top-level domains* (gTLDs), such as .com, .org or .net, or in the '*country code top-level domains* (ccTLDs), such as .au (Australia), .nz (New Zealand), .sg (Singapore), .uk (United Kingdom), etc. The use of the domain name system has grown rapidly over the last seven years and the growth in domain name registrations is expected to continue, particularly in light of the introduction of competition among domain name registration authorities (i.e. registrars) for the gTLDs.

The *domain name system* (DNS) which is the central system for routing traffic on the internet has assumed a key role in electronically enabled business. It both facilitates the ability of consumers to navigate the internet to find the websites they are looking for, and it facilitates businesses' ability to promote an easy-to-remember name or word which may, at the same time, serve to identify and distinguish the business itself (or its goods or services), and to specify its corresponding online, internet location.

As commercial activities on the internet have increased, domain names have acquired increasing significance as business identifiers and, as such, have come into conflict with the system of business identifiers that existed before the arrival of the internet and that are protected by intellectual property rights, namely trademarks and other rights of business identification, geographical indications and the developing field of personality rights. There is thus a tension between domain names and intellectual property rights: the DNS is largely privately administered and gives rise to registrations that result in a global presence accessible from anywhere in the world. The intellectual property rights system is publicly administered on a territorial basis and gives rise to rights that are exercisable only within the territory concerned.

Examples of bad practices include the deliberate, bad faith registration as domain names of well-known organizations, people and other trademarks in the hope of being able to sell the domain names to the owners of those marks, or simply to take unfair advantage of the reputation attached to those marks.

WIPO Internet Domain Name Process

On an international scale, WIPO set up the WIPO Internet Domain Name Process in the mid 1990s to make recommendations to the Internet Corporation for Assigned Names and Numbers (ICANN) which was established for the technical management of the domain name system. The Final Report in 1999 recommended the adoption of a number of improved, minimum 'best practices' for registration authorities ('registrars') registering domain names in the gTLDs, which are intended to reduce the tension that exists between domain names and intellectual property rights. These include:

- The use of a formalized agreement clearly setting forth the rights and obligations of the parties
- The collection and availability of accurate and reliable contact details of domain name holders as an essential tool for facilitating the protection of intellectual property rights on a borderless and otherwise anonymous medium.

Managing eGovernance Issues

At a managerial level, the factors that create obstacles to achieving and sustaining gains made in electronically enabled business include difficulty in setting priorities, roles and responsibilities, and difficulty in risk taking and approval processes (Accenture, 2000). However, it is clear that most issues are simply variations on the same basic theme: effective decision making and swift follow-through in the midst of constant market change.

The key factor in determining success, in other words, is governance that is not the rigid, conservative decision making of industrial age boardrooms, but rather is characterized by responsiveness and flexibility, and by an ability to adapt, in a matter of days or weeks, to the ever-changing electronically enabled business environment. There are three imperatives to setting up a governance system:

1. Build a robust governance model that links strategy and implementation
2. Align the model with strategic intent
3. Adapt the model (ensure flexibility) for each stage of the company's maturity.

Build a Robust Governance Model

Effective eGovernance enables decision **making** as well as decision **follow-through** in the new economy. It requires an increased capacity to anticipate and effectively deal with the multiple consequences of a decision within the complex system of a business using the internet for profit-making transactions. An imperative of effective eGovernance is the creation of a robust model that links strategy and implementation.

Governance is made up of three elements: **leadership, organizational structure** and **process alignment**. To be successful, an eGovernance model needs all three pillars firmly in place: without them a company's business is at risk.

1. *Leadership.* Leadership provides clear and consistent direction. It can be enhanced significantly in an electronically enabled business if it is provided not only by the CEO and other senior executives, but also by the members of the Board.
2. *Organizational structure.* Organizational structure creates the infrastructure to support the implementation of decisions. In the current world economy the primary issue in organizational design is how best to handle an established organization's move into electronically enabled business. Should the company establish a separate line of business to pursue the new strategic intent, or perhaps spin off the entity, or even maintain both the electronically enabled business and the established business? Such decisions about structure are extremely important in developing a successful business model for electronically enabled business.
3. *Process alignment.* Process alignment defines the activities that turn ideas into action. The alignment of key processes, including finance, technology and marketing, is a challenge for both established companies moving into electronically enabled business and those that are purely e-enabled. For the established company, process alignment concerns the critical links between the core processes of the organization and those of the electronically enabled business. For the purely e-enabled company, it concerns the links between core processes that are outsourced and those handled in house. In either case, lack of alignment spells trouble in product and/or service delivery.

Align the Governance Model with Strategic Intent

Alignment of strategy and governance is not a new concept and certainly does not solely apply to eGovernance. However, the major difference is that in the electronically enabled world of business, this alignment must be achieved quickly and adjusted often and the issues of direction and implementation must be addressed concurrently to achieve a competitive speed to market. In addition, and most challenging to traditional modes of business thinking, eGovernance models are temporary.

As a company learns more about electronically enabled business and adjusts its strategy accordingly, adjustments in its governance model must follow and must do so quickly. There are a number of different approaches to aligning the governance model with strategic intent and these include: establishing an electronically enabled business presence, creating the value chain for the electronically enabled business, reinventing the business for the eWorld, and creating a new electronically enabled business.

- *Establish an electronically enabled business presence.* With this strategic focus the goal is to protect the current franchise by creating the web-based capabilities required to remain competitive. This typically entails giving customers access to information about the existing menu of products and services through an internet channel. For example, an established business could expand its customer access to its products and services by simultaneously developing a website and a call centre. Although this option involves little risk and requires little change, the potential for creating significant new value is also low.
- *Create the value chain for electronically enabled business.* Here the strategic goal is to capture value by using the internet to automate the current business model, primarily to reduce costs and increase revenue but also to extend and deepen customer relationships. Because the electronically enabled business will be dependent on existing lines of business and in fact may also be perceived to be competitive with them, leadership and decision-making processes must firmly support electronically enabled business. In these cases, an integrated approach to organizational structure makes sense. Existing lines of business, with established customers, products and services, benefit from new channels for access and delivery. Therefore, electronically enabled business is integrated into the existing lines of business.
- *Reinvent the business for the eWorld.* A strategy of reinventing the current business model and finding new sources of value creation is possible using the internet to develop and deliver new products to existing customers as well as to new ones. Current products and services may or may not be phased out, but the focus is on the bundling of products and services that is made possible by the internet. Although the potential gains associated with using this strategy are huge, it is also the most threatening to the existing organization because it leverages existing customer relationships with new, alternative products.
- *Create a new electronically enabled business.* A company can explore the potential of new products and services, tap into new customer groups and adopt totally different business models relatively unencumbered by the existing organization using this strategy.

The issue for established companies remains whether to spin off these new eBusinesses or keep them within the company fold. Reinventing a traditional business and creating a new electronically enabled business also has distinct implications for each pillar of eGovernance. The reinvention strategy requires strong leadership with experience in the eEconomy, performance measures must be clear, with full agreement among the electronically enabled business leadership and the established organization's executive leadership and the Board. Finally, an explicit process must be defined for identifying and addressing issues concerning the focus and execution of the electronically enabled business so that the eGovernance model can change as the company adjusts its strategy.

Ensure Flexibility for an Evolving Electronically Enabled Business

An electronically enabled business must have the ability to change itself in order to keep up with the business environment in which it operates. In terms of eGovernance, this means that executives need to consider not only what model is appropriate for today, but also how to build adaptability into that model so that it can evolve along with the company and meet new marketplace challenges.

This change in a company's governance model typically tracks its life cycle maturity curve. Each of the three main phases of the curve – launch, growth and maturity (Fig. 9.3) – has a different business objective with a corresponding eGovernance form. For example, the **launch phase** equates to an emergence governance form, the **growth phase** equates to a scaling up of governance and the **mature phase** equates to a structured governance form. The correct eGovernance model responds to the needs of the enterprise as it proceeds along the maturity curve.

1. *Launch phase.* The business's needs revolve around raising money and building a sustainable market presence and brand. Senior executives must prioritize and balance critical tasks because the decision-making and leadership challenges are significant. The Board of Directors must take a very active role, complementing and filling gaps in the executive management team. This Board is likely to be heavy with venture capitalists who can keep the flow of cash going while the executive team typically focuses on the market and on the product or service.

2. *Growth phase.* In this stage, the business should have established a market presence and will be growing rapidly. The major concerns for the business will be about retaining satisfied customers by making sure that the underlying processes do not buckle under the weight of increased volume, recruiting and retaining appropriately skilled personnel, staying competitive in a maturing marketplace, and converting revenues to profits. The composition of the Board during this phase should change to include leaders with experience in efficient business management (e.g. process excellence) and in reengineering. In addition, the focus on alliance evaluation, management and change requires Board expertise in these areas and possibly a Board committee focused on them.

3. *Maturity phase.* As the company reaches maturity and growth flattens, the major concerns shift from operational excellence to reinvention. The company must find new engines of growth to satisfy its broader investor base. This implies changing the Board composition to secure the capabilities to sense the market, current and future, and to generate appropriate new ideas. New Board members should include an executive with experience in restructuring and reinvention and there should be a greater number of external members to enable the company to think outside the box.

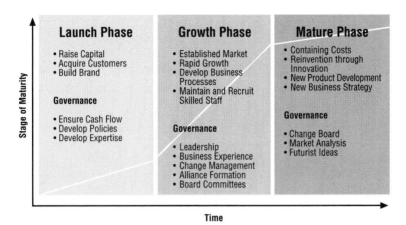

Fig. 9.3. Electronically enabled business life cycle governance curve

Summary

Good governance of an organization is, if not *the* most important, then at least one of the most important components of a successful and well run organization. This chapter has looked at the basic principles of good governance for, and within, an organization and has briefly described the regulatory environment in terms of international business and trade governance issues.

The chapter has also covered eGovernance issues in some detail – that is, governance associated with managing the internet, managing the activities that can occur when people and organizations use the internet as a communication mechanism and business tool, and the issues associated with corporate governance of an electronically enabled business.

There are numerous difficulties associated with maintaining good governance online – mainly because of the almost instantaneous copying and transmission capabilities afforded by the internet and other digital technologies. By far the most difficult of all tasks is to manage a business's or a person's intellectual property (IP) in all its forms. Such IP is a valuable business and/or personal asset and should be protected as far as it is possible to do so. The chapter describes the efforts of the World Intellectual Property Organization and other similar bodies to create regulatory mechanisms that will ensure the protection of IP online.

Finally, the chapter suggests a three-point plan to assist managers in managing eGovernance issues within an electronically enabled business. This consists of creating a governance model that is robust, aligning that model with the strategic intent of the business and ensuring that it is flexible enough to change with the ever-changing business environment.

A Smart Thinker who steered a major internet company to the top through adherence to good governance is Tim Koogle (see Smart Thinker).

Smart Thinker – *Tim Koogle*

Ex-CEO of Yahoo!

Tim Koogle ('TK') was president and CEO of Yahoo! Inc. from 1995 to May 2001 and was then on the Yahoo! Board of Directors until 2003 when he stepped down. Under his leadership as CEO (1995 – 2001) and as Chairman, a role he assumed in 1999, he helped build Yahoo! into a leading internet web portal and media company with over 145 million registered users and annual net revenues of more than one billion dollars. During his tenure, Business Week named him one of The Top 25 Executives of the Year in 1999 and 2000.

When asked what had made Yahoo! such a success he commented that it was a combination of things – strategy, branding and execution. For example in determining their strategy, they identified that the one key thing was branding because Yahoo! has always been a consumer-based franchise and media business from the very beginning, making branding, content, and distribution always important. In terms of execution, a global vision was adopted from the start, a good team was assembled, and the culture was established along with the right infrastructure for managing the business and growing it aggressively.

Currently involved in a new generation of startup companies, Tim's four major things to look out for in predicting the success of a company are:

1. Is the company a leader?
2. Is the company addressing a big market that's getting bigger?
3. Is the company well run?
4. Does the company have a fundamentally strong business model?

Source: ZDNet (www.zdnet.com), Forbes.com (www.forbes.com),
Ecommerce News (www.ecommercenews.com)

The Review and Critical Thinking Questions below have been provided for you to revise and reflect on the content of this chapter.

Review Questions

1. What is corporate governance and what are the major principles as outlined by the OECD?
2. Describe the two main governance models and discuss why different countries adopt different models.
3. What are the seven components of good governance and why are they so necessary?
4. What are the three major eGovernance arenas and describe with examples what they mean operationally?
5. List the governance issues associated with electronically enabled business and briefly describe each issue using an example.
6. What is WIPO and what does it do?
7. What are the major cyber law pressures – describe and discuss.
8. What does intellectual property cover – describe each major category briefly.

Critical Thinking Questions

1. What are the potential eGovernance issues associated with a state-owned telecommunication organization being privatized – use an example company to illustrate the issues.
2. Why is intellectual property so difficult to manage online – give examples of problem areas and how they might be addressed.
3. What did Tim Koogle do 'right' in terms of governance during his tenure as CEO of Yahoo?
4. Compare and contrast the downfall of Enron, WorldCom and Parmalat - what have turned out to be the major governance issues in their respective collapses?
5. Discuss, in terms of the three major eGoverance issues discussed in the chapter and with specific examples, the main issues facing a retail pharmacy when it starts doing business online.
6. Create an eGovernance strategy for a mythical electronically enabled agribusiness of your own – give reasons for each of your major strategies, and the way in which you would ensure they were carried out.

Final Thoughts – Future Proofing

The idea that agribusiness – or any business for that matter – is a balancing act for managers with the need to keep up to date while at the same time innovatively moving forwards, was highlighted in the Introduction. The thesis being that due to the complex, dynamic and competitive global business environment that agribusinesses now exist in, the adoption of modern business practices has to be backed by innovative business processes and technologies. This book has been about the 'E' issues in agribusiness, and has covered the what, why and how of electronically enabling an agribusiness to ensure that it can indeed exist in the current global business environment. However, no matter how good the technology and processes, there is another perhaps even more potent problem looming on the horizon for agri-industries in general – a problem that the 'E' component of business may well provide an answer to.

The issue is that worldwide, young people are not wanting to enter the agribusiness arena. Why is it so - the agribusiness sector is a complex, dynamic, innovative, interesting, capitally intensive, and risky industry to be in; what are the reasons behind this lack of interest in the sector and how can the 'E' factor halt the slide?

The problem is four-fold: Firstly there is the *'Image Problem'* – that is, the word 'agri' attached to anything has the connotation of manual hard work in difficult conditions – almost certainly related to production, almost certainly related to working outside in the elements, and almost certainly associated with financial hardships related to variable climatic conditions (consider the media images of dying animals or bleached skulls on the cracked and sun dried earth), or to disease (the burning piles of animal carcases in the recent foot and mouth outbreak in the UK will stay with people for many years to come).

Secondly there is the *'Image Problem'* – that is, academic versus non-academic. As a result of the above, those that have a modicum of academic capability steer away from agri-anything because the vision is of a life of manual labour in production, rather than white collar or knowledge work. Conversely, those who actively choose agri-anything may be regarded as non-academic and therefore unlikely to amount to much in a knowledge-driven global society.

Thirdly there is the *'Image Problem'* – that is, urban versus rural. 'Urban' signifying all that is bright, glamorous, upmarket and full of opportunities, and 'rural' signifying exactly the opposite with the additional negative being that rural locations in most countries are by definition, geographically distant from major commerce centres with associated poor services and service delivery, and limited opportunities.

Finally there is the *'Image Problem'* – that is, the generational issue of older people who have grown up with the images described above and who want a better life for their children, so actively encourage them to do other things.

Image or branding is everything, and while many of the agri-image related issues outlined have been true of the sector at one time or another, the real point is that it is not so today:

1. The agri- prefix is not the domain of the production end of the industry chain alone – it covers everything from production to retail in all of the food and fibre sectors, the demand for which continues to grow daily.

2. As a result, agribusinesses are some of the largest most complex and profitable global companies in the world.

3. In the past ten years everything in the business world has changed due to the internet and other ICTs. The effect on all components of agri-industry chains has been – as for other industry sectors – profound. Nothing will ever be the same again – the image needs to change.

4. There is little doubt that the challenges for the agri-food sector worldwide over the next 5–10 years will continue to mount and innovators will be needed to provide creative, profitable and sustainable solutions.

Given that these systemic issues in the agri-food industry exist, the questions are 'how does the agri-industry sector future-proof itself in terms of gaining the people it needs in agribusinesses' and 'how does the sector address the image issue and the knowledge acquisition process that develops the core competencies necessary to maintain the sector?' Is it simply a matter of better media – a good spin doctor perhaps? Unlikely, the situation is deeply ingrained in culture and as such, needs to be addressed through long term strategies that embrace a positive outlook for the sector.

Given that this book has been discussing the 'E' issues in agribusiness, it is not surprising that at least one (albeit of many) potential solutions would be an 'E' solution – that is, to make use of the ICT focus to '*pull*' or entice young people into the industry on the twin carrots of innovation and high tech tools, and to prime them to the possibilities of such opportunities through a '*push*' mechanism of innovative education – using those same tools.

An example would be to use the internet which, as has been said many times, breaks down the barriers associated with distance, to actively create interest by marketing the industry sector online to the general public – most especially high school and tertiary level students. The idea would then be to tap into this growing interest by providing internet-based or eLearning opportunities using online learning management systems such as Blackboard (www.blackboard.com) (one of many). Such eLearning allows a far greater reach to be created by educationalists and training organizations for delivery of innovative educational material - although it is acknowledged that creating good learning environments online is an art in itself (Bryceson, 2004).

In addition – as an extension to this basic 'use-the-software' eLearning approach, the use of *Virtual Environments (VE)* is being trialled in agribusiness education in Australia as a way to deliver a more interactive and dynamic learning experience (Bryceson, 2005 – www.nrsm. uq.edu.au/agribusiness).

Virtual Environments are computer-generated environments which simulate reality sufficiently for the human senses to experience it. They are environments which are partially or totally based on computer generated sensory inputs. VEs are at the core of most successful computer games (a hugely influential popular culture – so why not make use of media that attracts young people, to educate them), and encompasses a range of interactive computer environments from text-oriented online forums and multiplayer games, to complex simulations that combine audio, video, animation, and three-dimensional graphics.

The advantage of VEs is that they can immerse people in an environment that would normally be unavailable due to cost, safety, or perception restrictions and thus educational applications make use of the fact that they can allow learning through experience of many types of simulated tasks (Vince, 2004). VEs can also be advanced computer-assisted training systems (Julien *et al.*, 2003) developed around the internet as a delivery mechanism. This is because VEs can represent any three-dimensional world that is either a real system such as buildings and landscapes (Palamidese *et al.*, 1993) or an abstract system such as stock market behaviour, information flows, or any other business processes that can be modelled. Developed as an interactive system such environments allow the trialling of ideas, scenarios and situations without cost to the user – with the results of good and bad decision-making capable of being modelled, visualised, assessed and ranked (if required).

The Virtual Reality Agribusiness World (VAG) (Bryceson 2005) is a prototype of an animated, interactive, virtual reality environment that models an agri-industry sector supply chain and involves a game of strategy related to that chain. The wish in developing VAG was to take producers / professionals / students / school children, on a journey of discovery about agribusiness in all its complexity, as well as providing some thought provoking information about the integration of agribusiness with environmental and social issues to promote sustainability. In addition the idea was to develop something that was immediately engaging, interesting and fun for all concerned, that also delivered learning outcomes, used technology innovatively, and which could be loosely based around the Learning Communities concept of Wenger (1998). That is, the VAG concept was to use virtual reality techniques to:

1. Visualize supply and value chain issues.
2. Deliver innovative interactive and immersive educational experiences online via the use of 3D virtual reality gaming technologies.
3. Create and visualize business decision making scenarios and the related information flows both as input into, and as a result of, decision making.
4. Show and evaluate the impact and effects of decision making and information flows across components in an industry chain.

In terms of the 'E' technologies involved, VAG is designed around a three tier information architecture with a modular, object orientated programming approach to enable scalability onto the internet, scalability in this case meaning the ability to incorporate multiple supply chains, multiple players and multiple issues (physical, economic, environmental and social). Customised XML schema and game application programming interfaces (API) have been used to provide ultimate generic flexibility of the design to allow non-programmers to create and use detailed physical, economic, environmental and social variable definition in order to develop new scenarios and game plays. In addition, information flows through the communication schema and 'events' of all types – (e.g. decisions, market changes, climatic events such as drought starting or stopping, hazards etc. – either randomly timed or controlled) are possibilities.

Research continues in migrating VAG from CD to the internet to enable a massive multiplayer online game to be developed with the aim of both actively pushing the high tech, innovative image of agribusiness while at the same time providing a very different learning environment for the industry and students of all ages.

Back to reality, and the bottom line is that the agri-industry sector needs not only to be aware of the 'E' issues it faces in business, but also how it can make use of these very same tools to generate interest in itself as the employer of choice over the next five to ten years. In other words, the sector needs to be proactive in its adoption of 'E' technologies in order to stay current and it needs to be future proofing itself by actively marketing itself in the medium of its future managers in the most innovative of ways, in order to compete for the brightest and best of future employees.

Appendix 1

Technologies associated with the eLandscape

An excellent book that covers the technologies associated with the eLandscape in detail is Perry and Schneider's (1999) book 'Electronic Commerce' and the associated Online Companion. Good encyclopedias of terms can be found at the online dictionaries Webopedia (www.pcwebopedia.com), Wikipedia (http://en.wikipedia.org), HowStuffWorks (www. howstuffworks.com) and Jargon Buster (www.wiredmedia.co.uk/docs/64.html).

Local Area Network (LAN). A computer network that spans a relatively small area. Most LANs are confined to a single building or group of buildings. However, one LAN can be connected to other LANs over any distance via telephone lines and radio waves. A system of LANs connected in this way is called a *wide-area network* or WAN.

Most LANs connect workstations and personal computers (PCs). Each *node* (individual computer) in a LAN has its own Central Processing Unit (CPU) with which it executes programs, but it is also able to access data and devices anywhere on the LAN. This means that many users can share expensive devices, such as laser printers, as well as data. Users can also use the LAN to communicate with each other, by sending email or having online discussion via chat sessions. There are many different types of LANs; *Ethernets* are the most common for PCs. Most Apple Macintosh networks run on Apple's AppleTalk network system, which is built into Macintosh computers.

The following characteristics differentiate one LAN from another:

1. **Topology:** The geometric arrangement of devices on the network. For example, devices can be arranged in a ring or in a straight line.
2. **Protocols:** The rules and encoding specifications for sending data:

   • *Transmission Control Protocol (TCP)* – TCP enables two hosts to establish a connection and exchange streams of data. TCP guarantees delivery of data and also guarantees that packets will be delivered in the same order in which they were sent.
   • *Internet Protocol (IP) Technologies* – Technologies enabling the delivery of electronic services over a network using the internet protocol to encode and decode messages. Essentially involves delivering packets of data that include a delivery address.

Protocols also determine whether the network uses a peer-to-peer or client/server architecture.

   • **Peer-Peer (P2P).** This is a type of network in which each workstation has equivalent capabilities.
   • **Client/Server Architecture**. This type of network has some computers that are dedicated to serving the others.

LANs are capable of transmitting data at very fast rates, much faster than data can be transmitted over a telephone line - but the distances are limited, and there is also a limit on the number of computers that can be attached to a single LAN (www. pcwebopedia.com).

Wide Area Network (WAN). A computer network that spans a relatively large geographical area. Typically, a WAN consists of two or more local-area networks (LANs).

Computers connected to a wide-area network are often connected through public networks, such as the telephone system. They can also be connected through leased lines or satellites. The largest WAN in existence is the Internet.

Virtual Private Networks (VPNs). Smaller, privately-used networks within larger networks that offer the same characteristics, features, and benefits of legitimate, hard-wired private networks.

The Internet. An electronic communications and information network that is global in reach. It has grown through the interconnection of computer networks, both local and regional. It was developed in the 1960s with no centralized control or organization and has become one of the most powerful mediums for information exchange in the world. It is often referred to as 'the Net' or 'the Web'.

Intranets. A private network that runs on TCP/IP and is accessible only to authorized users. An intranet's web sites are the same as any other web sites but are not typically for public use.

Extranets. Extended private networks that are accessed via the public internet, that enable privileged users to securely share information.

Media. Objects on which data can be stored e.g. CD, Diskette etc. On a computer network the term 'media' also refers to the cables connecting devices together which can be twisted-pair wire, coaxial cables, or fibre optic cables. Some networks do without connecting media altogether, communicating instead via radio waves. Recently the **USB Memory Stick** *(or Flash Memory Card)* – is superceeding CDs as a storage device. They are small, easy-to-use plug-in devices that are compatible with most computers and can be used to store or move data, music and video clips from computer to computer. They are currently the size of a pocket knife and can be worn on a necklace. Currently available in several capacities, such as 128MB, 256MB, 512MB, 1GB, 2GB and 4GB.

Modem (*Modulator/Demodulator*). A device that enables computer signals to travel over phone lines. Modems come in different speeds. For internet use, a V90 standard modem with a fast 56 kilobit per second (56 Kb) is effective. There are also special modems for access to faster ISDN or broadband connections.

Mobile phone. A wireless device which behaves as a normal telephone while being able to be used over a wide area. Most current mobile phones use a combination of radio wave transmission and conventional telephone circuit switching. Mobile phone manufacturers include: Alcatel, LG, Motorola, Nokia, Panasonic, Philips, Sanyo, Siemens, and Sony Ericsson. The mobile phone has, in less than twenty years, gone from being a rare and expensive piece of equipment used only by businesses to a pervasive low-cost personal item. With high levels of mobile telephone penetration almost everywhere in the world, a mobile phone culture has evolved, where the mobile phone has become both a key social tool and a business tool. A history of the mobile phone can be found at: www.radio-electronics.com/info/cellulartelecomms/celldev/cellulardev.php.

Personal Digital Assistant (PDA). A handheld device that combines computing, telephone/fax, and networking features which is often used for information storage retrieval for personal or business use. Many PDAs incorporate handwriting and/or voice recognition features. PDAs are also called palmtops (Palm.com), Blackberry (www.blackberry.com), handheld computers, and pocket computers – see www.mobiletechreview.com for additional information.

Business Information Systems

Office Automation Systems

These systems facilitate everyday information processing tasks in offices and business organizations. They provide individuals with effective ways to process personal and organizational business data, to perform calculations and to create documents. Examples include:

- Desktop publishing packages (e.g. Adobe)
- Word processors (e.g. MSWord)
- Spreadsheets and personal note-taking systems (e.g. MS Excel)
- Electronic document management systems (EDMS) (e.g. Laserfiche)

Their impact on an organization is major – they act as tools for creating documents and presentations, and as communication tools for analysing information to provide help in implementing decisions.

Communication Systems

Ideally, communication systems help people work together by enabling them to exchange or share information in many different forms and at a distance, in many instances. Examples include:

- Telephone and video conferencing which provide oral communication and support decision making
- Email, fax, internet, intranet, extranet and virtual private networks which provide electronic document and message communication and which allow information gathering and information sharing
- Electronic calendars allow electronic scheduling of meetings, group information sharing and joint decision making.

However, technology may in fact hinder rather than help communication if the social context, the form and content of communication, and the time, place and direction of communication taking place is not understood. There are some issues that should be particularly taken note of in relation to email, fax, the internet, intranets and extranets.

These technologies filter out the social context of messages resulting in the danger of misinterpreting the message since there are no visual keys and tonal inflection to convey additional meaning. Additionally:

- New power relationships can develop – high level managers find it easy to obtain information directly from people without going through intermediary managers
- Privacy and confidentiality problems can arise because of company policies, and because nothing is ever completely deleted from a system
- Information overload can become a problem (100 emails a day take a lot of time and effort to read and reply to!)
- Intranets are private communication networks that are only accessible by authorised employees, contractors and sometimes, customers
- Extranets are also private networks but are directed more at customers or other company offices.

Transaction Processing Systems (TPS)

A transaction is a business event that generates or modifies data stored in an information system. A TPS collects and stores data about transactions (i.e. payments, credits and invoicing) and sometimes controls decisions made as part of a transaction.

Bill presentation and payment includes **Electronic invoicing** where invoices are presented by email or through a system that provides an online view of an account; **Clearance** where transmission, reconciliation and confirmation of payment orders that match invoices before settlement takes place; **Settlement** which is the process of recording the debit and credit positions of the parties involved in transferring funds and may also include 'Net' or 'Gross' settlements; **Collections** which is when one account is credited and another debited. A collection is assumed to be completed once authorisation from a Financial Service provider (ie bank) for the charge is confirmed.

These systems create a database that can be accessed directly, making direct people-to-people communication unnecessary. Example TPSs include:

- Point of Sale systems for sales transactions
- Credit Card payment processing systems
- Tracking systems of customer contacts during the sales cycle and for tracking movement of a work process in a factory
- Airline reservation systems
- Systems to allow the withdrawal of money from an ATM
- Online catalogue ordering systems and systems to download video clips

The technologies for electronic payments have now advanced to be mature and well understood. Secure online payment protocols are being implemented. These include **SSL (Secure Socket Layer)**, **Secure Electronic Transactions standards** and **IOTP** (Internet Open Trading Protocol) http://security.ittoolbox.com; **SWIFT** (Society for Worldwide Interbank Financial Telecommunications) http://www.swift.com; **Fedwire** and **CHIPS** (Clearing House Interbank Payment System) www.chips.org.

Management and Executive Information Systems (MIS and EIS)

MISs provide information for an organization's managers and generate information for monitoring performance, maintaining coordination and providing background information about the organization's operation.

EISs convert TPS data into information for monitoring performance and managing an organization thus providing readily accessible interactive information. Both MISs and EISs may incorporate email and other communication methods in the presentation of computerised data. These systems provide easy ways to analyse different types of monitoring and performance information. Examples include systems to:

- Summarise weekly sales reports by product and region
- Determine the planning of purchases based on a production schedule
- Flexibly access the corporate financial plan by line item

Managerial Decision Support Systems (DSS)

Decision Support Systems (DSSs) are interactive information systems that provide information and tools for analytical work in semi-structured and unstructured situations (Sauter, 1997, Holsapple and Whinston, 2002). They help people make decisions and evaluate scenarios by providing information, models and analysis tools that enable a clear rationale for explaining a decision. Examples include systems to analyse market data, sales data and the shopping habits of customers, as well as land management options, production and credit risk assessment, financial monitoring and quality assurance systems and environmental hazard assessment (www.dssresources.com).

Expert Systems

Expert systems are more sophisticated computer programs than DSSs, generally using artificial intelligence programming techniques to solve problems with expert knowledge embedded within their core. This is obtained from recognised experts in the field and coded so that a computer can apply it. Once set running on a problem, there is little human input. This reliance on the knowledge of a human domain expert is a major feature of expert systems (Turban *et al.*, 2005). The general relationship between MIS, DSS and Expert systems are shown in Fig. 1 below.

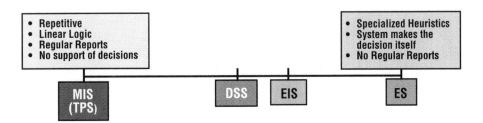

Fig. 1. The relationship between MIS (Managerial Information Systems) / TPS (Transaction Processing Systems), DSS (Decision Support Systems) and Expert Systems

Knowledge Management Systems (KM)

KM systems e.g. OpenText's Livelink (www.opentext.com) are communication systems designed to facilitate the sharing of knowledge rather than just information. The computer applications underlying KM systems are built on technologies such as intranets, electronic mail, groupware, EDMS, search engines and databases. Functions supported by these technologies include codifying knowledge and providing organized ways for people who have knowledge to share it with others.

Appendix 2

Table of Internet Advertising Jargon

Information sourced from: www.webcertain.com; www.tamingthebeast.net; www.e-channel. com.au/Resources_Jargon+Buster.html

Terminology	Description
Ad Clicks	Number of times users click on an ad banner.
Ad Click Rate	Sometimes referred to as 'click-through' – this is the percentage of ad views that resulted in an ad click.
Ad Views (Impressions)	Number of times an ad banner is downloaded and presumably seen by visitors. If the same ad appears on multiple pages simultaneously, this statistic may understate the number of ad impressions, due to browser caching. Corresponds to net impressions in traditional media. There is currently no way of knowing if an ad was actually loaded. Most servers record an ad as served even if it was not.
Banner	An ad on a Web page that is usually "hot-linked" to the advertiser's site.
Button	Button is the term used to reflect an Internet advertisement smaller than the traditional banner. Buttons are square in shape and usually located down the left or right side of the site. The IAB and CASIE have recognized the following sizes as the most popular and most accepted on the Internet:

Standard Internet Ad Sizes

468 x 60	Full banner
392 x 72	Full Banner/ Vertical Navigation Bar
234 x 60	Half Banner
125 x 125	Square Button
120 x 90	Button #1
120 x 60	Button #2
88 x 31	Micro Button
120 x 240	Vertical Banner

Terminology	Description
CASIE	CASIE stands for the Coalition for Advertising Supported Information and Entertainment. It was founded in May 1994 by the Association of National Advertisers (ANA) and the American Association of Advertising Agencies (AAAA) to guide the development of interactive advertising and marketing.
Click through	The percentage of ad views that resulted in an ad click.
CPC	Cost-per-click is an Internet marketing formula used to price ad banners. Advertisers will pay Internet publishers based on the number of clicks a specific ad banner gets. Cost usually runs in the range of US$ 0.10-0.20 per click.
CPM	CPM is the cost per thousand impressions for a particular site. A Web site that charges US$15,000 per banner and guarantees 600,000 impressions has a CPM of US$25 (US$15,000 divided by 600).
Hit	Each time a Web server sends a file to a browser, it is recorded in the server log file as a "hit". Hits are generated for every element of a requested page (including graphics, text and interactive items). If a page containing two graphics is viewed by a user, three hits will be recorded – one for the page itself and one for each graphic. Webmasters use hits to measure their server's work load. Because page designs vary greatly, hits are a poor guide for traffic measurement.
IAB	IAB stands for the Internet Advertising Bureau. The IAB is a global non-profit association devoted exclusively to maximizing the use and effectiveness of advertising on the Internet. The IAB sponsors research and events related to the Internet advertising industry.
Interstitial	Meaning in between, an advertisement that appears in a separate browser window while you wait for a Web page to load. Interstitials are more likely to contain large graphics, streaming presentations, and applets than conventional banner ads, and some studies have found that more users click on interstitials than on banner ads. Some users, however, have complained that interstitials slow access to destination pages.
Page	All Web sites are a collection of electronic 'pages'. Each Web page is a document formatted in HTML (Hypertext MarkUp Language) that contains text, images or media objects such as RealAudio player files, QuickTime videos or Java applets. The 'home page' is typically a visitor's first point of entry and features a site index. Pages can be static or dynamically generated. All frames and frame parent documents are counted as pages.

Terminology	Description
Jump Page	A jump page, also known as a 'splash page', is a special page set up for visitors who clicked on a link in an advertisement. For example, by clicking on an ad for Site X, visitors go to a page in Site X that continues the message used in the advertising creative. The jump page can be used to promote special offers or to measure the response to an advertisement.
Page Views	Number of times a user requests a page that may contain a particular ad. Indicative of the number of times an ad was potentially seen, or 'gross impressions'. Page views may overstate ad impressions if users choose to turn off graphics (often done to speed browsing).
Sticky	'Sticky' sites are those where the visitors stay for an extended period of time. For instance, a banking site that offers a financial calculator is stickier than one that doesn't because visitors do not have to leave to find a resource they need.
Unique Users	The number of different individuals who visit a site within a specific time period. To identify unique users, Web sites rely on some form of user registration or identification system.

Appendix 3

XML (eXtensible Markup Language)

1. **XML is a method for putting structured data in a text file.** Structured data includes spreadsheets, address books, configuration parameters, financial transactions, technical drawings, etc. Programs that produce such data often also store it on disk, for which they can use either a binary format or a text format. The latter allows if necessary, to look at the data without the program that produced it. XML is a set of rules, guidelines, or conventions for designing text formats for such data, in a way that produces files that are easy to generate and read by a computer, are unambiguous, and avoid common pitfalls, such as lack of extensibility, lack of support for internationalisation and platform-dependency.

2. **XML looks a bit like HTML but isn't HTML.** Like HTML, XML makes use of tags (words bracketed by '<' and '>') and attributes (of the form name = 'value'), but while HTML specifies what each tag and attribute means (and often how the text between them will look in a browser), XML uses the tags only to delimit pieces of data, and leaves the interpretation of the data completely to the application that reads it. XML also permits the definition of tags for many business-related fields such as tags for dates, names, addresses, prices etc., i.e. if you see '<p>' in an XML file, don't assume it is a paragraph. Depending on the context, it may be a price, a parameter, a person, a p... It is thus possible to produce an electronic document or online form that can be displayed to the user and can also be read by applications which can easily pick out the key data fields.

3. **XML is text, but isn't meant to be read.** XML files are text files, but they are not meant to be read by humans. They are text files, because that allows experts (such as programmers) to more easily debug applications, and in emergencies, they can use a simple text editor to fix a broken XML file. The rules for XML files are much stricter than for HTML. A forgotten tag or an attribute without quotes makes the file unusable, while in HTML such practice is often explicitly allowed, or at least tolerated. The official XML specification does not allow applications to try to second-guess the creator of a broken XML file; if the file is broken, an application has to stop right there and issue an error.

4. **XML is a family of technologies.** XML 1.0 – the specification that defines what 'tags' and 'attributes' are, but around XML 1.0 there is a growing set of optional modules that provide sets of tags and attributes, or guidelines for specific tasks and there are other modules under development at all times.

   - *Xlink* - describes a standard way to add hyperlinks to an XML file;
   - *XPointer and XFragments* – syntaxes for pointing to parts of an XML document (an Xpointer is a bit like a URL, but instead of pointing to documents on the web, it points to pieces of data inside an XML file);
   - *CSS*, the style sheet language, is applicable to XML as it is to HTML;

- *XSL* is the advanced language for expressing style sheets, based on *XSLT,* a transformation language that is often useful outside XSL as well, for rearranging, adding or deleting tags and attributes;
- *DOM* is a standard set of function calls for manipulating XML (and HTML) files from a programming language;
- *XML Namespaces* is a specification that describes how you can associate a URL with every single tag and attribute in an XML document. What that URL is used for is up to the application that reads the URL though (RDF, W3C's standard for metadata, uses it to link every piece of metadata to a file defining the type of that data);
- *XML Schemas 1 and 2* help developers to precisely define their own XML-based formats.

5. **XML is verbose, but this is not a disadvantage.** Since XML is a text format, and it uses tags to delimit the data, XML files are nearly always larger than comparable binary formats. That was a conscious decision by the XML developers. The advantages of a text format are evident (see 3 above), and the disadvantages can usually be compensated at a different level. Disk space isn't as expensive as it used to be, and programs like *zip* and *gzip* can compress files very well and very fast. Those programs are available for nearly all platforms and are usually free. In addition, communication protocols such as modem protocols and HTTP/1.1 (the core protocol of the web) can compress data on the fly, thus saving bandwidth as effectively as a binary format.

6. **XML is new, but not that new.** Development of XML started in 1996. Before XML there was SGML (widely used for large documentation projects) developed in the early 1980s, and an ISO standard since 1986, and of course HTML, the development of which started in 1990. The designers of XML simply took the best parts of SGML, guided by the experience with HTML, and produced something that is no less powerful than SGML, but vastly more regular and simple to use.

7. **XML is licence-free, platform-independent and well-supported.** By choosing XML as the basis for a project, you buy into a large and growing community of tools and engineers experienced in the technology. Opting for XML is a bit like choosing SQL for databases: you still have to build your own database and your own programs/procedures that manipulate it, but there are many tools available and many people that can help you. And since XML is licence-free, you can build your own software around it without paying a thing. The large and growing support means that you are also not tied to a single vendor for addons either.

Further useful information on XML can be found at the World Wide Web Consortium (W3C) (www.w3.org/). XML Tutorials can be found at: www.w3schools.com/xml/default.asp

References

Accenture (2001) *Reality Online: No Pain, No Gain Outlook Special Edition: Part II: The Internet Integration Lifecycle.* Available from: www.accenture.com/Global/Research_and_Insights/Outlook/By_Alphabet/OnlineAchieved.htm (accessed 15 November 2005)

Accenture Outlook Special Edition (2000) *Governance at Speed – The eSeries III* [online]. Available from: www.accenture.com/xd/xd.asp?it=enWeb&xd=ideas\outlook\special2000_3\se3_contents.xml (accessed 8 November 2005)

Afuah, A. and Tucci, C. (2003) *Internet Business Models and Strategies: Text and Cases*, McGraw Hill, USA.

AIM (1999) *Radio Frequency Identification (RFID): A Basic primer.* AIM Inc White Paper Document No: 1.11, 1999-09-28. Available from: www.aimglobal.org/technologies/rfid/resources/papers/rfid_basics_primer.asp (accessed 8 November 2005)

Allaire, J. (2005) *Building New Business Models on the Web* [online]. Available from: www.macromedia.com/ (accessed 8 November 2005)

Allee, V. (2000) Reconfiguring the Value Network. *Journal of Business Strategy*, July/August, 36-39.

Alli, I. (2003) *Food Quality Assurance: Principles and Practices.* Taylor & Francis (CRC Press), USA.

Alter, S. (1999) *Information Systems – A Management Perspective.* Addison Wesley Longman, USA.

Amanor-Boadu, V., Amanor-Boadu, Y. and Dyer, G. (2001) Challenging cognitive barriers to rural revitalization in an internet economy. In: *Proceedings of the IAMA Conference*, 27-28 June 2001. Sydney, Australia.

Amit, R. and Zott, C. (2001) Value Creation in e-Business. *Strategic Management Journal* 22, 493-520.

Anderson, D. and Lee, H. (2000) The Internet-Enabled Supply Chain: From the First Click to the Last Mile. *ASCET* [online], vol. 2., 4 January 2000. Available from: www.ascet.com/documents.asp?d_ID=199 (accessed 8 November 2005)

Anderson, D. and Lee, H. (2001) New Supply Chain Business Models – The Opportunities and Challenges. *ASCET* [online], vol. 3, 5 January 2001. Available from: www.ascet.com/documents.asp?grID=146&d_ID=426 (accessed 8 November 2005)

ArctiSoft (2003) *Overview of PKI White Paper* [online]. Available from: www.articsoft.com/pki.htm (accessed 8 November 2005)

Armstrong, J.S. (2001) Standards and Practices for Forecasting. In: Armstrong, J. S. (ed.) *Principles of Forecasting: A Handbook for Researchers and Practitioners.* Kluwer Academic Publishers, Norwell, MA. Available from: www-marketing.wharton.upenn.edu/forecast/standardshort.pdf (accessed 8 November 2005)

Arntzen, B.C., Brown, G.G., Harrison, T.P. and Trafton, L. (1995) Global Supply Chain Management at Digital Equipment Corporation. *Interfaces*, January-February 1995.

Australian Food Standards Agency (2005) *The Australia New Zealand Food Standards Code* [online] (up to Amendment 81, 2005). Available from: www.foodstandards.gov.au/standardsdevelopment/gazettenotices/index.cfm (accessed 11 November 2005)

Ballenger, N. and Blaylock, J. (2003) Consumer-Driven Agriculture. *Amber Waves* [online] 1, (2). Available from: www.ers.usda.gov/AmberWaves/April03/Features/ConsumerDrivenAg.htm (accessed 8 November 2005)

Baily, M.N. and Lawrence, R.Z. (2001) Do we have a new E-conomy? *American Economic Review,* 91(2).

Barney, D. (2005) Get Ripped! 20 Steps to a Lean, Mean Supply Chain Machine. *CSCO Magazine* [online] June 1 2005. Available from: www.cscomagazine.com/ (accessed 8 November 2005)

Barney, J.B. (2001) Resource-based theories of Competitive Advantage. A 10 year retrospective on the Resource Based View. *Journal of Management* 27, 625-641.

Barton, D., Coombes, P., and Wong, S. (2004) Asia's governance challenge. *The McKinsey Quarterly*, Number 2, 54-62.

Beamon, B.M. and Bermudo, J.M. (2000) A hybrid push/pull control algorithm for multistage, multiline production systems. *Production, Planning & Control*, vol. 11(4), 349-356.

Berne Convention for the Protection of Literary and Artistic Works [online], (1886) Available from: www.wipo.int/treaties/en/ip/berne/trtdocs_wo001.html (accessed 8 November 2005)

Berry, M.J. and Linoff, G.S. (2004) *Data Mining Techniques*, 2nd edn. John Wiley & Sons, New York.

Berson, A., Smith, S. and Thearling, K. (2000) *Building Data Mining Applications for CRM*, McGraw Hill Publishing. New York.

Betts, M. (2003) Management Dashboards Becoming Mainstream. *Computerworld*, 14 April 2003. Report no. 37401.

Bhatt, G.D. and Emdad, A.F. (2001) An Analysis of the Virtual Value Chain in Electronic Commerce. *Logistics Information Management*, vol. 14, no. 1/2, 78-84.

Bierce, P. and Kennerson, A. (2002) *Insights on effective outsourcing – When not to outsource* [online]. Available from: www.outsourcing-law.com/articles/what-should-not-be-outsourced.asp (accessed 8 November 2005)

Bitcom (2005) *B2E Portals* [online]. Available from: www.bitpipe.com/tlist/B2E-Portals.html (accessed 8 November 2005)

Blossom, J. (2005) *Models for success: 2005 ushers in an era of major shifts in Content Business Models*. News Commentary, Shore Communications, 3 January 2005. Available from: www.shore.com/commentary/newsanal/items/2005/20050103preview.html (accessed 8 Nov 2005)

Boehlje, M. (1999) Structural Changes in the Agricultural Industries: How do we measure, analyze and understand them? *American Journal of Agricultural Economics*, vol. 81(5), 10-28.

Boehlje, M., Dooley, F., Akridge, J. and Henderson, J. (2000) E-Commerce and Evolving Distribution Channels in the Food and Agribusiness sector. In: *Proceedings of the IAMA Conference – Advances in Technology*. Chicago, 2000.

Boey, P., Grasso, M., Sabatino, G. and Sayeed, I. (1999) Architecting e-Business Solutions, *Distributed Enterprise Architecture*, vol. II, no. 7. Cutter Consortium, Executive Report.

Borck, J. (2001) Develop a supply chain strategy – End-to-end integration and collaborative commerce capabilities are key. InfoWorld, 20 April 2001.

Bovet, D. and Martha, J. (2000) *Value Nets – Breaking the Supply Chain to Unlock Hidden Profits*. John Wiley & Sons, New York.

Brighan, E.F. (1995) *The Bar Code Manual*. Thompson Learning, New York.

Brooke, M. (2005) RFID Strategy – RFID: The Lean Machine. *Industry Week Article No: 10562* [online] 26 July 2005. Available from: www.industryweek.com/ (accessed 8 November 2005)

Brooking, A. (1999) *Corporate Memory – Strategies for Knowledge Management*, International Thompson Business Press. London.

Brooks, P. (1997) March of the Data Marts. *Data Base Management Systems*, March 1997.

Brown, S.A. (1997) *Revolution at the Checkout Counter: The Explosion of the Bar Code*. Harvard University Press, Boston, MA.

Brown, S. and Eisenhardt, K.M. (1997) The art of continuous change: Linking complexity theory and time-paced evolution in relentlessly shifting organizations. *Administrative Science Quarterly* 42 (1), 1-34.

Brown, S. and Eisenhardt, K.M. (1998) *Competing on the edge: Strategy as Structured Chaos*. Harvard Business School Press, Boston, MA.

Bryceson, K.P. and Marvenek, S. (1998) The role of satellite data and precision farming technologies for on-farm monitoring in Australia. In: *Proceedings of the First International Conference on Geospatial Information in Agriculture & Forestry*, 1-3 June 1998, Florida.

Bryceson, K.P. (2003) eBusiness Impacts on the Peanut Industry in Queensland – a Case Study. *Queensland Review*, vol 10 (1), 103-122.

Bryceson, K.P. (2004) ESCIE – A New Model using Social Constructivism and the Concept of 'Ba' as a Theoretical Framework. *Effective Teaching & Learning Conference.* Brisbane, November 2004.

Bryceson, K.P. and White, D.H. (eds) (1993) *Proceedings of The Drought and Decision Support Workshop*, March 1992. Bureau of Resource Sciences, Canberra. ACT.

Bryceson, K.P. and Pritchard, A. (2003) The Impact of eBusiness Technologies on the Grains Value Chain. *Final Report to Grains R&D Corporation,* Canberra, ACT, Australia.

Bryceson, K.P. and Cover, M. (2004) A Market and Value Chain Analysis of the Australian Guar Industry. *RIRDC Report,* no. 04, 165.

Bryceson, K.P. and Kandampully, J. (2004) The Balancing Act: 'E' Issues in the Australian Agri-Industry Sector. In: *Proceedings of the McMaster World Congress on The Management of Electronic Business* 14-16 January. Hamilton, Ontario.

Bryceson, K.P. (2005) VAG – A Game of Agribusiness Supply Chain Strategy. *Proceedings of the ASCILITE 2005 Conference* December 4-7, Brisbane, Australia.

Buhr, B.L. (2000) Information Technology and Changing Supply Chain Behaviour. *American Journal of Agricultural Economics* 82(5), 1130-1132.

Burgess, P. (2001) *The Electronic Business Environment: Exploding the Hype. A Consultancy Report on the 'Dot Com' Phenomenon, and Current Trends in eBusiness.* The University of Queensland, Australia.

Burton-Jones, A. (1999) *Knowledge capitalism.* Oxford University Press, UK.

Busch, J.S. and Hantusch, N. (2000) I don't trust you, but why don't you trust me? *Dispute Resolution Journal*, vol. 55 no.3, August/October 2000, 56-65.

Cavaye, J. (2001) Rural Community Development, New Challenges and Enduring Dilemmas. *Journal of Regional Analysis and Policy* 31(2), 109-124.

CDC (2005) *Salmonella enteritidis* [online]. Centres for Disease Control and Prevention, USA Department of Health and Human Services. October 2005. Available from: www.cdc.gov/ (accessed 8 November 2005).

CFSAN (2005) Centre for Food Safety and Applied Nutrition, HACCP Overview. USA Food and Drug Administration. Available from: www.cfsan.fda.gov/~comm/haccpov.html (accessed 11 November 2005)

Chapman, C.B. and Ward, S.C. (2002) *Managing Project Risk and Uncertainty.* J. Wiley, Chichester, UK.

Chapman, R.L., Soosay, C. and Kandampully, J. (2002) Innovation in logistic services and the new business model: a conceptual framework. *Managing Service Quality* 12 (6) 358-371.

CIES (2004) Implementing Traceability in the Food Supply Chain. *The Food Business Forum* [online] January 2005. Available from: www.ciesnet.com/ (accessed 11 November 2005)

Clarke, J.B. (1907) *Essentials of Economic Theory.* Macmillan, New York.

Clemmons, R.M. (2002) *The Complete Idiot's Guide to Knowledge Management: Chapter 8: Communities of practice – the killer application.* CWL Publishing Enterprises. Madison, WI.

Coase, R.H. (1937) The Nature of the Firm. *Economica* 4, 386-405.

Coat, J. (1993) *The Principles of Cyberspace Innkeeping: Building Online Communities* [online]. Available from: http://gopher.well.sf.ca.us:70/0/Community/innkeeping (accessed 8 November 2005)

Coles, J.W., McWilliams, V.B. and Sen, N. (2001) An examination of the relationship of governance mechanisms to performance. *Journal of Management* 27(1), 23-50.

Colkin, E. (2002) Where was IT? *Information Week,* Issue 895, 1 July 2002, 20-23.

Collins, J. (2003) Zebra Unveils RFID Label Maker. *RFID Journal*, September 2003.

Collins, R., Dunne, T. and O'Keefe, M. (2002) The 'Locus of Value': a hallmark of 'chains that learn'. *Supply Chain Management: An International Journal Research Note.* 7(5), 318-321.

Collis, D. and Montgomery, C.A. (1997) *Corporate Strategy – Resources and Scope of the Firm.* Irwin, Chicago.

Collits, P. (2003) The Regional Divide and the Future of Small Towns. *Proceedings of the 'From Strength to Strength' Regional Development Conference* [online], Timaru, New Zealand, September 2003. Available from: www.regdev.govt.nz/conferences/2003/collits/background.html (accessed 11 November 2005)

Concise Finance Encyclopedia (2004) [online]. Available from: www.finance-encyclopedia.com/term/due_diligence?c=0256520225) (accessed 8 November 2005)

Connor, D.R. (1998) How to create a nimble organization. National Productivity Review 17 (Autumn).

Coombes, P. and Watson, M. (2000) Three surveys on Corporate Governance. *McKinsey Quarterly* no. 4, 74-79.

Coombes, P. and Watson, M. (2001) Corporate Reform in the Developing World. *McKinsey Quarterly* special edn., no. 4, 89-93.

Corbett and Associates (2000) Ten Years of Outsourcing Practice: Tactical, Strategic, and Transformational. *CIO Magazine, 10 Years that Shook IT* [online] 1 October 1999. Available from: www.corbettassociates.com/firmbuilder/articles/19/50/591/ (accessed 8 November 2005)

Corbett, M. (2001) *Primer on Outsourcing.* [online]. Available from: www.firmbuilder.com/firmbuilder/articles/19/48/388/ (accessed 8 November 2005)

Courtney, H. (2001) Making the most of uncertainty. *McKinsey Quarterly,* Strategy no. 4, 38-48.

Courtney, H.G., Kirkland, J. and Viguerie, S.P. (2000) Strategy under Uncertainty. *McKinsey Quarterly,* Strategy no. 3.

Corritore, C.L., Wiedenbeck, S. and Kracher, B. (2001) The Elements of Online Trust. *Chi 2001,* March 31 – April 5, 504-505.

Cox, A. (1999) Power, Value and Supply Chain Management. *Supply Chain Management,* vol. 4(4), 167-175.

Cox, A. (2001) Managing with Power: Strategies for improving value appropriation from supply relationships. *Supply Chain Management* vol. 37, no. 2, 42-47.

Coyne, K.P. and Subramaniam, S. (1996) Bringing discipline to strategy *McKinsey Quarterly,* Strategy no. 4, 14-25.

CRC (2003) Australian Biosecurity Cooperative Research Centre website. Available from: www1.abcrc.org.au/pages/About.aspx?MenuID=25 (accessed 11 November 2005)

Credit Lyonnais Securities Asia (CLSA) (2002) *CLSA Ltd. Report on Emerging Markets* [online]. Available from: www.clsa.com/ (accessed 8 November 2005)

CXO Media Inc. (2001) Outsourcing. *CIO Magazine.* Available from: www.cio.com/summaries/outsourcing/index.html (accessed 11 November 2005)

Dalton, J.P., Manning, H. and Gardiner, K. (2001) Managing Content Hypergrowth. *The Forrester Report* [online] January 2001. Available from: www.vignette.com/Downloads/FORRESTER_MANAGING_CONTENT.pdf

Darwin, C. (1859) *The Origin of the Species.* New Edition (1998), Oxford University Press, Oxford.

Das, T. and Teng, B.S. (1998) Between Trust and Control: Developing Confidence in Partner Cooperation in Alliances. *The Academy of Management Review,* vol. 23:3, July.

Das, T.K. and Teng, B.S. (2001) Trust Control and Risk in Strategic Alliances: An integrated Framework. *Organization Studies,* 22(2), 251-283.

Dasgupta, K. and Chandramouli, A. (2001) Peer-to-Peer Computing in the Enterprise. *EBizQ* [online] 27 August 2001. Available from: www.ebizq.net/topics/dev_tools/features/2319.html (accessed 8 November 2005)

Dash, J. (2001) Outsourcing Wave hits Japanese markets. *Computerworld* [online]. Available from: www.computerworld.com/managementtopics/management/outsourcing/story/0,10801,62813,00.html (accessed 9 November 2005)

Davenport, T. (1997) *Information Ecology.* Oxford University Press, UK.

Davis, P.O. (2004) Credit Risk Measurement:Avoiding Unintended Results, Part 1. *The RMA Journal* April 2004, 86-89.

Deck, S. (2001) *What is CRM?* [online]. The CRM Research Centre, CIO.com. Available from: www.cio.com/research/crm/edit/crmabc.html accessed 11 November 2005)

Distefano, J. (2000) The Decisioning Frontier: Information Strategy. *DM Review* [online] July 2000. Available from: www.dmreview.com/master.cfm?NavID=198&EdID=2356 (accessed 9 November 2005)

Ditsa, G. (2003) Executive Information Systems use in organisational contexts: An Explanatory User Behaviour Testing. In: *Information Management: Support Systems and Multimedia Technology,* 109-155.

DMReview (2003) Industry Implementations. *DM Direct Newsletter* [online] 26 September 2003. Available from: www.dmreview.com/article_sub.cfm?articleId=7452 (accessed 9 November 2005)

Dodgson, M. (1991) Technology, learning, technology strategy and competitive pressures. *British Journal of Management* 2/3, 132-149.

Dove, R. (2000) Knowledge Work and Trust – The Key Relationship in Relationship Management. *Paradigm Shift International* [online]. Available from: www.parshift.com/Essays/essay058.htm (accessed 9 November 2005)

Drucker, P. (1995) *Managing in a Time of Great Change.* Truman Talley, New York.

Drucker, P. (1999) *Management Challenges for the 21st Century.* Harper Business, New York.

Drucker, P. (2001) Survey: The Near Future – Part 1 and 2. *CFO.com* [online] 2 November 2001. Available from: www.cfo.com/article/1,5309,5637,00.html (accessed 9 November 2005)

Dubinski, A.J., Yammarino, F.J. and Jolson, M.A. (1995) An examination of the linkages between personal characteristics and dimensions of transformational leadership. *Journal of Business and Psychology* 9(3) 315-335.

Dunne, A.J. (2004) *The Learning Organisation: a new imperative for Australian Agribusiness?* Agribusiness Perspectives Papers 2004, Paper 62. Melbourne University, Australia.

Dunne, A.J. (2001) Supply Chain Management: Fad, Panacea or opportunity? *Occasional Paper,* 8 (2), 1-40. School of Natural and Rural Systems Management, The University of Queensland, Gatton, Queensland, Australia.

ECCC (2003) *Tracking and Tracing Food Products in Canada.* Whitepaper on the Can-Trace Initiative. www.can-trace.org/About/docs/TrackingAndTracingInitiativeWhitePaper.pdf (accessed 9 November 2005)

Edwards, J., Lehtoranta, A. and Sjöström, A. (2002) Mobile internet: A solution still in search of a problem? *Focus E-Zine* Issue 8. Cap Gemini Ernst & Young, USA.

Edwards, J.B. (2002) How are Payoff-Conscious CEOs Changing IT Trends? *Journal of Corporate Accounting and Finance* 13:5, July/August, 33-38.

Eisenbach, R., Watson, K. and Pillai, R. (2001) Transformational leadership in the context of organizational change. *Journal of Organizational Change Management* 12:2, 80-89.

Electronic Transactions Bill (1999) Australian Government. Available from: www.aph.gov.au/library/pubs/bd/1999-2000/2000BD064.htm (accessed 15 November 2005).

Elmasri, R. and Navathe, S. (2000) *Fundamentals of Database Systems,* 4th edn. Benjamin Cummins, Redwood City, CA.

ENVI Application Challenge (2005) [online]. Research Systems Inc. Available from: www.rsinc.com/envi/applications_challenge/index.asp - one (accessed 11 November 2005)

Euro-Lex Portal (2005) *Regulation No 178/2002 of the European Parliament and of the Council* [online]. Available from: http://europa.eu.int/eur-lex/en/index.html (accessed 9 November 2005)

European Commission (2004) *Corporate Social Responsibility: National Public Policies in the European Union.* European Commission, Directorate-General for Employment and Social Affairs. Luxembourg.

European Framework of Quality Management (2004) *The EFQM Model.* Available from: www.efqm.org/Default.aspx?tabid=122 (accessed 17 November 2005)

European Guide to Good Practice Knowledge Management (2004) *CWA 14924 Series* [online]. Available from: www.cenorm.be/cenorm/businessdomains/businessdomains/isss/cwa/knowledge+management.asp (accessed 9 November 2005)

European Union Commission (2002) *B2B Internet trading platforms: Opportunities and barriers – A first assessment.* Commission staff working paper [SEC (32002) 1217].

European Union Commission (2003) *Final Report of the Expert Group on B2B Internet Trading platforms* [online]. Available from: http://europa.eu.int/comm/enterprise/ict/policy/b2b/wshop/fin-report.pdf (accessed 9 November 2005)

Fahy, J. and Smithee, A. (1999) Strategic Marketing and the Resource Based View of the Firm. *Academy of Marketing Science Review* [online] 99 (10). Available from: www.amsreview.org/articles/fahy10-1999.pdf (accessed 11 November 2005)

FAO (2005) *Understanding the Codex Alimentarius*. WHO and FAO, Rome. Available from: www.fao.org/documents (accessed 10 November 2005)

FDA Portal (2005) *Public Health Security and Bioterrorism Preparedness and Response Act of 2002* [online] (the Bioterrorism Act, Section 36). USA Food and Drug Administration. Available from: www.fda.gov/oc/bioterrorism/bioact.html (accessed 10 November 2005)

Federal Bank of Dallas (2005) *Innovation, Technological Change and the Economy*. Federal Bank of Dallas

Fensel, D. (2001) *Ontologies: Silver Bullet for Knowledge Management and Electronic Commerce*. Springer-Verlag, Heidelberg.

Ferreira, J., Schlumpf, E. and Prokopets, L. (2002) *Collaborative Commerce: Going Private to Get Results*. Deloitte Research Papers. Available from: www.deloitte.com (accessed on 11 November 2005)

Firestone, J.M. (1999) Defining the Executive Information Portal [Executive Information Systems Inc., White Paper No. 13, 31 July. Available from: http://www.dkms.com/papers/eipdef.pdf (accessed 11 November 2005).

Firestone, J.M. (2003) *Enterprise Information Portals and Knowledge Management*. KMCI Press/Butterworth-Heinemann , Burlington, MA.

Firmbuilder (2001) *The Disciplines of Outsourcing* [online]. Available from: www.corbettassociates.com/firmbuilder/articles/5/ (accessed 10 November 2005)

Forsyth, G. (2004) eCommerce in Supply and Demand Chains. Rural Industries Research & Development Corporation. *RIRDC Report* no. 04/106, Canberra, Australia.

Franklin, C. (2000) *How BlueTooth Works*. [online]. Available from: http://electronics.howstuffworks.com/bluetooth.htm (accessed 10 November 2005)

Franks, J.R. and Mayer, C. (1994) *Corporate control: A synthesis of the international evidence*. Institute of Finance and Accounting (IFA) Working Paper. London School of Economics, pp. 165-92.

Fritz, M. and Schiefer, G. (2003) Dynamic Market Monitoring for Agrifood Industries over the Internet. In: *EFITA 2003 Conference Proceedings* 5-9 July 2003. Debrecen, Hungary. pp. 131-137.

Frost, P.J., Moore, F.L., Louis, M.R., Lundberg, C.C. and Martin, J. (eds) (1985) *Organizational Culture*. Sage Publications Inc.

Fukuyama, F. (1995) *Trust: The Social Virtues and the Creation of Prosperity*. Penguin. London.

Fukuyama, F. (2001) The virtual handshake: E-commerce and the challenge. *Proceedings of the Seventh Americas Conference on Information Systems*, 689-695.

Gartner Group (1999) IT Expo, Brisbane, Australia.

Garvin, M. (2001) Challenges of Infrastructure Management. *Report to the Institute for Civil Infrastructure Systems*. USA.

Gaski, J.F. (1984) The Theory of Power and Conflict in Channels of Distribution. *Journal of Marketing,* vol. 59, 9-29.

Gaudin, S. (2002) The Emergence of Collaborative Commerce. *IT Management* [online] 2 December 2002. Available from: http://itmanagement.earthweb.com/it_res/print.php/1551061 (accessed 10 November 2005)

GeoSINC International (2002) *e-Readiness Guide for Developing Countries – How to Develop and Implement a National e-Readiness Action Plan*.

Ginns, D. (2002) *The Agri-food sector in Australia; Where is it going?* Agribusiness Perspectives Papers 2002, Paper 54. Melbourne University, Australia.

Gluck, F.W., Kaufman, S.P. and Walleck, A.S., (1980) Strategic management for competitive advantage. *Harvard Business Review*, July/August, 154-161.

Gluck, F.W., Kaufman, S.P. and Walleck, A.S. (1982) The Four Phases of Strategic Management, *Journal of Business Strategy*, Winter, pp. 9-21.

Gluck, F.W., Kaufman, S.P., Walleck, A.S., McLeod, K. and J. Stuckey (2000) Thinking strategically. *The McKinsey Quarterly, Strategy Anthology,* Spring 2000.

Golan, E. (2004) Traceability in the US food supply: An Overview. *USDA Briefing Notes* [online] January 12, 2004. Available from: www.ers.usda.gov/briefing/traceability/overview.htm (accessed 10 November 2005)

Golan, E., Krissoff, B., Kuchler, F., Nelson, K., Price, G. and Calvin, L. (2003) Traceability in the US Food Supply: Dead End or Superhighway? *Choices Magazine* 18(2), 36-40.

Golan, E., Krissoff, B., Kuchler, F., Calvin, L., Nelson, K. and Price, G. (2004) Traceability in the US Food Supply: Economic Theory and Industry Studies. AER-830, USDA/ERS. March 2004.

Goleman, D. (1998) What makes a Leader? *Harvard Business Review* 76(6), 92-102.

Goleman, D. (2000) Leadership gets results. *Harvard Business Review,* March/April 2001, 78-90.

Gouvea, A., Laouchez, J.-M. and Lindenboim, C. (2002) Brazilian Boardrooms. *McKinsey Quarterly* no. 2, 6-12.

Griffin, J. (2000) Information Strategy: Integrating to Achieve E-Business Success. *DM Review* [online] October 2000. Available from: www.dmreview.com/master.cfm?NavID=198&EdID=2733 (accessed 10 November 2005)

Gulati, R. (1998a) Alliances and Networks. *Strategic Management Journal*, 19, 293-317.

Gulati, R. (1998b) The architecture of cooperation: managing coordination costs and appropriation concerns in strategic alliances. *Administrative Science Quarterly,* December.

Gunnlaugsdottir, J. (2002) An International Standard on Records Management: An Opportunity for Librarians. *Libri*, vol. 52, 231-240.

Hachi, R. and Lighton, J. (2001) The Future of the Networked Company. *McKinsey Quarterly*, no. 3, 26-39.

Hagel, J. (1996) Spider versus Spider – Are webs a new strategy for the information age? *McKinsey Quarterly* no. 1, 4-18.

Hagel, J. and Armstrong, A.C. (1997) Net Gain: Expanding markets through virtual communities. *McKinsey Quarterly* no. 1, 140-153.

Hagel, J. and Singer, M. (2000) Unbundling the Corporation. *McKinsey Quarterly* no. 3, 148-161.

Haisten, M. (2000) Real-Time Data Warehouse: Putting the E in E-BI. *DM Review* [online] June. Available from: www.dmreview.com/master.cfm?NavID=198&EdID=2320 (accessed 10 November 2005).

Hajibashi, M. (2001) *E-Marketplaces: The Shape of the Economy*. Ascet vol 3, 15 April.

Hall, C. (2000) Enterprise Information Portals: hot air or hot technology? Cutter Consortium Press March 2000.

Hall, M. (2001) Outsourced to the Core. *ComputerWorld* [online] March 26, 2001. Available from: www.computerworld.com/managementtopics/management/story/0,10801,58566,00.htm (accessed 10 November 2005)

Ham, S. and Atkinson, R.D. (2001) A Third Way Framework for Global E-Commerce. *Technology and New Economy Project* [online] Progressive Policy Institute. Available from: www.ndol.org/documents/global_ecommerce.pdf (accessed 10 November 2005)

Handy, C. (1993) Balancing corporate power: A new federalist paper. *The McKinsey Quarterly* no. 3, 159-182.

Hannon, P. and Atherton, A. (1998) Small Firm Success and the Art of Orienteering: the Value of Plans, Planning and Strategic Awareness in the Competitive Small Firm. *Small Business and Enterprise Development* 5 (2), 102-119.

Hausman, C. (1999) The changing role of farmers, corporations, governments and the WTO in maintaining trust in a Global Trade Environment. In: *IAMA Congress Proceedings*, June 1999, Florence, Italy.

Hausman, W.H. (2002) Supply chain performance metrics. In: Billington, C., Harrison, T., Lee, H. and Neale, J. (eds) *The Practice of Supply Chain Management.* Kluwer, Boston.

Hay McBer Corporation (1996) *Four Circle Model of Leadership.* Hay McBer Inc.

Heffernan, W., Hendrickson, M. and Gronski, R. (1999) Consolidation in the Food and Agriculture System. *Report to the National Farmers Union,* 5 February 1999.

Hildebrand, C. (2000) What is EDI? *Darwin Magazine,* 1 October 2000.

Hildreth, P.J. and Kimble, C. (2002) The Duality of Knowledge. *Information Research* [online] 8 (1), paper no. 142. Available from: http://informationr.net/ir/8-1/paper142.html (accessed 10 November 2005)

Hillson, D.A. (1999) Developing effective risk responses. In: *Proceedings of the 30th Annual Project Management Institute Seminars and Symposium.* 11-13 October 1999. Philadelphia, USA.

Hillson, D.A. (2002a) Using the Risk Breakdown Structure (RBS) to understand risks. In: *Proceedings of the 33rd Annual Project Management Institute Seminars and Symposium (PMI 2002),* 7-8 October 2002. San Antonio, USA.

Hillson, D.A. (2002b) What is risk? Towards a common definition. *InfoRM* Journal of the UK Institute of Risk Management, April 2002, 11-12.

Hines, P. (2004) *The lean way to enhanced performance and profits* [online]. Available from: www.laa.asn.au/__data/page/347/Peter_Hines_Presentation_402a.pdf (accessed 10 November 2005)

Hoffman, I. (2002) *The web site audit check list* [online]. Available from: http://www.ivanhoffman.com/audit.html (accessed 10 November 2005)

Hoffman, N.P. (2000) An Examination of the 'Sustainable Competitive Advantage' Concept: Past, Present, and Future. *Academy of Marketing Science Review* [online] 00(04). Available from: www.amsreview.org/articles/hoffman04-2000.pdf (accessed on 14 November 2005)

Hogg, T., Huberman, B. and Franklin, M. (2000) Protecting Privacy while sharing Information in Electronic Communities. In: *Proceedings of the 10th Conference on Freedom and Privacy – Challenging Assumptions.* ACM, vol. 10, 73-75.

Hollander, E.P. and Offerman, L.R. (1990) Power and Leadership in Organizations. *American Psychologist,* February, 179-188.

Holsapple, C. and Singh, M. (2000) Toward a Unified View of Electronic Commerce, Electronic Business, and Collaborative Commerce: A Knowledge Management Approach. *Knowledge and Process Management* 7(3), 151-164.

Holsapple, C.W. and Whinston, A.B. (1996) (Revised edn 2002) *Decision Support Systems: A Knowledge-Based Approach.* International Thompson Publishers, Cambridge, Massachusetts.

Hooker, N.H., Helig, J. and Ernst, S. (2001) What is unique about E-Agribusiness? In: *Proceedings of the IAMA Conference,* 27-28 June 2001. Sydney, Australia.

Hoque, F. (2000) *e-Enterprise – Business Models, Architecture and Components.* Cambridge University Press and SIGS Books, UK.

Hoschka, T.C. (2002) A market for the well governed. *McKinsey Quarterly* no 3, 26-28

Huang, H., Kesar, C., Leland, J. and Shachat, J. (2003) Trust, the Internet and digital Divide. *IBM Systems Journal,* vol. 42, no. 3.

Huberman, B.A., Franklin, M. and Hogg, T. (1999) Enhancing Privacy and Trust in Electronic Communities. In: *Proceedings of the 1st ACM Conference on Electronic Commerce (EC99)* ACM Press, NY, 78-86.

Huston, G. (2002) Telecommunications Policy and the internet. *OnTheInternet* [online] January/February 2002. Available from: www.isoc.org/oti/articles/1201/huston.html (accessed on 14 November 2005)

IBM Global (2000) *From idea to IPO. Building a successful and supportive eBusiness infrastructure* [online]. Available from: www-1.ibm.com/services/us/its/pdf/ngipo.pdf (accessed 10 November 2005)

Ickis, J.C. and Brenes, E.R. (1999) Building trust in American Central American Agribusiness. In: *IAMA Congress Proceedings,* June 1999. Florence, Italy.

International Chamber of Commerce (2002) *A Global Action Plan for eCommerce,* (3rd edn) [online]. Available from: http://www.iccwbo.org/home/electronic_commerce/word_documents/3rd%20Edition%20Global%20Action%20Plan.pdf

International Corporate Governance Network (1999) *Statement on OECD Global Corporate Governance Principles.* [online]. Available from: www.icgn.org/documents/globalcorpgov.htm (accessed 10 November 2005)

Investopedia (2005) *Dictionary of Finance and Investment.* Available from: www.investopedia.com (accessed 17 November)

IP Australia (2005) *An Introduction to IP.* Available from: www.ipaustralia.gov.au/ip/introduction.shtml (accessed 15 November 2005)

Ishaya, T. and Macaulay, L. (1999) The Role of Trust in Virtual Teams. *Journal of Organizational Virtualness,* 1(1) 140-157.

ISO (2001a) *ISO 15489:2001 Information and Documentation – Records Management.* ISO Central Secretariat.

ISO (2001b) *ISO Standards Compendium ISO 9000 Quality Management.* ISO Central Secretariat.

ISO (2004) *ISO 9000:2000 Standards*, edn. 3, pp. 23. ISO Central Secretariat.

ISO (2005) *ISO 17799 Services Portal.* Available from: www.iso17799software.com/ (accessed 18 November 2005)

iSource Editorial (2003) The Time for CPFR is Now. *Supply Demand Chain Executive Magazine,* 4 February. Available from: www.isourceonline.com/article.asp?article_id=3579 (accessed 15 November 2005)

Jabali, T. (2003) Relationship Dynamics: The Saviour of CRM. *Intelligent Enterprise Magazine* [online] 10 October 2003. Available from: www.intelligententerprise.com/ (accessed 11 November 2005)

Jaffe, B. (2001) Be it bricks or clicks, a buck is still a buck. *eWeek.com* [online] 12 March 2001. http://www.eweek.com/article2/0,1759,464437,00.asp

James, D. and Wolf, M.L. (2000) A Second Wind for ERP. *The McKinsey Quarterly* no. 2, 100-108.

Jenson, M.C. (2003) *A Theory of the Firm: Governance, Residual Claims and Organisational Forms.* Harvard University Press.

Johansen, A. and Sornette, D. (2000) *The NASDAQ Crash of April 2000: Yet another example of log-periodicity in a speculative bubble ending in a crash* [online] 3 May 2000. Available from: www.edpsciences.org/articles/epjb/abs/2000/18/b0225/b0225.html, (accessed 14 November 2005)

Jones, N. and Basso, M. (2003) Mobility Makes Business Processes Fast and Responsive. *Strategy, Trends and Tactics* [online] 18 April 2003. Gartner. Available from: www4.gartner.com/ (accessed 10 November 2005)

Julien, T.U. and Shaw, C.D. (2003) Firefighter Command Training Virtual Environment. TAPIA, 03, Oct 15-18, Atlanta, Georgia, USA. *ACM* 1-58113-790.

Juniper Research (2005) Revenue from Mobile Sports Video-Clips & Text-Updates to double in 2005. *Press Release 6th April 2005.* Available from: www.juniperresearch.com (accessed on 16 November 2005).

Kahaner, L. (1997) *Competitive Intelligence – How to gather, analyse and use information to move your business to the top.* Simon & Schuster, Touchstone Books, USA.

Kaplan, R.S. and Norton, D.P. (1992) The Balanced Scorecard – Measures that Drive Performance. *Harvard Business Review* 70(1), 71-79.

Kaplan, R.S. and Norton, D.P. (1993) Putting the Balanced Scorecard to Work. *Harvard Business Review* 71(5), 134-147.

Kaplan, R.S. and Norton, D.P. (1996) Using the Balanced Scorecard as a Strategic Management System. *Harvard Business Review* 74(1), 75-85.

Keighley, T. (1993) Creating an empowered organization. *Training and Development in Australia.* December, 6-11.

Kirtland, A. (2003) Executive Dashboards. *Boxes and Arrows* [online] 24 November 2003. Available from: www.boxesandarrows.com/archives/executive_dashboards.php (accessed 10 November 2005)

Koch, C. (2002) The ABCs of ERPs [online]. the ERP Research Centre. Cio.com. available from: www.cio.com/research/erp/edit/erpbasics.html (accessed 10 November 2005)

Kojima, S., Nakanishi, H., Yoda, K. and Sora, H. (2004) Traceability System for Manufacturer Accountability. *Hitachi Review*, vol. 53 no. 5.

Kozlowski, M.A. (2001) Trust may be the most significant impediment to Digital Commerce. *Delphi Group Report on Group Trust,* July 2001.

Kuhn, H., Junginger, S. and Bayer, F. (2000) How Business Models influence eBusiness Applications. In: Stanford-Smith, B. and Kidd, P.T. (eds) *Proceedings of eBusiness and eWork 2000,* 18-20 October 2000. Madrid, Spain.

Kumar, K. and Van Hillegersberg, J. (2000) ERP Experiences and Evolution. *Communications of the ACM,* vol. 43, 4.

Lam, J. (1999) Enterprisewide Risk Management and the Role of the Chief Risk Officer. ERisk White Paper, 25 March 1999. Available from: http://erisk.com/Learning/Research/011_lamriskoff.pdf (accessed 17 November 2005)

Lam, J. (2000) *All Bets are On.* Harvard Business School Press.

Lambert, R. (1996) Data Warehousing Fundamentals: What You Need To Know To Succeed. *DM Review* [online] March 1996. Available from: www.dmreview.com/master. cfm?NavID=198&EdID=1313 (accessed 10 November 2005)

Langlois, M. (2001) Selected Issues in Business Process Outsourcing. *Strikeman Elliot Regional Meeting,* 15 May 2001. Toronto, Canada.

Lanner Group (2004) The WITNESS Suite. Available from: www.lanner.com/products/product_ overview.php (accessed on: 11 November 2005)

Lanvin, B. (2002) Reaping the Full Benefits of ICTs in the New Global Economy. *GIIC Annual Meeting* [online] 23 April 2002. Beijing. Available from: www.giic.org/events/beijing/bruno_ lanvin.ppt (accessed on 14 November 2005).

Laplant, M. and Levine, A. (1999) Web Content syndication: Issues on the Sell side. *Fastwater Rapids* 1(5).

Laseter, T.M., Houston, P.W., Wright, J.L. and Park, J.Y. (2000) Amazon your Industry: Extracting the Value from the Value Chain. *Business Strategy*, First Quarter 2000**.** Available at: www.boozallen. de/scm/downloads/602.pdf (accessed on 14 November 2005)

Lawrence, E., Corbitt, B., Tidwell, A., Fisher, J. and Lawrence, J.R. (2000) *Internet Commerce: Digital Models for Business,* 2nd edn. John Wiley & Sons, Australia.

Lawrence, F.B., Jennings, D.F. and Reynolds, B.E. (2003) *e-Distribution.* Thomson South Western, Ohio, USA.

Leclerc, A. (2000) Integrating the Enterprise. *Distributed Enterprise Architecture,* vol. III, no. 5. Cutter Consortium, Executive Report.

Lee, H. (2001) Ultimate Enterprise Value Creation using Demand Based Management. *Stanford Global Supply Chain Management Forum* SGSCMF-W1-2001.

Lee, H. and Whang, S. (1998) *Information Sharing in a Supply Chain.* Research Paper No 1549. Graduate School of Business, Stanford University.

Lengnick-Hall, M. and Moritz, S. (2003) The impact of eHR on the Human Resource Management Function. *Journal of Labour Research* 24:3, 365.

Leon, M. (2003) Dashboard Democracy. *Computer World,* 16 June 2003. Report No: 38766.

Leroux, N., Wortman, M.S. and Mathias, E. (2001) Dominant factors impacting the development of Business-to-Business (B2B) E-Commerce in Agriculture. In: *Proceedings of the IAMA Conference,* 27-28 June 2001. Sydney, Australia.

Lewicki, R.J. and Bunker, B.B. (1996) Developing and maintaining trust in working relationships. In Kramer, R.M. and Tyler, T. (eds) *Trust in Organizations.* Sage, Thousand Oaks, CA.

Lewis, C.D. (1998) *Demand Forecasting and Inventory Control: A Computer Aided Learning Approach.* Wiley & Sons, New York.

Lief, L. (1992) In: Applewhite, A., Evans, W.R. and A. Frothingham (eds) *And I Quote:The Definitive Collecton of Quotes, Sayings, and Jokes for the Contemporary Speechmaker* (Revised Edition). St. Martin's Press.

Litan, R. and Rivlin, A. (2001) The Economy and the Internet: What lies ahead. *Brookings Institute, Conference Report* no. 4.

Lothair Inc. and Norbridge Inc. (2001) Supply Chain Collaboration – Close Encounters of the Best Kind. *BusinessWeek Special Advertising Sections* [online] 26 March 2001. Available from: www.businessweek.com/ adsections

Louis, M.R. (1985) Perspectives on Organizational Culture. In: Frost, P.J., Moore, L.F., Louis, M.R., Lundberg, C.C. and Martin, J. (eds) *Organizational Culture*. Sage, Beverly Hills, CA.

Lucansky, P. and Burke, R. (2003) What is Lean Enterprise? *Supply Chain Planet* [online] 28 October 2003. Available from: www.supplychainplanet.com/e_article000203543.cfm (accessed 10 November 2005)

Luce, C. (2002) Three trends in enterprise information portals. *ZDNet Tech Update* [online] 19 March 2002. Available from: www.techupdate.zdnet.com/techupdate/stories/main/0,14179,2855580,00.html (accessed 10 November 2005)

Lummus, R.R. and Vokurka, R.J. (1999) Managing the Demand Chain Through Managing the Information Flow: Capturing Moments of Information. *Production and Inventory Management Journal*, 40 (1), 16-20.

Lundquist, E. (2001) Last Word on DotCom Flops. *ZDNet E-Commerce* [online] 12 February 2001. Available from: www.eweek.com/article2/0,1759,160545,00.asp (accessed 14 November 2005)

Ma, H. (2000) Of Competitive Advantage: Kinetic and Positional. *Business Horizon* 43, 53-64.

Machiavelli, N. (1513): *The Prince, Ch 6: Concerning New Principalities Which Are Acquired by One's Own Arms And Ability*. Medieval Source Book. Available from: www.fordham.edu/halsall/basis/machiavelli-prince.html (accessed 15 November 2005)

Macintosh, A., Filby, I., Kingston, J. and Tate, A. (1998) Knowledge Asset Road Maps. In: *Proceedings of The Second International Conference on Practical Aspects of Knowledge Management (PAKM98)*. Basel, Switzerland.

Mackson, J., Wright, G., Rachaputi, N.C., Krosch, S. and Tatnell, J. (2000) *Aflatoxin in Peanuts* [online]. Department of Primary Industries and Fisheries Note. Available from: www.dpi.qld.gov.au/fieldcrops/3027.html (accessed 11 November 2005)

Malcolm, A. (2005) *Food Safety Training and Management Manual* [online]. Available from: www.foodhazardcontrol.com/ (accessed 11 November 2005)

Maney, K. (2001) 'First mover advantage' no longer an advantage. *USA Today, Marketplace, Cyberspeak* [online] 18 July 2001. Available from: www.usatoday.com/tech/columnist/2001-07-18-maney.htm (accessed 10 November 2005)

Marinos, G. (2005) The Information Supply Chain: Achieving Business Objectives by Enhancing Critical Business Processes. *DMReview*, April 2005.

Martin, C. (1999) Creating the Agile Supply Chain. *ASCET*, vol. 1. January 1999.

Martin, R., Miller, P. and Mayes, S. (2003) The 'New' Economy: Myths, Realities and Regional Dynamics. *Proceedings of High-Tech Business: Clusters, Constraints and Economic Development*, Robinson College, Cambridge, May 2003. Available from: www.cbr.cam.ac.uk/news/pdf/Martin.pdf (accessed 15 November 2005)

Maza, J. (2001) Corporate Portals: The Value to your Enterprise. *DMReview*, February 2001.

McAmis, D. (2001) Bringing Business Intelligence to the Portal, *The Intranet Journal* [online]. Available from: www.intranetjournal.com/articles/200104/pt_04_25_01e.html (accessed 10 November 2005)

McGrath, R.G. (2003) Strategy Development in Turbulent Times: Entrepreneurial Leadership. *13th International Food and Agribusiness Management Association (IAMA) Forum*. June 23, 2003. Cancun, Mexico

McGrath, R.G. and MacMillan, I. (2000) *The Entrepreneurial Mindset*. Harvard Business School Press, MA.

McKean, J.D. (2001) The importance of traceability for public health and consumer protection. *Rev. Sci. Tech. Off. Int. Epiz.* 20 (2), 363-371.

McLean, B. (2001) ENRON – Is Enron Overpriced? *Fortune* [online] 5 March 2001. Available from: www.fortune.com/fortune/specials/2003/1027/enron.html (accessed 10 November 2005)

McLean, B. (2002) ENRON – Monster Mess. *Fortune* [online] 20 January 2002. Available from: www.fortune.com/fortune/investing/articles/0,15114,369754,00.html (accessed 11 November 2005)

McLellan, T. (1995) *Data Modeling: Finding the Perfect Fit: An Introduction to Data Modeling* [online]. Available from: www.islandnet.com/~tmc/html/articles/datamodl.htm (accessed 10 November 2005)

Mehr, R.I. and Hedges, B.A. (1963) *Risk Management in the Business Enterprise.* Richard D. Irwin, Inc. Homewood, IL

Mercy, J.-B. (2001) A Better Understanding of the Enterprise Information Portal Market. *Internet Journal* [online] October 2001. Available from: www.intranetjournal.com/articles/200110/eip_10_03_01a.html (accessed 10 November 2005)

Miller, J. (2003) High Tech and High Performance: Managing appraisal in the Information Age. *Journal of Labour Research.* 23:3, 409.

Mintzberg, H. (1990) Strategy formation: schools of thought. In: Frederickson, J.W. (ed.) *Perspectives on Strategic Management.* Harper & Row, New York.

Mintzberg, H. (1994) *The Rise and Fall of Strategic Planning.* Prentice-Hall, Englewood Cliffs.

Monks, R.A.G. and Minow, N. (1999) *Power and Accountability* [online]. Available from: www.lens-library.com/power/ (accessed 10 November 2005)

Montgomery, C.A. (ed.) (1995) *Resource-based and evolutionary theories of the firm: Towards a Sustainable Synthesis.* Kluwer Academic, Boston/Dordrecht.

Morecroft, J.D. (1994) Executive Knowledge, Models, and Learning. In: Morecroft, J.D. and Sterman, J.D. (eds) *Modeling for Learning Organizations.* Productivity Press, Portland, pp. 3-28.

Mougayar, W. (2002a) *From Horizontal to Peripheral* [online]. Available from: www.mougayar.com/ (accessed 14 November 2005)

Mougayar, W. (2002b) *The Open Corporation* [online]. Available from: www.mougayar.com (accessed 14 November 2005)

Narasimhan, R. and Das, A. (1999) Manufacturing Agility and Supply Chain Management Practices. *Production and Inventory Management Journal,* 40 (1), 4-10.

National Archives of Australia (2000) *The Dirks Methodology – Users Guide* [online]. Available from: www.naa.gov.au/recordkeeping/dirks/dirksman/part1.html - fn2 (accessed 10 November 2005)

National Occupational Health and Safety Commission (2003) [online]. Commonwealth of Australia. Available from: www.natindex.nohsc.gov.au/ (accessed 10 November 2005)

New Economy Index (1998) *A Report from the Progressive Policy Institute,* 1-54. Washington, DC, USA. Available from: www.neweconomyindex.org/index_nei.html (accessed 30 May 2006)

Newell, R. and Wilson, G. (2002) A Premium on Good Governance. *McKinsey Quarterly,* no. 3.

Newton, D. (2000) Supply Chain Learning for Australian Agribusiness. *Agriculture, Fisheries and Forestry Report.* Commonwealth of Australia.

NFIS (2002) National Food Industry Strategy, Australia [online]. Available from: www.nfis.com.au/ (accessed 10 November 2005)

Ng, E. (2005) Types of B2B E-Business Models commonly used: An Empirical Study on Australian Agribusiness firms. *Australasian Agribusiness Review* 13, Paper 1. University of Melbourne, Australia.

NHS (2005) Developing the KM Environment. *Specialist Library of Electronic Health* [online]. Available from: www.nelh.nhs.uk/knowledge_management/km2/processes.asp (accessed 10 November 2005)

Niccolai, J. (2001) *Sabre, EDS get approval for giant outsourcing deal* [online]. Available from: www.computerworld.com/industrytopics/travel/story/0,10801,61843,00.html (accessed 10 November 2005)

Noddle, A. (1999) Global Food Retailing and Consumer Trust. In: *IAMA Congress Proceedings,* June 1999. Florence, Italy.

NOIE (2002) *Advancing Australia – The Information Economy Progress Report.* Commonwealth Government of Australia, Canberra, ACT.

Nonaka, I. (1991) The Knowledge-Creating Company, *Harvard Business Review,* vol. 6 no. 8, 96-104.

Nonaka, I. (1994) A Dynamic Theory of Organisational Knowledge Creation. *Organisational Science* 5/1, 14-37.

Nonaka, I. and Konno, N. (1998) The Concept of 'Ba': Building a Foundation for Knowledge Creation. *California Management Review* 40 (3) Spring, 40-54.

Nonaka, I. and Takeuki, H. (1995) *The Knowledge-Creating Company: How Japanese Companies Create the Dynamics of Innovation.* Oxford University Press, New York.

Norris, M., West, S. and Gaughan, K. (2001) *E-Business Essentials – Technology and Network Requirements for the Electronic Marketplace.* Wiley-BT Series. John Wiley & Sons Ltd, Chichester, UK.

Norton, D. and Kaplan, R. (1992) The Balanced Scorecard: Measures that drive performance. *Harvard Business Review* 70(1), 71-79.

Oak, H. (2002) A Primer on Wireless Application Protocol (WAP) *Builder.com*, 3 Jul 2002. Available from: http://builder.com.com/5100-6387-1045252.html (accessed 14 November 2005)

O'Keefe, M. (1998) Establishing Supply Chain Partnerships: Lessons from Australian Agribusiness. *Supply Chain Management,* 3(1), 5-9.

O'Leary, D. (2000) *Enterprise Resource Planning Systems: Systems, Life Cycle, Electronic Commerce, and Risk.* Cambridge University Press, Cambridge, England/New York.

Organization for Economic Co-operation and Development (OECD) (2004) *Principles of Corporate Governance* [online]. Available from: www.oecd.org/dataoecd/32/18/31557724.pdf (accessed 10 November 2005)

Osterwalder, A. and Pigneur, Y. (2002) An e-Business Model Ontology for modeling e-Business. In: *Proceedings of the 15th Bled Electronic Commerce Conference. E-Reality: Constructing the Economy* June 17-19, 2002. Bled, Slovenia, pp. 1-11.

Outsourcing Research Council (1999) *The 1999 Outsourcing Study* [online]. Available from: www.corbettassociates.com/firmbuilder/articles/19/42/ (Member requirement), (accessed on 14 November 2005)

Outsourcing Research Council (2002) *The 2002 Outsourcing Study* [online]. Available from: www.diamondcluster.com/Go/OutsourcingStudy/GlobalIT-OutsourcingStudy.pdf (accessed 14 November 2005)

Owen, H. (1994) *The Millennium Organization.* An occasional paper from the Open Space Institute of Canada.

Palamidese, P., Betro, M. and Muccioli, G. (1993) The Virtual Restoration of the Visir Tomb. *Proceedings of the IEEE Conference on Visualization '93*, 420-423. IEEE Computer Society Press, Los Alamitos, California, USA.

Paris Act of the Berne Convention (1971) [online]. Wipo Database of Intellectual Property. Available from: www.wipo.org/treaties/en/ip/berne/pdf/trtdocs_wo001.pdf (accessed 11 November 2005)

Peat, B. and Webber, D. (1997) *Introducing XML/EDI – The Ebusiness fundamentals.* XML/EDI, 29 August 1997.

Peterovic, O., Kittl, C. and Teksten, R.D. (2001) Developing Business Models for eBusiness, *International Conference on Electronic Commerce,* 31 October-4 November 2001. Vienna.

Peterson, M.J. (2003) *CPFR®: The 21st Century's Most Powerful Process for Consumer Satisfaction* [online]. Available from: www.cpfr.org/CPFR_Process_for_Consumer_Satisfaction.doc (accessed 11 November 2005)

Phan, D. (2003) E-business development for competitive advantages: a case study. *Information and Management,* vol. 40, 581-590.

Pidgeon, N. (2000) How the ethernet works [online]. Available from: www.computer.howstuffworks.com/ethernet.htm (accessed 11 November 2005)

Plant, R. (2000) *eCommerce: Formulation of Strategy.* Prentice Hall.

Polyani, M. (1966) *The Tacit Dimension.* Doubleday & Co. Garden City, New York.

Porter, M.E. (1980) *Competitive Strategy: Techniques for Analyzing Industries and Competitors.* The Free Press, New York.

Porter, M.E. (1985) *Competitive Advantage, Creating and Sustaining Superior Performance.* The Free Press, New York.

Porter, M.E. (1990) *The Competitive Advantage of Nations.* The Free Press, New York.

Porter, M.E. (1996) What is strategy? *Harvard Business Review* 96(6), 61-78.

Porter, M.E. (1998) On Competition. *Harvard Business Review Book Series.*

Porter, M.E. (2001) Strategy and the Internet. *Harvard Business Review* 79(3), 63-78.

PPI (2002) The 2002 *State New Economy Index: Benchmarking Economic Transformation in the States* [online]. Progressive Policy Institute, Strategic Indicators of the New Economy. Available from: www.neweconomyindex.org/states/2002/index.html (accessed 14 November 2005)

Prahalad, C.K. and Hamel, G. (1990) The Core Competence of the Corporation. *Harvard Business Review,* May-June, 1990, pp 79-91.

PricewaterhouseCoopers (1999) *Electronically enabled business regulation: The Borderless Economy in a World of Borders* [online]. Available from: www.pwcglobal.com/fr/pwc_pdf/pwc_policy.pdf

Pritchard, B. and McManus, P. (eds) (2000) *Land of Discontent: The Dynamics of Change in Rural and Regional Australia,* University of NSW Press, Kensington, Sydney, Australia.

Rappa, M. (2005) Data Privacy , Available from: http://digitalenterprise.org/privacy/privacy.html (accessed 14 November 2005)

Quinn, J.B., Julien, F. and Negrin, M. (2000) Outsourcing Strategy: Minimising Strategic Risk. *Global Focus* 12 (3) 99-112.

Rachaputi, N.C., Wright, G.C., Krosch, S., Tatnell, J. and Mackson, J. (2001) Management Practices to Reduce Aflatoxin in Peanut. In: *Proceedings of 10th Australian Agronomy Conference,* Hobart. Available from: www.regional.org.au/au/asa/2001/1/b/rachaputi.htm (accessed 11 November 2005).

Read, C., Ross, J., Dunleavy, J., Schulman, D. and Bramante, J. (2001) *ECFO: Sustaining Value in the New Corporation.* John Wiley & Sons, NY.

Reason, T. (2001) Financial Portals: Behind the Green Door, CFO.com. *Tools and Resources for Financial Executives,* 15 September 2001.

Reddy, R. (2001) Taming the Bullwhip Effect. *Intelligent Enterprise,* 13 June 2001.

Redmond, P. (2002) Measuring the Costs and Value of Enterprise Mobility. *Strategy, Trends and Tactics* [online]. Available from: www4.gartner.com (accessed 10 November 2005)

RFID News (2005a) *The Controversial ISO 11784/85 Standard: The "Standard" with a blemish.* Available from: www.rfidnews.com/iso_11784.html (accessed 16 November 2005)

RFID News (2005b) *The Controversial ISO 11784/85 Standard: A short discussion.* Available from: www.rfidnews.com/iso_11784short.html (accessed 16 November 2005)

Rhodes, C. (1996) Researching Organizational Change and Learning: A Narrative Approach. *The Qualitative Report,* 2 (4) December 1996.

Robert, P.C., Rust, R.H. and Larson, W.E. (eds) (1994) Site Specific Management for Agricultural Systems. *ASA, CSSA, SSSA Second International Conference,* March. Minneapolis, USA.

Roberts, B. (2003) Portal takes off: Jetnet, American Airlines' employee web page, aims to become a one-stop destination. *HR Magazine* [online] February 2003. Available from: www.findarticles.com/p/articles/mi_m3495/is_2_48/ai_97873163 (accessed 14 November 2005)

Robson, A., Phinn, S., Wright, G. and Fox, G. (2004) Combining Near Infrared Spectroscopy and Infrared Aerial Imagery for Assessment of Peanut Crop Maturity and Aflatoxin Risk. *Proceedings of the 4th International Crop Conference,* 26 September-1 October 2004. Brisbane, Australia.

Ruggles, R. (1998) The State of the Notion *California Management Review* 40 (3) 40-54, Spring.

Ryan, J. (2000) The Internet and e-Business Opportunities. *The Technology Guide Series* [online]. Available from: www.techguide.com.(Accessed 14 November 2005)

Sacht, J. (2003) E-HR Strategy: *An Electronic (E) Human Resource (HR) Strategy is attainable by small and medium sized business* [online]. WorkINfo.com Free Paper. Available from: www.workinfo.com/free/Downloads/58.htm (accessed 14 November 2005)

Samarajiva, I.R. (1998) Trust and privacy in cyberspace: A view from an Asian vantage point. *Infoethics '98 Conference Ethical, Legal and Societal Challenges of Cyberspace.* Available from: http://www.unesco.org/webworld/infoethics_2/eng/papers/paper_11.htm (accessed 14 November 2005)

SAP (2003) Gateway to the Future. *Look@SAP* [online]. Available from: www.sap-si.com/company/look_at_sapsi/archive/3_2003/3_03E_14_1.pdf (accessed 14 November 2005)

Sauter, V. (1997) *Decision Support Systems- A Managerial Approach.* John Wiley & Sons, New York.

Say, P. (2001) ECRM Deployment Options. *Internet.com* [online] 22 January 2001. Available from: www.clickz.com/b2b_mkt/b2b_mkt/article.php/835691 (accessed 14 November 2005)

Schein, E. (1992) *Organizational Culture and Leadership,* 2nd edn. Jossey-Bass, USA.

Schiefer, G. (2003) New technologies and their impact on Agriculture, Environment and the Food Industry. In: *EFITA 2003 Conference Proceedings*, 5-9 July. Debrecen, Hungary, pp. 3-11.

Scott, J. (2000) The e-Business Hat Trick: Adaptive Enterprises, Adaptable Software, Agile IT Professionals. *Cutter IT Journal,* 13, (4) 7-12, April.

Schroeder, C.M., Naugle, A.L., Schlosser, W.D., Hogue, A.T., Angulo, F.J. and Rose, J.S. (2005) Estimate of illnesses from *Salmonella* Enteritidis in eggs, United States, 2000. *Emerging Infectious Diseases* [online]. Available from: www.cdc.gov/ncidod/EID/vol11no01/04-0401.htm (accessed on 11 November 2005).

Sculley, A. and Woods, W.W. (2001) *B2B Exchanges: The Killer Application in the Business-to-Business Internet Revolution.* Basic Books. New York, NY.

Seely-Brown, J. and Duguid, P. (1998) Organizing Knowledge. *California Management Review* 40 (3), 90-111.

Senge, P. (1990) *The Fifth Discipline: The Art and Practice of the Learning Organization.* Doubleday Inc, NY.

Senge, P. (ed.) (1994) *The Fifth Discipline Fieldbook: Strategies and Tools for Building a Learning Organisation.* Doubleday Inc, NY.

Shapiro, C. and Varian, H.R. (1998) *Information Rules: A Strategic Guide to the Network Economy.* Harvard Business School Press, Cambridge, MA.

Shephard, J. (2001) Is there a First Mover Advantage in Converting to EBusiness? *AMR Research* [online]. Available from: www.amrresearch.com/Content/View.asp?pmillid=14889 (accessed 14 November 2005)

Shore, D. (2001) Growing your Business. *Price Waterhouse Report* September/October.

Simons, D. and Zokaei, A.K. (2005) Application of the lean paradigm in red meat processing. *British Food Journal,* 107 (4), 192-209.

Simons, D., Francis, M., Bourlakis, M. and Fearne, A. (2003) Identification of the determinants of Value in the UK Red Meat Industry. *Journal of Chain & Network Science,* 3(2), 109-121.

Sims, D. (2000) What is CRM? *CRMGuru* [online] 21 March 2000. Available from: www.crmguru.com/content/features/sims01.html (accessed 14 November 2005)

Smith, T. (2004) A Focus on Animal Electronic Identification. *USDA Food Safety Research Information Office Fact Sheet 12* [online]. Available from: www.nal.usda.gov/fsrio/research/fsheets/fsheet12.htm (accessed 14 November 2005)

Solvik, P. (2001) Transforming your company with Internet Business Solutions. *CISCO Systems iQ* [online] 29 June 2001. Available from: http://newsroom.cisco.com/dlls/tln/newsletter/2001/june/part2.html (accessed 14 November 2005)

Stace, D. and Dunphy, D. (2001) *Beyond the Boundaries,* 2nd edn. McGraw Hill, Australia.

Stata, R. (1989) Organizational Learning – the Key to Management Innovation. *Sloan Management Review*, Spring.

Sterman, J.D. (2000) *Business Dynamics: Systems Thinking and Modeling for a Complex World.* McGraw-Hill, Boston.

Storer, C. (2001) Inter-Organizational Information Feedback Systems in Agribusiness Chains: A Chain Case Study Theoretical Framework. In: *Proceedings of the 2001 International Agribusiness Management Association World Food & Agribusiness Symposium*, 25-28 June 2001, Sydney.

Strassman, P. (1995) *The Politics of Information Management.* Information Economics Press, New Canaan.

Stricker, S., Sumner, D.A. and Mueller, R.A.E. (2003) Wine on the web in a global market: A Comparison of e-Commerce readiness and use in Australia, California and Germany. In: *EFITA 2003 Conference Proceedings,* 5-9 July. Debrecen, Hungary, pp. 253-263.

Sullivan, A. (2001) *Groups to offer global e-commerce conduct code* [online]. Reuters article on Privacy Exchange. Available from: www.privacyexchange.org/news/archives/bpi/buspriv0105. html (accessed 14 November 2005)

Sullivan, B. (2001) Content Management for the Masses. 28 June.

Sullivan, J. (1998) *The Ten Tenets of the 21st Century HR.* [online] Article 29, Gately Consulting. Available from: http://ourworld.compuserve.com/homepages/gately/sullivan.htm (Accessed 14 November 2005)

Sullivan, J. (2003) *E-HR (Electronic HR) a walk through a 21st century HR Department.* [online] Article 76, Gately Consulting. Available from: http://ourworld.compuserve.com/homepages/ gately/pp15js76.htm (accessed 14 November 2005)

Sundstrom, F.J., van Deynze, A. and Bradford, K. (2002) *Identity Preservation of Agricultural Commodities.* Agricultural Biotechnology in California Series: Publication No: 8077

Swanson, E.B. and Ramiller, N.C. (1993) Information Systems Research Thematics: Submissions to a New Journal, 1987-1992. *Information Systems Research* 4(4).

Tai, G. (2001) *Balance your Outsourcing* [online]. Available from: www.itworldcanada.com/a/ Network-World/26a8b617-3a1e-41f2-860c-fbeb227bf7a9.html (accessed 14 November 2005)

Tapscott, D. (2001) Rethinking Strategy in a Networked World (or Why Michael Porter is Wrong about the Internet), *Strategy+Business,* 24, 34-41.

Tapscott, D., Ticol, D. and Lowry, A. (2000) *Digital Capital: Harnessing the Power of Business Webs.* Harvard Business School Press.

Tellis, G.J. and Golder, P.N. (1996) First to Market First to Fail? Real causes of enduring market leadership. *Sloan Management Review* Spring, 65-74.

Thearling, K. (1999) *An Introduction to Data Mining.* [online]. White Paper. Thearling Corporation. Available from: http://www3.primushost.com/~kht/text/dmwhite/dmwhite.htm (accessed 14 November 2005)

Theuvsen, L. (2003) Selling food on the Internet: Chances and strategies of small and medium sized manufacturers. In: *EFITA 2003 Conference Proceedings.* 5-9 July, Debrecen, Hungary, pp. 206-211.

Thompson, R. (2000) Internet Intelligence Infrastructures for the Information Economy. *DM Review,* December.

Thompson, S.J., Hayenga, M. and Hayes, D. (2000) E-Agribusiness. In: *Proceedings of the IAMA Conference* 4 July 2000. Chicago, USA.

Todd, B. (2000) *From Plate to Paddock: Turning the Tables.* Agriculture, Fisheries and Forestry, Commonwealth of Australia.

TRIPS Agreement, The (1994) *Agreement on Trade-Related Intellectual Property Rights* [online]. Available from: www.wto.org/english/tratop_e/trips_e/t_agm0_e.htm (accessed 14 November 2005)

Turban, E., King, D., Lee, J. and Viehland, D. (2004) *Electronic Commerce 2004: A Managerial Perspective.* Prentice Hall, New Jersey.

Turban, E.W., Aronson, J.E. and Liang, T.P. (2005) *Decision Support Systems & Intelligent Systems,* 7th edn. Pearson/Prentice Hall, New Jersey.

Turner, C. (2000) *The Information E-conomy. Business Strategies for competing in the Digital Age.* Kogan Page Ltd, London.

Tushman, M. and Romanelli, E. (1985) Organizational Evolution: A Metamorphosis Model of Convergence and Reorientation. In: Cummings, L.L. and Staw, B.M. (eds) *Research in Organizational Behavior.* JAI Press, Greenwich, CT 7, pp. 171-222.

Tyner, R. (2001) *Sink or Swim: Internet Search Tools and Techniques V 5.0* [online]. Available from: www.lboro.ac.uk/library/sink.html (accessed 14 November 2005)

Tyson, J. (2000) *How Internet infrastructure works* [online]. Available from: http://computer.howstuffworks.com/internet-infrastructure1.htm (accessed 14 November 2005).

Ulrich, W. (1999a) *The Essential Information Integration Strategy* [online]. Tactical Strategy Group. Available from: www.systemtransformation.com/integration_strategy.htm (accessed 14 Noveber 2005)

Ulrich, W. (1999b) *Synchronize EAI with Tactical and Strategic Initiatives* [online]. Tactical Strategy Group. Available from: www.eaiindustry.org/docs/SynchronizeEAIwithTacticalandStrategic.pdf (accessed 14 November 2005).

United Kingdom Food Standards Agency (2005) *Guidance notes on the Food Safety Act 1990* [online] (Amendment) Regulations 2004 & General Food Regulations 2004. Available from: www.food.gov.uk/foodindustry/guidancenotes/foodguid/generalfoodsafetyguide (accessed 14 November 2005)

United Nations Commission on International Trade Law (UNCITRAL) (1996) *Model Law on Electronic Commerce with Guide to Enactment* [online]. (with Additional Article 5*bis* as adopted in 1998). Available from: www.jus.uio.no/lm/un.electronic.commerce.model.law.1996/doc (accessed 14 November 2005)

UQ (2005) *Opportunities for Queensland Tropical Fruits and Related Services in China.* Final Report to Queensland Department of Primary Industries and Fisheries, 4 November. Brisbane, Australia.

US Department of Labor and Education (2000) *What Work Requires of Schools: A SCANS Report for America 2000.*

US Government (2001) *The Budget and Economic Outlook: An Update.* Congressional Budget Office, August 2001.

Van Der Spek, R. and De Hoog, R. (1995) A Framework for Knowledge Management Methodology. In: *Knowledge Management Methods: Practical Approaches to Managing Knowledge,* vol. 3 of 3, Schema Press, Arlington, Texas, pp. 379-398.

van Hemert, N. (2005) E-Business and the Dutch Flower Industry – A survey for strategic opportunities. In: *Proceedings of the IAMA Conference,* 30 April 2005, Chicago. USA.

Varangis, P. and Larson, D.F. (1996) *Dealing with Commodity Price Uncertainty.* Policy Research Working Paper 1667. World Bank, Washington, D.C.

Varangis, P., Hess, U. and Bryla, E. (2003) Innovative Approaches for Managing Agricultural Risks. In: Scott, N. (ed.) *Agribusiness and Commodity Risk Strategies and Management.* Incisive RWG Ltd, London.

Varian, H. and Shapiro, C. (1999) *Information Rules: A Strategic Guide to the Network Economy.* Harvard Business School Press, Boston, USA.

Varon, E. (2002) Portals finally get down to business, *CIO,* 16:5, 1-4.

Vaxelaire, R. (2004) Traceability: responding to Consumer Needs. *14th International Food and Agribusiness Management Association (IAMA) Forum.* 12-15 June. Montreux, Switzerland. Available from: www.ifama.org/conferences/2004Conference/Forum/Vaxelaire.ppt (accessed on 16 November 2005)

VICS (2004) *CPRF – An Overview* [online]. Available from: www.vics.org/committees/cpfr/CPFR_Overview_US-A4.pdf (accessed 14 November 2005)

Vince J. (2004) *Introduction to Virtual Reality.* Springer, New York.

von Krogh, G. and Roos, J. (2004) (Eds.) *Managing Knowledge: Perspectives on Co-operation.* Idea Group Inc.

Waide, P. (2001) *Australian E-tailing: First-mover Disadvantage.* Internet.com, 22 February 2001.

Watson, W. (2000) *The Net Effect: eHR and the Internet* [online]. Research Report. Available from: www.watsonwyatt.com/asia-pacific/pubs/wwwired/articles/2000_11_01.asp (accessed on 14 November 2005)

Watson Wyatt (2005) *The Human Capital Index® (HCI) Report* September 2005. Available from: www.watsonwyatt.com/research/resrender.asp?id=w-822&page=1 (accessed 16 November 2005)

Webber, J. (2000) Second-Mover Advantage? – An investigation into the marketing strategies and successes of Internet companies that are second-to-market. *Internet Marketing*. Spring 2000.

Webmergers.com (2000) *Speed Kills* [online]. Available from: www.webmergers.com (accessed 14 November 2005)

Wenger, E. (1998) Communities of practice: learning as a social system. *The Systems Thinker*, 9(5) June/July.

Wenger, E. and Snyder, W. (2000) Communities of practice the organizational frontier. *Harvard Business Review*. January-February, 139-145.

Whitaker Associates (2001) *The Impact of Clicks on Bricks: VET Facilities Planning in an information age*. Formal Report to Department of Public Works and Services , PMG/Programs/ Education Facilities, Research Group for NSW Department of Education and Training (TAFE). Available at: www.flexiblelearning.net.au/clicks/res_libr/finarepo/media/pdf/full.pdf (accessed 14 November 2005)

White and Case LLP (2002) *Global Privacy Law: A Survey of 15 Major Jurisdictions* [online]. Available from: www.whitecase.com/report_global_privacy.pdf. (accessed 14 November 2005)

White, A. (2000) The Rise and Fall of the Trading Exchange 'shhh – don't tell my competitor!' *ASCET* [online], vol. 3, April 2001. Available from: www.ascet.com/documents.asp?d_ID=488 (accessed 14 November 2005)

White, D.H., Tupper, G. and Mavi, H.S. (1999) *Climate variability and drought research in relation to Australian agriculture* [online]. LWRRDC Occasional Paper CV01/99. Canberra, ACT. Available from: www.acslink.aone.net.au/asit/compend98.htm (accessed 14 November 2005).

Wikipedia (2005) *Free Online Encyclopedia*. Available from: http://en.wikipedia.org/ (accessed 17 November 2005)

Wilkinson, J. (2003) *Agriculture to Agribusiness*. Australian Government Briefing Paper 3/2003.

Wilson, R. (2000) The six principles of viral marketing. *Web Marketing Today* [online] 70. Available from: www.wilsonweb.com/wmt5/viral-principles.htm (accessed 14 November 2005)

WIPO (2005) *Intellectual Property on the Internet: A Survey of Issues*. Available from: www.wipo. int/copyright/ecommerce/en/html/#3a (accessed 16 November 2005)

Womack, J. and Jones, D. (1996) *Lean Thinking*. Simon and Schuster, New York.

Womack, J.P., Jones, D.T. and Roos, D. (1990) *The Machine that changed the world*. Rawson Associates, New York.

Woods, J. and Jimenez, M. (2002) *Balance optimization and synchronization-focused SCM*. Gartner Research Note: Strategic Planning Assumption, 30 September 2002.

World Intellectual Property Organization (WIPO) (2000) *Primer on Electronic Commerce and Intellectual Property Issues*. [online]. Available from: www.wipo.int/copyright/ecommerce/en/ index.html# (accessed 14 November 2005)

World Intellectual Property Organization (WIPO) (2005) *Intellectual Property on the Internet: A Survey of the Issues* [online]. Available from: www.wipo.int/copyright/ecommerce/en/ip_survey/ ip_survey.html (accessed 14 November 2005)

World Trade Organization (2005) Website Section on Agricultural Trade. www.wto.org/english/ tratop_e/agric_e/agric_e.htm (accessed 14 November 2005)

Yee, A. (2001) *Integrating Your e-Business Enterprise*. QUE Publishing, USA.

Yee, A. (2002) *Best practices for successful EAI* [online]. Available from: www.ebizq.net/topics/eai/ features/1712.html (accessed 14 November 2005)

Yee, A. (2004) *Building integration solutions with deployment in mind* [online]. Available from: www. ebizq.net/topics/dev_tools/features/4177.html (accessed 14 November 2005).

Yuzdepski, J. (2000). The WAP Primer. *Wireless Review*, April 1, 2000. Available from: http:// wirelessreview.com/mag/wireless_wap_primer/index.html (accessed 16 November 2005)

Zeeuw, de A. (1999) Realities of the WTO negotiating process. In: *IAMA Congress Proceedings*, June 1999, Florence, Italy.

Zoufaly, F. (2002) *Issues and Challenges Facing Legacy Systems*. Developer.com. Management Article No: 1492531, 1 November 2002.

Useful URLs

Many URLs are included in the actual text of each chapter as they pertain to the discussion. The following are an alphabetical list of URLs where additional useful information has been obtained in putting together this book. All were 'live' as of 14 November 2005.

NOTE: there are a plethora of websites on all the subjects discussed in this book – the reader is advised to do a search for themselves if the following list does not provide them with information they are looking for.

Australian Accounting Standards Board (AASB): www.aasb.com.au

Accounting Standards Board (UK): www.asb.org.uk

Australian Department of Agriculture Fisheries and Forestry: www.daff.gov.au

Australian Food & Fibre industries: www.dfat.gov.au/facts/affaoverview.html

Australian Food News: www.ausfoodnews.com.au/flapa/laws.htm

Australian Government National Office for the Information Economy: www.noie.gov.au

Australian National Occupational Health and Safety Commission: www.nohsc.gov.au

Australian National Food Industry Strategy: www.nfis.com.au

Australian Risk Portal: www.riskmanagement.com.au

BBBOnLine (Better Business Bureau): www.bbbonline.com

BitLaw (technology law online): www.bitlaw.com

Biodiversity Organisation: www.biodiv.org/biosafety

Business Plan Site: www.bplans.com

Centre for Business Planning: www.businessplans.org

CIO Magazine Outsourcing Centre : www.cio.com/research/outsourcing

CIO Supply Chain Research Centre: www.cio.com/research/scm

Codex Alimentarius Commission: www.codexalimentarius.net

Computerworld: www.computerworld.com

Corporate Governance Network (Corporate Governance resources online)**:** www.corpgov.net

Data Mining Resources: www.thearling.com

Deloitte Consulting: www.deloitte.com

Digital Governance: http://digitalenterprise.org/governance/gov.html

DMReview (business intelligence and analytics articles): www.dmreview.com

EBusiness Gateway: www.ebusiness-gateway.co.uk

Economic Intelligence Unit (EIU): www.eiu.com

EDI Standards at the Uniform Codes Council: www.uc-council.org

EGovernance Newsletter: www.governance.co.uk/index.htm

eReadiness metrics: www.infodev.org/ereadiness/methodology.htm

Ernst & Young Trust Centre: www.ey.com/global/content.nsf/Malaysia/Ernst_&_Young_Trust_ Services

European Food Safety Authority: www.efsa.eu.int

European Union Corporate Governance Codes: www.ecgi.org/codes/all_codes.php

Find Articles.com (an online database of articles): www.findarticles.com

FindLaw (an online law encyclopedia): http://news.findlaw.com/legalnews/scitech/tech

Food and Agricultural Association of the United Nations (FAO): www.fao.org

Food Traceability Forum: www.foodtraceabilityforum.com

Fortune.com (an online business magazine): www.fortune.com/fortune

Grains Research & Development Corporation (GRDC) (Australia): www.grdc.gov.au

Global Corporate Governance Forum: www.gcgf.org

Global EAN Party Information Register (GEPIR): www.gepir.org

Harvard Business Review: www.hbr.com

Human Resources Guide: www.hr-guide.com

HTML Primer: http://archive.ncsa.uiuc.edu/General/Internet/WWW/HTMLPrimer.html

HTML Tutorial: www.w3schools.com/html

How Stuff Works (an online encyclopedia): www.howstuffworks.com

Information Economy site: www.sims.berkeley.edu/resources/infoecon

Institute of Biosecurity: www.bioterrorism.slu.edu

Intelligent Enterprise (data mining information): www.intelligententerprise.com

International Chamber of Commerce: www.iccwbo.org

International Accounting Standards Board (IASB): www.iasb.org/index.asp

ISO – International Organization for Standardization: www.iso.org

Internet.com (an online business magazine): www.internet.com

Investopedia (an online encyclopedia): www.investopedia.com

IP Australia: www.ipaustralia.gov.au

IT Papers: www.itpapers.com

Jargon Buster: www.wiredmedia.co.uk/docs/64.html

Meat and Livestock Australia (MLA): www.mla.com.au

Mobile Phone History: www.radio-electronics.com/info/cellulartelecomms/celldev/cellulardev.php.

Modem Information: www.computerhope.com/modemsta.htm

MRO online marketplace directory: www.b2business.net/eMarketplaces/Major_Markets/MRO_&_ Sourcing

NASDAQ Stock Exchange: www.nasdaq.com

OnFarm software guide: www.rirdc.gov.au/farmersguide/listings/bs.htm

PCWebopedia.com (an online encyclopedia): www.pcwebopedia.com

PKI Resource Centre: http://csrc.nist.gov/pki

Privacy Organisation: www.privacy.org

Product Surety Centre (resources on product safety): www.productsurety.org

Management Portal (online management resources): www.valuebasedmanagement.net

Product Surety Centre (useful information on food related traceability):www.productsurety.org

RFID Journal: www.rfidjournal.com

RFID Technologies at AIM: www.aimglobal.org/technologies/rfid

Rural Industries Research & Development (RIRDC) (Australia): www.rirdc.gov.au

Société Générale de Surveillance (SGS) (verification and certification company): www.sgs.com

Supply Chain Resource Page: www.createcom.8m.com

The ABC of ERP: www.cio.com/research/erp/edit/erpbasics.html

The ABC of CRM: www.cio.com/research/crm/edit/crmabc.html

The Internet Society: www.isoc.org

The Outsourcing Centre (outsourcing information): www.outsourcing.com

The Risk Jigsaw (business risk information): www.erisk.com

Total Quality Management (TQM information): www.gslis.utexas.edu/~rpollock/tqm.html

Truste (not for profit online trust/privacy certification company): www.truste.org

UK Department for Environment Food and Rural Affairs:www.defra.gov.uk

United States Department of Agriculture: www.usda.gov

USA Food & Drug Administration: www.fda.gov

Wikopedia (an online encyclopedia): http://en.wikipedia.org

World Health Organization (WHO): www.who.int/foodsafety

World Wide Web Consortium3: www.w3.org

XML Tutorials: www.w3schools.com/xml/default.asp

Index

A

Aflatoxin 192, 200
agile supply chain 125–126, 134–135
agri-food 99
agri-industry 1, 5, 18, 20, 25, 32, 34, 40–41, 59,
    78, 99–100, 114, 137, 142, 149–150, 152,
    171, 173, 192, 210, 326–327
agribusiness 1, 4, 10, 23, 25, 27, 29, 32, 37,
    40–43, 45, 48, 50, 56, 60, 77–78, 99–100,
    103, 117, 125, 127, 130, 135–138, 142,
    149–150, 159, 170, 211–212, 219, 250,
    258, 323, 325, 326–327
Alliance for Global Business (AGB) 305
alliances 38, 137, 140, 145, 161, 182, 226, 260
Anglo-Saxon 290–291
anthrax 172
antitrust legislation 301–302
applicable law 306, 309
AQIS 173, 175–176, 194
artificial neural networks 83
Asia Pacific Economic Cooperation (APEC) 304,
    307
Aspergillus parasiticus 192
asset management 203, 261
asset management systems 240–243, 245
attributes 5, 66, 181, 186–188, 201, 206, 208,
    262, 287, 339–340
auction 35–36, 38, 42, 47–48, 51, 56, 86, 215,
    227–228
auction house 147
audit 95, 116, 144, 157, 178, 182, 186, 235–236,
    265, 286, 293–294
audit programme 157
Australia 4–5, 43, 46, 48, 74, 77, 84, 86, 114,
    123, 127, 130, 136, 141, 146, 148, 151,
    160, 169, 173, 175, 178–179, 181–182,
    187, 189, 192–197, 199–201, 220, 223,
    236, 238, 256, 259, 293, 297, 301,
    303–304, 312–313, 317, 326
Australian Standards 194, 196
authentication 104, 143
authorization 104, 144
auto-ID 184
automation 35, 58, 81–82, 101, 114–115, 135,
    236, 245, 253, 262
Avian Influenza 174

B

B2B 24, 29, 35, 43, 51, 55–60, 120, 126, 128,
    135, 145–148, 170, 211, 213, 253, 264,
    296, 300
B2C 24, 29, 35, 43–47, 49, 50, 56–57, 60, 128,
    147, 211, 253, 264, 296, 300, 307
Balanced Scorecard Approach 13
balancing act 10, 325
bandwidth 21, 340
barcode 181–184
barriers 7, 46, 59, 120, 173, 204, 213, 301, 326
barriers to change 274
BCP 162–167
Beauregard 314
bench strength 262
Berne Convention 312
best practice 110, 235
Better Business Bureau Program 144
bidding and auctioning 47
bio-safety 172
biosecurity 171–174, 180, 190, 194, 201
biotechnology 174–175, 189
bioterrorism 171–172, 176
Board of Directors 2, 277, 284, 286–287, 320,
    322
bonding factors 203, 213
Borderless Economy, The 300
bots 67–68
bovine spongiform encephalitis (BSE) 171–172,
    174, 188, 195, 199
BPay 18
brand 144, 153, 214
Brazil 86, 141, 223
Brochureware 45, 51
browsers 21, 78
bullwhip effect 111–113
business case 251, 261, 271, 277–278, 280
business components 264
business continuity plans 162
business impact analysis 166
business model 11–13, 24, 25, 27, 33, 43,
    45–46, 56, 60, 84, 97, 99, 100–101,
    120, 126, 147–148, 228, 233, 253, 257,
    275–276, 300, 318–319, 322
business process 14–15, 82, 117, 131, 136, 166,
    267–268, 312

LIBRARY

DUCHY COLLEGE
LIBRARY